Advances in Science, Technology & Innovation

IEREK Interdisciplinary Series for Sustainable Development

Advances in Science, Technology & Innovation (ASTI) is a series of peer-reviewed books based on important emerging research that redefines the current disciplinary boundaries in science, technology and innovation (STI) in order to develop integrated concepts for sustainable development. It not only discusses the progress made towards securing more resources, allocating smarter solutions, and rebalancing the relationship between nature and people, but also provides in-depth insights from comprehensive research that addresses the **17 sustainable development goals (SDGs)** as set out by the UN for 2030.

The series draws on the best research papers from various IEREK and other international conferences to promote the creation and development of viable solutions for a **sustainable future and a positive societal** transformation with the help of integrated and innovative science-based approaches. Including interdisciplinary contributions, it presents innovative approaches and highlights how they can best support both economic and sustainable development, through better use of data, more effective institutions, and global, local and individual action, for the welfare of all societies.

The series particularly features conceptual and empirical contributions from various interrelated fields of science, technology and innovation, with an emphasis on digital transformation, that focus on providing practical solutions to **ensure food, water and energy security to achieve the SDGs.** It also presents new case studies offering concrete examples of how to resolve sustainable urbanization and environmental issues in different regions of the world.

The series is intended for professionals in research and teaching, consultancies and industry, and government and international organizations. Published in collaboration with IEREK, the Springer ASTI series will acquaint readers with essential new studies in STI for sustainable development.

ASTI series has now been accepted for Scopus (September 2020). All content published in this series will start appearing on the Scopus site in early 2021.

Disha Mishra • Rishikesh Singh •
Puja Khare
Editors

Biochar-Based Nanocomposites for Contaminant Management

Synthesis, Contaminants Removal, and Environmental Sustainability

Springer

Editors
Disha Mishra
CSIR—Central Institute of Medicinal
and Aromatic Plants
Lucknow, Uttar Pradesh, India

Rishikesh Singh
Department of Botany
Panjab University
Chandigarh, India

Puja Khare
CSIR—Central Institute of Medicinal
and Aromatic Plants
Lucknow, Uttar Pradesh, India

ISSN 2522-8714 ISSN 2522-8722 (electronic)
Advances in Science, Technology & Innovation
IEREK Interdisciplinary Series for Sustainable Development
ISBN 978-3-031-28872-2 ISBN 978-3-031-28873-9 (eBook)
https://doi.org/10.1007/978-3-031-28873-9

This Springer imprint is published by the registered company Springer Nature Switzerland AG
The registered company address is: Gewerbestrasse 11, 6330 Cham, Switzerland

Preface

Anthropogenic activities and accelerated pace of industrialization significantly enhanced environmental pollution and become a serious global concern due to its numerous carcinogenic and mutagenic effects on various life forms. The release of large amount of organic and inorganic pollutants pose a serious threat to the human body, food safety, farming, and fishing activities. To combat these problems, low-cost, sustainable, and easily adaptable solution is needed. Recently, biochar has emerged as low-cost, green, and sustainable material for the removal of pollutants from various environmental matrices. The abundance of biomass as a raw material for biochar production has added ease to their efficient production, as biomass source and pyrolysis conditions play a major role in deciding the properties and performance of biochar. However, surface engineering or the production of biochar-based nanocomposites through various methods such as ultrasonication, metal impregnation, microwave, electrochemical, or addition/coating of functional entities has helped in increasing adsorptive sites, surface area, uniform pore distribution, tolerance toward wide pH range, removal of heavy load of contaminants, and improving mechanical stability. These composite materials have immense potential for the remediation of heavy metals, antibiotics, pesticides, personal care products, oils, dye, and nutrients from the soil, sediments, water, or air components of the environment.

Currently, the progression toward biochar research has generated many environmental and economic benefits. Many countries across the globe are now focusing more on interdisciplinary research agendas for biochar applications. Thus, generation of engineered and functional biochar composites through various physiochemical processes has engendered more advanced biochar-based composite materials with improved characteristics, and thus, provides extended application as per the choice of the pollutants. The designed and engineered biochar composites have demonstrated various kinds of chemical bonding like surface complexation, π–π interaction, electrostatic interaction, hydrogen bonding, as well as different processes like partitioning, complex adsorption, Fenton process, and photocatalytic degradation for the removal of hazardous pollutants. Even the recent research on the immobilization of indigenous microorganisms on the biochar templates has rapidly enhanced the microbial degradation of pollutants.

Current book has been compiled through the contribution of various authors and provides holistic knowledge about the ongoing and future research works on the biochar-based composite materials and their extended application for the removal of pollutants. This is the first attempt created toward the creation of a platform for the identification of current and future research needs through an intense literature review. The book will be helpful for a wide variety of readers for the identification of knowledge gaps and designing future frameworks to stimulate advanced research about biochar-based composite materials. This book has been thematically classified for synthesis and application of biochar-based nanocomposites. The chapters were arranged as per their information about the biochar-based nanocomposites.

Overall, this book contains 11 chapters which are grouped in three major Parts, viz. (i) Biochar-based nanocomposites: An Introduction (Part 1) providing a holistic outlook on the topic, (ii) Synthesis of biochar-based heterostructures/composites and their characteristics (Part 2) which contains total four chapters (Chap. 2 to Chap. 5) covering different physical, chemical, and biological approaches of biochar-based nanocomposite synthesis, and (iii) Application of biochar-based nanocomposites for remediation of emerging contaminants from the environment (Part 3) which also contains total six chapters (Chap. 6 to Chap. 11) providing a detailed overview of different application scenarios of biochar-based nanocomposites for the contaminants management from different environmental matrices. Detailed overview on different chapters is as followed:

Chapter 1, 'Biochar-based nanocomposite materials: Types, characteristics, physical activation, and diverse application scenarios,' which has been written by Jaiswal et al., discussed the basic introduction of biochar-based composites by briefly covering their production routes, types, and characteristics. They have further provided an idea about the application of these composites in agriculture, energy, catalysis, and as sustainable industrial materials.

Chapter 2, by Sarmah et al., from India, have presented extensive knowledge about the physical production routes and the conversion process into nanocomposites of biochar through an extensive literature survey. The discussion about the nanoscale engineering of biochar materials will provide new insight into the in situ and cost-effective production mechanism and their efficient applications. In Chapter 3, authors, namely Borgohain et al., from India, have highlighted the importance of biochar-based hydrogel composite system, their properties, detailed classification, and mechanism of action for the removal of a wide range of contaminants. They have also included the importance of biochar-based hydrogels for the simultaneous remediation and degradation of pollutants from aquatic and soil ecosystems.

Chapter 4, authored by Verma and Singh, discussed the detailed methods of production including chemical and biological routes for the production. They have provided an overview of the main methods involved in the production of nano-biochar composite structures and their possible application in wastewater treatments, soil amendments, fuel additives, sensors, and enzyme biocatalytic reactions. In Chapter 5, entitled 'Comparative investigation of biochar-based nanocomposites over pristine biochar: An overview', Mishra et al. from India have presented the comparison between the pristine and modified biochar structures, including the comparison on the basis of their upgraded properties and performances toward the removal of pollutants. Chapter 6, by Mahour and Srivastava, have especially focused on the remediation of inorganic contaminants from the soil through the application of biochar-based nanocomposites. They have concluded that biochar-based nanocomposites have emerged as sustainable, revolutionary, and cost-effective solutions for industrial applications as well.

In Chapter 7, Kumar et al., from Canada and India, have emphasized the potential of biochar-based nanocomposites for the removal of organic contaminants from various environmental matrices. In Chapter 8, Ramola et al., from China and India, have focused on the role of biochar-based composites for the removal of dyes. They have elucidated the preparation and mechanism of biochar-supported photocatalysts for the in situ degradation of dyes. Chapter 9, by Jain et al., have provided insight into regeneration, toxicity, and future challenges for the preparation of biochar-based nanocomposites. Further, they elaborated on transport, adsorption, toxicity, and risk load in various regimes like water, soil, and air. Chapter 10, written by Mayank Singh, has provided an idea about the usage of biochar-based nanocomposites as a sustainable solution for agricultural applications. The ultimate chapter of the book, i.e., Chapter 11, developed by Lodhe et al., from India, highlighted the production of specific types of waste-derived biochar-based nanocomposites for the wastewater treatment, fuel cells, and other extended applications. They have outlined the importance of various doped nanocomposites and their characterization for efficient applications.

Overall, the current book has included sufficient knowledge, latest research work, theoretical and practical concepts of biochar composite systems in the form of a detailed literature review, figures, and tables which could be used for drawing the attention of various research communities around the globe. The book will further pave the way forward for the advanced, engineered, and goal-oriented production of biochar-based nanocomposites. Furthermore, the book will be equally beneficial for students, research scholars, academicians, environmental scientists, agriculturists, industries, early career researchers, non-governmental organizations (NGOs), government institutions, and especially those working in areas of pollutant remediation through green and sustainable options. We hope that chapters of this book will provide deep insights into the production and application of biochar-based composite systems.

Disha Mishra
Central Institute of Medicinal
and Aromatic Plants (CIMAP)
Lucknow, India
e-mail: mishra.disha1@gmail.com

Rishikesh Singh
Panjab University
Chandigarh, India
e-mail: rishikesh.iesd@gmail.com

Puja Khare
Central Institute of Medicinal
and Aromatic Plants (CIMAP)
Lucknow, India
e-mail: pujakhare@cimap.res.in

Contents

Biochar-Based Nanocomposites: An Introduction

Biochar-Based Nanocomposite Materials: Types, Characteristics, Physical Activation, and Diverse Application Scenarios

Ravikant Verma, Swapnamoy Dutta, Arvind Kumar, Tulsi Satyavir Dabodiya, Naveen Kumar, Karthik Selva Kumar Karuppasamy, B. Sangmesh, Ajeet Jaiswal, and Krishna Kumar Jaiswal

Abstract

Biochar activation has been used to augment the physicochemical properties, such as an increase in specific surface area, porosity, and surface functional groups, while improving efficiency for different applications. The physical activation method has generally driven the activation and modification of biochar. The physical activation method involved simply heating the feedstock to a higher temperature in the presence of activating agents such as CO_2, inert atmosphere, steam, or pyrolysis gases that caused the loss of gases and volatile components of the feedstock and formed a porous matrix with enhanced specific surface area. Physical activation has been predominantly controlled by different parameters such as temperature, duration of heating, degree of activation, quality of precursors, and activating agent. Biochar-based nanocomposites have been considered advanced materials and exhibit distinctive properties with enhanced pore size, functional groups, surface-active sites, etc. Generally, it has been distinguished based on the nanomaterial, viz., oxide/hydroxide biochar, magnetic biochar, and functional nanoparticles-coated biochar. The chapter has deliberated mechano-chemical strategies such as steam activation, microwave, magnetic modifications, ball milling, and heat treatment methods for fabricating biochar-based nanocomposites with improved properties. It has promised different applications in agro-environment arena with several other applications (such as energy, catalysis, and biomedical). Also, the pros and cons of the different techniques have been explored for future applications.

Keywords

Biochar-based nanocomposites • Energy and environment • Engineered biochar • Pyrolysis • Surface activation

R. Verma · A. Kumar · N. Kumar
Department of Ecology and Environmental Sciences, Pondicherry University, Puducherry, 605014, India

S. Dutta
Bredesen Center for Interdisciplinary Research and Graduate Education, University of Tennessee, Knoxville, TN 37996, USA

A. Kumar
Department of Earth and Atmospheric Sciences, University of Alberta, Edmonton, AB T6G 2E3, Canada

T. S. Dabodiya
Department of Chemical and Material Engineering, University of Alberta, Edmonton, AB T6G 2E3, Canada

K. S. K. Karuppasamy · B. Sangmesh · K. K. Jaiswal (✉)
Department of Green Energy Technology, Pondicherry University, Puducherry, 605014, India
e-mail: kkjindia.get@pondiuni.ac.in

A. Jaiswal
Department of Epidemiology and Public Health, Central University of Tamil Nadu, Thiruvarur, 610005, Tamil Nadu, India

1 Introduction

Carbon-rich solid biochar is the by-product generated in the process of thermochemical transformation of residual biomasses into bio-oils and syngas. It mainly comprises a carbon skeleton, a lesser amount of heteroatom functional groups, water molecules, and mineral constituents (Mohan et al. 2006). The unique physicochemical properties of biochar lead to many valuable technological applications, including carbon sequestration, decline of greenhouse gas releases, waste management, renewable energy generation, soil amendments, and environmental remediation (Kuppusamy et al. 2016). The particle size of carbon-rich solid biochar ranges from µm to cm with remarkable physical properties, which vary as per the pyrolysis techniques used and efficiently applied for different applications

D. Mishra et al. (eds.), *Biochar-Based Nanocomposites for Contaminant Management*, Advances in Science, Technology & Innovation, https://doi.org/10.1007/978-3-031-28873-9_1

(Sangani et al. 2020). The further decrease in the particle size from μm to nm or less and the increase in the surface/volume ratio simultaneously upsurge the adsorption potential, the surface energy, and, consequently, the contaminant removal efficiency (Naghdi et al. 2019). Biochar of nano-size also possesses a small hydrodynamic size, higher negative zeta potential, higher oxygen-containing functional group, and more carbon defects. The higher oxygen content and carbon defect properties of nano-biochar increase reactive oxygen species (ROS), improving adsorption capacity and catalytic activity, respectively (Chausali et al. 2021).

The carbon in biochar remains stable for many years; however, upon aging, the carbon defects arise due to slow aerial oxidation, which incorporates peripheral hydroxyl and carboxyl groups (Vaccari et al. 2011). Raw carbon nanoparticles (rCNPs) formed during carbonization methods, called biochar dust, perform a key role for the retention and controlled release of soil nutrients including ammonium and nitrate ions for growing plants (Saxena et al. 2014). Biochar nanocomposites (i.e., nano-biochar) have revolutionized the new research area of biochar nanotechnology, and the nanomaterial produced by the methods has better physicochemical and surface properties. Thus, it can help to succeed for four integrated goals, i.e., remediation of pollutants, management of wastes, carbon sequestration, and renewable energy generation (Sohi 2012; Jaiswal et al. 2022). Nano-biochar has exceptional pollutant absorption, enzyme immobilization, and soil mobility properties compared to biochar, indicating the novelty of nano-biochar as a potential waste management substitute (Chausali et al. 2021).

Biochar materials comprise C (40–70%), O (10–45%), H (1–5%), N (0–3%), S (<1%), and ash. The ash components of biochar can be quantified via dry combustion technique, which depends on the quality of the feedstock. Many other trace elements, like Na, Mg, K, Ca, P, etc., could also be present and play a crucial role to treat metal-contaminated wastewater (Tan et al. 2017). The availability of minerals in biochar varies conferring to the types of biomass and the ranges of pyrolysis temperatures; e.g., in rice husk, the Si content is much higher (∼8%) compared to rapeseed (∼1%), wheat straw (∼4.5%), and softwood (0%), while the maximum content of K has been found in rapeseed (∼17%) compared to other biochars derived from biomass mentioned above. Similarly, at the increased temperature from 550 to 750 °C for rapeseed, the Si content increases from 1.1 to 2.8%, while the K content decreases from 17.1 to 5.9%. The atomic ratios present in biochar are an important factor in deciding aromaticity (H/C ratio) and hydrophilicity (O/C ratio) (Fan et al. 2004). Fixed carbon in biochar is the amount of carbon remaining after subtracting ash and volatile carbon; however, the fixed carbon contents, the pH values, and the extent of acidic functional groups enhance with the rise in pyrolysis temperature from 300 to 600 °C. In contrast, the adsorption capacity, yield, and the number of functional groups declined with increasing pyrolysis temperature (Sun et al. 2017). In this chapter, we have discussed the physical activation methods applied to biochar activation under different categories, such as gaseous modifications (using steams, CO_2, air, or ozone), thermal modifications (using conventional heat or microwave irradiations), and some advanced modifications (using ultrasound waves, plasma treatment, or electrochemical approaches). The physical and chemical characteristics of biochar, such as surface area, surface functional group, mesopore, micropore, pore volume, number of aromatics, polarity, organic components, combustion, and ash content, have also been discussed as an outcome of activation methods (Sajjadi et al. 2019).

2 Nano/-Functional Biochar

Nano-biochar is an advanced nanostructured material characterized by significant physicochemical differences, such as smaller hydrodynamic radius, higher surface area, larger pore volume, higher negative zeta potential, greater oxygen-containing functionalities, and carbon defects (Oleszczuk et al. 2016). It can produce reactive oxygen species to increase adsorption capacity in comparison to their macro-biochar and have analogous possessions to carbon black (Lyu et al. 2018; Qin et al. 2018). The feedstock, such as macro-biochar, has directly driven properties of the ensuing nano-biochar. In contrast, more than 98% of the engineered nano-biochar has been produced through mechanical grinding. It has been reported that the synthesis of nano-biochar is from peanut shell biochar by sonication and grinding (Liu et al. 2018). They have also reported nano-biochar interaction chemistry by studying the comparative structural properties of hematite/goethite-associated peanut nano-biochar and defined that the association of peanut nano-biochar with goethite formed larger interlocking hetero-structures compared to the hematite hetero-structure of the peanut nano-biochar (Bartoli et al. 2022). The surface area of nano-biochar differs with multiple factors, including pyrolysis temperatures and pore formation on the material surface due to the devolatilization of biomass materials. The eradication of functional groups of alkyl and ester during the pyrolysis may interpret for greater surface area (Huang et al. 2020; Tomczyk et al. 2020). Nano-biochar prepared at lower temperatures (300–400 °C) confers lower surface area (5.6–47.2 $m^2\ g^{-1}$), while a considerably more extensive range of surface area (342–430 $m^2\ g^{-1}$) has been reported at higher temperature (450–600 °C) (Li et al. 2020).

Moreover, apart from the biomass composition and the preparation methods, several other notable factors such as hemicellulose-rich biomass also govern the surface area, which produces a high surface area at high temperature compared to lignin-rich biomass (Chen et al. 2011). Similarly, it has been described that a notable shrinkage in surface area of nano-biochar (141.3–93.28 $m^2 g^{-1}$) was apparent in comparison to the macro-biochar at 400 °C. However, macro-biochar possesses a smaller surface area compared to nano-biochar at 700 °C. Magnetically modified wheat straw nano-biochar using the ball mill method has shown a significant increase in surface area (47.2–296.3 $m^2 g^{-1}$) with escalating temperature in comparison with the pristine magnetic nano-biochar (2.93–198.6 $m^2 g^{-1}$) (Li et al. 2020). Fixed carbon, yield, and volatile matter are directly influenced by feedstock, and fixed carbon increases as the power range increases. In contrast, volatile matter decreases compared to higher heating rates and longer residence time. These results are the consequences of the lignocellulosic functional groups. Wallace et al. (2019) have reported an approximate weight percentage of gross biomass for fixed carbon and volatile matter for softwood and hemp to be 19.5% and 18.8%, as well as 67.3% and 69.6%, respectively. Softwood biochar in macro-scale, e.g., dendro-biochar, comprises volatile components (19.7%) and fixed carbon (63.8%). However, there are no significant changes in the range of fixed carbon and volatile matter concerning the size of the particles (Mayakaduwa et al. 2016).

The water retention capacity (water holding capacity) is an essential characteristic of biochar and nano-biochar that depends on the functional groups exists on the nano-biochar's surfaces, the porosity, and the coordination of the pores rather than the size of the particles; i.e., as the temperature of pyrolysis increases, it diminishes the functional groups on the surface of biochar, and the hydrophobicity of the nano-biochar increases (Alam et al. 2018). The value of the zeta potential is influenced by the types of feedstock as well as the size of the nanoparticles; this characterization has been used to understand stability, aggregation, and electro-kinetics in water (Song et al. 2019). Biochar and nano-biochar have a negative charge on the surface oxygen molecule due to the deprotonation of functional groups on the surface (Yi et al. 2015).

Mechanical strength is the key property of the material, but it is inversely dependent on porosity; i.e., highly porous nano-biochar has low strength. The hardness value for biochar synthesis from softwood (0.35 GPa) and hemp (0.24 GPa) has not shown much variation with power applied (2100, 2400, and 2700) at 2700 watts. However, softwood (5.64 GPa) and hardwood (5.95 GPa) have shown a higher Young's modulus with increased pyrolysis temperature, which could increase the fixed carbon and decrease the components of residual volatiles (Wallace et al. 2019). The quasi-elastic behavior of nano-biochar plays a crucial role as a filler to add flexural properties to the biocomposites (Wallace et al. 2019). The elemental compositions of nano-biochar fluctuate with the types of feedstock and the pyrolysis temperature (Ramanayaka et al. 2020). The aromaticity, stability, and polarity of nano-biochar can be described using the ratio of H/C and O/C. Nano-biochar possesses lower aromaticity but a greater content of oxygen-containing functional groups and mineral component in comparison with the macro-biochar (Song et al. 2019). The alkalinity of nano-biochar is generally influenced by the quality of feedstock, preparation method, soluble organic compound, surface functional group, inorganic alkalis, and carbonates. Wood-derived nano-biochar has more alkalinity (>8). At the same time, pinewood and chipped wood show a neutral pH due to the loss of the carboxyl group and the relatively low ash content (Fidel et al. 2017). Similarly, pyrolysis temperature of goethite-modified biochar increases from 300 to 600 °C, and alkalinity varies from 6.40 to 7.28 with increasing ash content (Liu et al. 2018).

3 Preparation of Biochar and Nano/-Functional Biochar

Biochar preparation involves several vital techniques including hydrothermal carbonization, pyrolysis, torrefaction, gasification, and flash carbonization. These techniques are mainly classified as thermochemical conversion methods (Gururani et al. 2022). Specifically, the process to prepare biochar is to be selected on the basis of specific criteria including the quality of biomass and process parameter (e.g., temperature, heating rate, and residence time) to control the appropriate yields and physicochemical states of the biochar (Vijayaraghavan 2019). Each process has unique conditions and environments for biochar preparation. For example, the pyrolysis process has been carried out at 250–900 °C, maintaining an oxygen-free atmosphere. In the pyrolysis process, solid, liquid, and gaseous products can be produced by decomposing lignocellulosic components such as lignin, cellulose, and hemicellulose. This technique generally involves a set of reaction pathways such as depolymerization, fragmentation, and crosslinking under particular heat conditions and pressure. This technique's preparation of biochar mainly includes using various bioreactors, such as self-propelled reactors, bubbling fluidized beds, paddle furnaces, and agitated sand rotating kilns (Yaashikaa et al. 2020). The yield of biochar has been effectively impacted with the variable parameter of temperature. Generally, during the pyrolysis process, temperature intensification produces syngas that reduces biochar yield. The process of pyrolysis has been further classified into two main sections: slow pyrolysis and fast pyrolysis. Slow pyrolysis is mainly considered in biochar preparation, where the temperature range is maintained between 300 and 700 °C with the residence time <2 s (Funke and Ziegler 2010). Gasification is another thermochemical technique used to prepare biochar as a by-product formation; however, the yield of biochar obtained is much lower. This technique mainly uses syngas (containing CO, CO_2, CH_4, and H_2) as gaseous products by decomposing carbonaceous material. This process involves two main steps: drying the biomass to eradicate the maximum moisture content and then the oxidation and combustion reaction of the gasifying agent.

Torrefaction is an evolving method of biochar preparation under the lower rate of heating, which denotes to the mild pyrolysis. Different decomposition processes have been used to eliminate the O_2, CO_2, and H_2O of biomass using an inert atmospheric air without oxygen at 300 °C (Zhang et al. 2017; Bakraoui et al. 2020). The torrefaction method alters the properties of the biomass, e.g., particle size, surface area, heating rate, energy density, and moisture content. The torrefaction method can be carried out in different approaches, such as:

- **Steam torrefaction**: In this, the biomass has been treated using steam with a temperature not higher than 260 °C and a residence time of 10 min.
- **Wet torrefaction or hydrothermal carbonization**: It comes from the interaction of the biomass with water at 180–260 °C with the residence time 5–240 min.
- **Oxidative torrefaction**: It has been carried out by treating the biomass with oxidizing agent, such as the gases used in the combustion process, to generate thermal energy. The thermal energy has been applied to develop the necessary temperature.

Apart from torrefaction, pyrolysis, and gasification, hydrothermal carbonization process is another technique to be performed at the lower temperatures (180–250 °C) and with lower processing costs. In this process, the temperature slowly increases in the closed reactor, where the mixture of biomass and water can be maintained (Zhang et al. 2017). Generally, the hydrolyzed product undergoes a set of reactions, viz., dehydration, fragmentation, and isomerization and finally generates 5-hydroxymethylfurfural and its derivatives as intermediates. Subsequently, condensation, polymerization, and intramolecular dehydration reactions also form hydrochars (Bakraoui et al. 2020). However, in the hydrothermal carbonization process, the complex nature of high molecular weight lignin creates difficulties in char formation. When lignin breaks down through dealkylation and hydrolysis, it generates catechols, phenols, syringols, etc., as phenolic products (Jain et al. 2016; Kumar et al. 2022). Like the pyrolysis reaction, lignin components that cannot dissolve in the liquid phase have been converted to hydrochars.

4 Characteristics of Biochar and Nano/-Functional Biochar

Usually, it is always recommended to analyze the structural features, physicochemical properties, and elemental composition in studies, as it is extremely useful in gaining a concrete understanding of the prepared material (Chen et al. 2019). It is well designated in studies that factors such as structure, inclusion, lignin, and cellulosic components of various biomass are useful in understanding the structure of biochar (Lehmann 2007). Different aspects must be adequately evaluated, such as the shape and size of the pores, the structure of aromatic carbon, such as morphological characteristics, pH, density, existing functional group, particle size distribution, water retention capacity, surface area, and physicochemical parameters to make impact (Narzari et al. 2017). However, these parameters are highly influenced by the factors involved in the synthesis, including temperature, pressure, rate of heating, residence time, pre-and post-treatment techniques, etc. Impact of the carbonization temperature on pore sizes and structure of the aromatic carbon has been clearly observed, in which the rise in temperature enhances the pore sizes.

Although the microporous structure of the biochars surface severely damages at exceeded temperature of a specific limit (more than 700 °C), similarly, when the temperature is maintained at nearly 800 °C, the structure of the carbon skeleton of biochar loses its stability (Lian et al. 2011; Huang et al. 2014). Moreover, the elements also provide minute details to differentiate the prepared biochar. Generally, C, H, O, N, P, S, Na, Mg, Si, K, and Ca are present in biochar, where the amount of aromatic carbon (almost 60%) is maximal (Zhao et al. 2016). Fatty acids, phenols, alcohols, esters, and compounds analogous to humic acid and fulvic acid are carbon compounds in the biochar (Chen et al. 2021). The phosphates (P) and nitrogen (N) is also found on the surface of the biochar, but in a much smaller amount. N is mainly observed as the biochar surface's heterocyclic C–N structure (Hue 2020). With increasing carbonization temperature, the P amount fluctuates, especially decreasing. This distinction can be affected due to high pH and phosphates, as Mg and Ca form in the process of carbonization. The amount of Na, K, Ca, and Mg-like elements varies mainly depending on the feedstock from which the biochar has been prepared. Biochar produced from cattle manure, woody, and herbaceous plants has been found to have the highest contents of these four elements (viz., Na, K, Ca, and Mg). In the biochar, Na and K as low-valence metal ions are more predominant than Mg, Al, and Ca as high-valence metal ions (Silber et al. 2010; Chen et al. 2019).

Carbonization and pyrolysis processes are mainly adapted to produce biochar, and in those processes, inorganic minerals such as phosphates and carbonates have been formed because biochar becomes alkaline. The temperature and quality of feedstocks for pyrolysis or carbonization affect the pH level of the biochar generated. For example, compared to non-legume biochar, the pH of legume biochar is superior. Similarly, the pH of biochar produced from woody or herbaceous plant is lower in comparison with the poultry manure biochar under the same carbonization and pyrolysis reaction conditions (Yuan et al. 2011; Chen et al. 2019). In addition, with the rise in temperature, the pH level increases under different pyrolysis and carbonization reaction conditions. This occurs mainly due to the disintegration of acidic functional groups, e.g., carboxyl, hydroxyl, phenolic, and the volatiles of organic acids (Chintala et al. 2014; Chen et al. 2019). Another important parameter is the biochar's specific surface area, which increases along with the rise in the temperature of pyrolysis up to a specific range. It generally varies between 1.5 and 500 $m^2\ g^{-1}$ (Al-Wabel et al. 2013). When the structure of internal pores of biochar gets filled with volatiles, tars, and other products, the specific surface area gets reduced. This is due to the thermal disintegration of biomass at relative lower temperature. These substances (volatiles, tars, etc.) break down into volatile gases and escape with increasing temperature, and the pores' size is reduced. However, the number of pore structure in the biochar rises with resultant to the higher specific surface area and more microporous structures. Specifically, nano-biochars comprise the surface areas between 5.6 and 47.2 $m^2\ g^{-1}$, when it is generated at 300–400 °C, but at elevated temperatures (450–600 °C), the surface area increases dramatically (342–430 $m^2\ g^{-1}$) (Ramanayaka et al. 2020).

In addition to the surface area, carboxyl, carbonyl, and hydroxyl functional groups also influence the capacity of biochar. Alkaline or oxygen-containing functional group in biochar enhances its hydrophobicity or hydrophilicity, adsorption capacity, ion exchange capacity, and buffering (Antón-Herrero et al. 2018). With increasing carbonization temperature, the carboxyl and hydroxyl functional groups as well as the C–H, C–O, and O–H bonds in biochar decline, but the amount of alkaline group increases. So, as the carbonization temperature increases by combining all the functional groups, their density and number reduction also affect the degree of aromatization as well as water holding capacity of biochar. In the case of nano-biochars prepared from pine wood have water retention capacities of 2.7 and 9.75 at 400 and 525 °C, respectively. With the rise in temperature, the water holding capacity of biochar decreases, but the degree of aromatization increases (Kinney et al. 2012). Moreover, biochar has a high boiling point, a higher degree of carboxylate esterification and aromatization structure, higher carbon content, poor solubility, higher stability, and strong resistance to physical, chemical, and biological disintegration (Leng et al. 2019). Available studies showed that compared to micro- or macro-biochars, nano-biochars have lower stability, which was observed in a zeta potential study in which zeta potential values were found to be less negative for nano- and micro-suspensions of biochar (1.73–13.4 mV) (Qian et al. 2016).

5 Biochar-Based Nanocomposites (BNCs)

Biochar consists of pyrolyzed biomass rich in solid carbon and deficient in oxygen. It can be prepared from agricultural residues, woody biomass, animal litter, algae biomass, and solid waste through different thermochemical routes (pyrolysis, carbonization, torrefaction, gasification, etc.). The characteristic properties of biochar have been inherited from the pyrolysis technique, temperature, and external synthesis conditions (Chi et al. 2021; Jaiswal et al. 2021a, b, c, d; Sharma et al. 2021). Recently, the scientific community has been trying to develop a biochar-based composite, i.e., a biochar-based nanocomposite (BNC). BNCs can be used as a potential material for removing aqueous pollutants (both organic and inorganic), which solves the problems of water contamination and carbon sequestration. This classification is on the basis of types of nanomaterials applied to functionalization of biochar, e.g., nanoparticles of metal oxide, metal hydroxide, magnetic metal oxide, chitosan, graphene, $g\text{-}C_3N_4$, graphene oxide, carbon nanotubes, double-layer hydroxides, ZnS nanocrystals, nanoscale zero-valent iron, etc. (Jaiswal et al. 2018; Rodriguez-Narvaez et al. 2019; Dutta et al. 2020). BNCs are classified according to their method of synthesis. All types of biochar-based nanocomposites are described as following.

5.1 Oxide/Hydroxide BNCs

The oxide and hydroxide types of BNCs have been synthesized by the target element enrichment method by applying bioaccumulation to produce metal salt pretreatment biomass and inserting metal oxide nanoparticles after pyrolysis. In the process of pretreatment of metal salts formation and target enrichment, the biomass was first treated with a specific metal impregnation before pyrolysis. However, in the insertion-based formation process, the nanometal oxide/hydroxide can be directly introduced into the pyrolyzed biomass biochar (Panahi et al. 2020). Biochar-based nanocomposites and engineered biochar with desired properties can be achieved by utilizing target elements that can be enriched through bioaccumulation in biomass (Wang et al. 2017). Subsequently, the in situ conversion of metal precursors to nanometal oxide/hydroxide can be enabled with enriched biomass. Various physicochemical properties have been improved by incorporating metal precursors (e.g., Fe, Zn,

Mn, Al, and Ca) into the feedstock, i.e., bioaccumulation to develop biochar integrated with nanometal oxide/hydroxide (Wang et al. 2017). Furthermore, the hyperaccumulation technique facilitates the bioaccumulation of the target elements and provides the desired concentration and the specific properties of the oxide/hydroxide BNCs. The pretreatment technique uses chemical reagents on the biomass before pyrolysis. During this process, the metal ions adhere to the surface or are absorbed by the interior of the biomass (Zhang et al. 2019). In addition, metal oxide nanocomposites can be prepared after the pyrolysis of the biomass with any technique, such as evaporation, heating, wet impregnation, and direct hydrolysates (Wang et al. 2017).

5.2 Magnetic BNCs

Biomass pretreatment with iron ions or co-precipitation with iron oxides can convert BNCs into magnetic material, mainly because of their separation from the aqueous medium. Here, the biochar can be separated from the solution when an external magnetic field is introduced. The impregnation of iron oxide on the BNCs decides the adsorption capacity of the BNCs. Initially, the biomass was co-precipitated using Fe ions, which allowed them to impregnate into the biomass, then pyrolyzed under different conditions (Hassan et al. 2020). A novel porous biochar matrix incorporated with nanometric Fe_2O_3 has been recently developed that possesses improved ferromagnetic and surface properties through a thermal pyrolysis process. Microwave heating is another method that could be used to prepare magnetic biochar. In addition, natural hematite ore can be processed during biomass pretreatment to produce magnetic biochar in a facile way, as they have been shown to exhibit stronger magnetic properties. Direct co-precipitation of $Fe^{3+}/^{2+}$ or Fe on biochar can be another way to produce BNC from pyrolyzed biomass (Siddiqui et al. 2020). Physicochemical properties, such as saturation magnetization and thermal stability, mainly depend on the charge content of Fe_2O_3 or another magnetic oxide. Interestingly, the chemical co-precipitation of Fe–O on biomass prior to pyrolysis shows superior physicochemical properties compared to the co-precipitation of Fe–O after pyrolysis.

5.3 Functionalized BNCs

Chitosan, graphene oxide, carbon nanotubes, nanoscale zero-valence iron, ZnS nanocrystals, layered double hydroxides, etc., can be used as inexpensive filler materials in a matrix of biochar. Enabling these functional materials improves the surface activity, excellent porosity, higher surface area, and thermal stability of biochar. These materials are coated with the biomass before pyrolysis (Dahal et al. 2019). Furthermore, functional composite biochar can be prepared by facile dip-coating procedures. Here, nanoparticle suspensions have been prepared with deionized water, and the biochar has been immersed in the suspension with the help of a drying oven for further pyrolysis. Annealing of graphene–pyrene derivatives has resulted in the production of graphene-coated biochar to coat the biochar surface through p–p interactions, increasing the surface area and improving thermal stability (Tan et al. 2016). Synthesis of multi-walled carbon nanotube (MW–CNT)-coated biochar can be performed by dip-coating of biomass in solutions of carboxyl-functionalized CNTs of varying concentrations prior to slow pyrolysis. It has been observed that the biochar's stability, surface area, and porosity improved with the addition of CNTs. Impregnation of nanoparticles after pyrolysis can lead to the preparation of biochar-based composites, which enriches the biochar with additional properties of the added functional nanoparticle. Chitosan, ZnS nanocrystals, Mg/Al hydrogel, and zero-valent iron are generally used for the synthesis, resulting in superior functional properties enriched by the synergy of functional nanoparticles and biochar (Panahi et al. 2020). It is essential that the selected target materials do not produce any undesirable effects on the nanomaterials.

6 Physical Activation of Biochar

The activation of biochar is necessary to advance its properties and applicability. The biochar's properties including specific surface area, porosity, and oxygen-containing functional group can be improved through the activation of biochar. Various routes can be adapted to activate biochar, such as chemical, physical, and mechano-chemical approaches (Panwar and Pawar 2022). In this section, we have deliberated specifically on the development of various physical activation techniques for biochar, which include gaseous activation (such as using steam, CO_2, ozone, and air), thermal activation (such as conventional, and microwave), plasma activation, ultrasound activation, electro/electrochemical modification, etc. In general, carbonization and activation are the two main steps involved in the physical activations of carbonaceous materials. Under a range of temperature of 600–900 °C and an inert gas atmosphere, biochar's complete carbonization occurs (Gao et al. 2020). In the procedure of pyrolysis or partial carbonization, it is necessary to carry out complete carbonization to eliminate the non-carbon species and produce biochar with a high percentage of carbon. The biochar's internal surface area can be increased through the carbon skeleton's partial gasification. Under the influence of an inert agent (N_2 or He) or partial oxidizing medium (such as steam, limited air, ozone, and CO_2), a thermal treatment requires to perform gasification (Seow et al. 2021). However, due to the high temperature and energy requirements, novel techniques such as plasma treatment, microwaves, ultrasonic irradiation, and electrochemical modification processes can be employed to improve biochar quality while keeping process costs affordable (Mariana et al. 2021).

Gaseous activation is a frequently used physical activation approach where an oxidizing agent has been used in conjunction with a temperature of over 700 °C for biochar activation. Biochar has been exposed to oxidizing agents, e.g., ozone, steam, CO_2, or air, and as a result, the inaccessible pores widen and open. Generally, oxidizing agent enters into the internal structure of carbon atoms and gasifies, making oxygen functional groups available and improving inner surfaces (Sajjadi et al. 2019). Physical activation by CO_2 or steam involves the injection of these gases at the end of the pyrolysis technique. Furthermore, the activation step can be performed separately from the prepared biochar. The minor sizes of the H_2O molecule in comparison to the CO_2 molecule facilitate the steam activation. In contrast to the air activation, the reaction among carbon atoms and steam displays an endothermic manner with easy to control and more suitable to gasify carbon atoms with higher surface activity (Alvarez et al. 2015).

The general reaction between steam and coal involves the chemisorption of water, gasification of coal, displacement of water as well as gasification through steam, carbon dioxide, and hydrogen. Mesopores and micropores have been developed in this gaseous steam and produce a wide range of pore size distribution. The steam-to-carbon mass ratio, temperature, and the time of activation are 3-key parameter which affects the steam activation technique (Lupul et al. 2015). By increasing the temperature of steam and time of treatment, the ordered structures can be obtained. The pore volume, radius, and surface area have been increased by the elimination of carbon atom from the surface. The use of low-temperature steam results in a reduced surface area. However, over-activation has also resulted in higher ash content and lower total surface area due to longer duration at high temperatures. Over-activation leads to the development of pore size that exceeds the preferred sizes, and the volume fraction of micropores declines, which becomes problematic for the application (Fan et al. 2004). Simple activation with carbon dioxide involves the Boudouard reaction. Carbon monoxide and surface rust have been formed on the carbon surface by dissociative chemisorption of CO_2 (Kovalskii et al. 2022). The incidence of oxygen in carbon dioxide enhances the development of oxygen-containing functionalities on the surface of biochar (Feng et al. 2016). Exposure of biochar to ozone is another form of activation. Firstly, oxidizing with ozone and exposing it to high temperatures in an inert atmosphere is a route to obtain the synthesized biochar with the greater surface area (Jimenez-Cordero et al. 2015).

The ozonation of the biochar forms enormous carboxyl, hydroxyl, phenolic, quinone, and ethers on the biochar. The treatment of ozone on the surface of carbon also generates carboxylic and carbonyl groups as acidic functional groups (Gümüş and Akbal 2017). These functional groups mainly help improve soil cation exchange capacity and similar applications. In recent studies, short-term treatment of biochar with ozone is also considered to create acidic functional groups. In the study, carbon oxidation increased with increasing ozone dose (Smith 2011). In addition to steam, ozone, and carbon dioxide, oxygen (air) is another gasifying agent with an economical approach to activating biochar. However, due to the high reactivity of oxygen, this process is complicated and complex compared to others. Cyclic oxygen chemisorption/desorption is an efficient method to achieve the preferred properties of biochar (Sajjadi et al. 2019). In this process, oxygen chemisorption occurs mainly at an adequate temperature in oxygen-containing air in the first cycle. Then desorption of chemisorbed oxygen occurs during the second cycle in an inert atmosphere at the higher temperature. The desorbed oxygen exists from the surface in the form of carbon dioxide and carbon monoxide. As a result, the size of the pores becomes more prominent, and more active sites can be generated (Py et al. 2003). The chemisorption process occurs at low temperatures, while desorption occurs at high temperatures.

Conventional thermal modification is another effective technique used for physical activation. Here, for a specific time (1–2 h), a higher temperature (600–1500 °C) has been applied to the biochar, due to which oxygen heteroatoms and hydrogen can be removed, and as a result, a well-ordered/condensed carbonaceous structure has produced. From experimental observations, it has been seen that the amorphous structure changes at the time of gas discharge and the conversion of disordered structure and amorphous into the system of superior aromatic rings (Kleber et al. 2015). However, the important detail that has to be mentioned is that the impact of thermal activation is more significant in the initial stages. Still, the subsequent release of heteroatoms gradually reduces the effect. A thermal activation process can be performed in the presence of hydrogen, CO_2, steam, and limited air as oxidants or under an inert atmosphere (nitrogen, argon). Under the influence of the inert atmosphere, the hydrophilic groups can be removed from the biochar, and the hydrophobicity of the carbon increases. Wide-ranging surface functionalities can be obtained through the modification in the carbonaceous matrix of the biochar using a thermal post-treatment. In addition to the traditional heat treatment, microwave-assisted activation of biochar has drawn attention, as biochar can be selectively treated to modify the functionalities using chemical reagents. Uniform temperature distribution on carbon substrates is a key benefit of microwave-assisted activation. Microwave-activated biochar exhibits larger surface areas due to the greater micropores compared to the conventional technique (Chu et al. 2017). Furthermore, the method can effectively reduce energy consumption as it involves a lower activation temperature with a shorter reaction period. However, microwave-assisted activation and microwave-assisted pyrolysis are different, as the second one does not include chemical reagents. Physical and chemical modifications of biochar can also be achieved through the use of ultrasonic activation. Compared with the conventional method, this method also employs a lower energy yield without noticeable modification in carbon contents. Ultrasonically activated biochar mainly possesses higher heating values, internal surface area, and rate of reaction (Wang et al. 2021).

Plasma activation is another recently familiarized green technique through which chemical surface transformation has been performed to obtain desired textural properties, especially suitable porous structures (combination of mesopores, micropores, and macropores) that can be easily achieved (Gupta et al. 2015). In this technique, high-energy electrons emitted by the plasma are used to penetrate and break chemical bonds on the surface in a short period (between a few minutes and 30 min) (Niu et al. 2017). In this technique, biochar has been generally exposed to ionized oxygen in a temperature range of approximately 15–30 °C for at least 0.1–5.0 min (maximum 1 h) (Lee 2012). Altering the physicochemical properties of biochar can be done using an electric field. Using the electro- or electrochemical modification technique, the insertion of chemical products and surface area can be improved (Jung et al. 2015). However, in electrochemical technique, HClO and ClO^- are the two strong oxidants that dramatically improve biochar's morphology and porosity. However, this technique is still being optimized for efficient use.

7 Diverse Applications of BNCs

Applications of biochar, nano-biochar, and biochar-based nanocomposites have been well-recognized in the environment, agriculture, energy, catalysis, and materials science (Fig. 1). In addition to the domestic application of nano-biochar in agriculture and treatment of environmental pollutants, it has value-added characteristics to be applied in various areas such as power generation, biology, and medical application. It can also be used in catalysts and in developing energy-efficient materials (Liu et al. 2022). In this section, we have deliberated the numerous applications of biochar-based nanocomposites as follows:

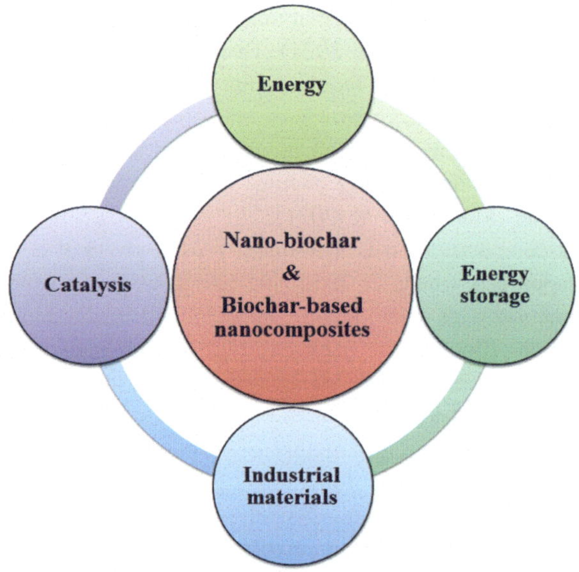

Fig. 1 Different applications of biochar and biochar-based nanocomposites

7.1 Applications of BNCs in Agriculture

Agricultural applications of biochar-based nanocomposites include disease management, plant growth enhancement, pesticide remediation/decomposition, fertilization, soil amendment, microbial growth management, and support. These materials can also help improve productivity, plant biomass, and soil performance. Interestingly, biochar has been demonstrated to improve soil fertility, thereby reducing the usage of chemical fertilizers, particularly nitrogen, which would further reduce nitrogen leaching into groundwater (Ding et al. 2010). In addition, it improves water retention capacity, soil fertility, and permeability. Additionally, biochar-based nanocomposites can improve charge density by transporting large amounts of nutrients that will enhance crop yields. The nanocarbon derivative of biochar can sequester pesticides due to better porosity, which enriches the adsorption sites in the soil. Similarly, biochar-based nanocomposites reduce desorption, leaching, pesticide mobility, and bioavailability of pesticides in soil layers (Chausali et al. 2021).

Biochar-based nanocomposites application to soil is effective in sequestering soil carbon and offers other benefits, such as neutralizing soil pH to improve soil quality, solidifying microbial growth in soil, and enhancing soil cation exchange capacity. Due to the beneficial properties of biochar-based nanocomposites, including alkaline nature, porosity, larger surface area, water holding capacity, organic carbon and minerals, etc., it has excellent agricultural utilization. It enriches the micro-ecological environment of the soil (Wang et al. 2020). BNCs promise diverse applications in the field of agriculture and environment with certain other applications and advances in the arenas of energy, catalysis, material science, biomedical, etc. (Chausali et al. 2021). Under natural environmental conditions, these physicochemical properties and inherent structures allow biochar to persist in soil for several years without abrupt decomposition (Das et al. 2021). As a result, the use of biochar has strongly affected the nutrient content, water content, soil bulk density, porosity, cation exchange capacity, etc. (Jain et al. 2019). Various investigations have identified the aspects of biochar in crop growth and development, suppression of pathogen infestation, alleviation of soil limiting factors, reduction of leaching, and removal of uptake of heavy metals/pesticides. The application of biochar in soil amendments has shown significant improvement in elemental composition and surface properties. From the studies, clarifications were found on the interactions between hydrogen ions existing in the soil and biochar-containing functional groups (e.g., carboxylic, hydroxyl, and phenolic groups) that increase the pH level in the soil. Mainly, the H^+ ions react with the carbonates, bicarbonates, and silicates present in the biochar and neutralize the soil's pH level (Ahmad et al. 2012).

Improving soil fertility and structure through biochar is crucial to increase crop yields. Uzoma et al. (2011) produced maize using biochar at 20 t ha^{-1} and 15 t ha^{-1} on sandy soil, and as a result, crop yield enhances by 98% and 150%, respectively. Biochar-based fertilizers have a significant impact on crop yields. In fact, studies have shown an evident growth in the production of peanut, soybean, and maize crops. In particular, the number of grains per plant and grain weight significantly improved (Chen et al. 2019). Biochar also helps minimize aluminum toxicity, enhancing productivity by maintaining water retention and improving the soil's cation exchange ability (Tan et al. 2017). Additionally, biochar can remove heavy metals, pesticides, and hazardous pollutants from agricultural soil. Biochar contains basic cations that help to increase the soil surface and cation exchange capability of the soil and provide the ability to adsorb more cations. The presence of elements (e.g., Ca, K, P, and N) in higher concentrations in biochar intensifies the concentration of soil nutrients. It has been adequately valuable for supplying these nutrients to the existing microbial communities in the soil (Bian et al. 2013). In addition to these factors, from the point of view of reducing the impact of greenhouse gases and supporting carbon sequestration, biochar can be enormously considered for agricultural use (Venegas et al. 2016).

7.2 Applications of BNCs in the Removal of Organic Contaminants from Wastewater

Biochar-based nanocomposites have excellent adsorption characteristics that help to remove organic contaminants from the aquatic environment. Various biochar-based nanocomposites interact with organic pollutants such as rhodamine B, methylene blue, acid orange 7, eriochrome black T, crystal violet, methyl orange, reagent blue 19, tetracycline hydrochloride, reagent red, norfloxacin, and ofloxacin have been reported. The biochar substrate influences the adsorption capacity of biochar-based nanocomposites (Jaiswal et al. 2021a, b, c, d; Zeghioud et al. 2022). The interaction kinetics between the functional group of the biochar-based nanocomposite and the organic contaminants involves a certain number of mechanisms involving hydrogen bonding, electrostatic interaction, ion exchange, hydrophobic interaction, etc. Furthermore, the coating of nanoparticles on the biochar structure can increase the adsorption sites, which has been observed with increased surface area. Graphene-coated biochar can interact with organic contaminants and graphene sheets on the biochar surface. Hydrophobic interactions between graphitic/hydrophobic sites on CNTs and organic pollutants influence the adsorption mechanism of organic contaminants (Zhang et al. 2012; Jaiswal et al. 2021a, b, c, d).

7.3 Applications of Biochar-Based Nanocomposites in Catalytic Applications

Nano-biochars and biochar-based nanocomposites can be effectively used in bio-based reactions as enzyme carriers and acid catalysts. The biocatalytic system of biochar-based nanocomposites improves the properties by steering efficiency by increasing the surface area and loading capacity and decreasing mass transfer resistance (Do Minh et al. 2020). Biochar-based nanocomposites effectively immobilize support material in enzyme systems due to their highly porous structure, further increasing the higher surface area-to-volume ratio and thus allowing an additional amount of surface functional groups (Mandal et al. 2021). In general, the inherent properties of biochar-based nanocomposites, such as pore characteristics and functional groups, enhance catalytic capacity and active sites, thus improving charge carrier separation (Liu et al. 2015). Biochar-based nanocomposites can also be used as a heterogeneous acid catalyst in esterification reactions.

Another, most serious concern that needs to be addressed worldwide is environmental pollution. In everyday life, every living organism in the world contributes in some way to increasing environmental pollution, especially humans, contributing as much as possible to pollution through the use of vehicles, the operation of machines, the use of plastic materials, etc. Among all the above mentioned, plastic waste has been considered the most important cause of various problems (Nielsen et al. 2020). Generally, $\sim 20\%$ of plastic waste has been recycled conventionally. Plastic has properties of carbon–hydrogen bonds that can be used effectively in the generation of energy through the pyrolysis process (Kalargaris et al. 2017). During heat treatment, the catalysts play an essential role in the improvement process (Bezerra et al. 2007). Associated with that, biochar-based materials have been considered a necessary catalyst in the pyrolysis process of plastic waste degradation. The biochar contents with silica, alumina, and aluminum hydroxide enhance the degradation process through bonding and intricate construction with plastic structures. Furthermore, biochar significantly contributes to removing contaminants during the pyrolysis process of plastics (Singh et al. 2021).

7.4 Applications of Biochar-Based Nanocomposites in Biofuels Productions

The extreme conditions of acidic nature in the anaerobic associated digestion process have resulted in hostile behavior on the microbes. The highly acidic conditions associated with anaerobic digestion show antagonistic effects on the microbial community. The amalgamation of biochar in the process results in a significant reduction in the generation of ammonia and carbon dioxide and also improves the activity of microbes in the soil. This is basically due to the biochar's alkalinity (pH level). On top of that, some of the alkali metals like Mg, Ca, and K associated with biochar can dissolve and improve biomass digestion, subsequently leading to better methane generation (Chiappero et al. 2021). In addition, it has been observed that the CO_2 goes through a conversion process, resulting in the product of carbonates and bicarbonates that increase the stability of the digestion and a stable alkaline pH. Due to the higher electrical conductivity nature of biomass, higher methane generation occurs in the system. In certain studies, it has been shown that the use of sulfonated-based biochar leads to an improved catalyst in the transesterification process and improves the effectiveness of biodiesel production (Jaiswal et al. 2021a, b, c, d; Zailan et al. 2021). In addition, using feedstocks with a high content of fatty acids (cooking oils) can enhance the production of biofuels.

7.5 Applications of Biochar-Based Nanocomposites as an Additive in Fuels

Biochar-based nanocomposites are helpful in fuel additives and play an efficient role in reducing harmful vehicle emissions. Biochar-based nanocomposite additives perform better than nanometallic additives due to the non-toxic nature of bio-nano-additives (Chausali et al. 2021). Researchers recently discovered that nanobiocarbon particles in a blend of diesel, biodiesel, and ethanol could improve engine performance and positively affect exhaust emissions (Weidong et al. 2020). Recently, nanoparticles from bagasse, i.e., sugarcane nano-biochar have shown excellent performance as a substitute for metal-based fuel additives (Ardebili et al. 2020).

Additives used in fuels for the operation of various automobiles and other industrial machinery play an essential role in reducing the emission of harmful gases and other particles into the atmosphere (Dey and Mehta 2020). Apart from nanometal additives, bio-nano additives have been preferred nowadays due to their non-toxic behavior during operation and emission (Saha et al. 2020). Due to the increasing pollution and harmful gases emitted during various operations of fuel-based automobiles and machinery, many researches have been focused on developing energy-efficient bio-additives and biofuels to reduce the emission of toxic gases and particles into the atmosphere. Several researches on nano-biochar particles in conjunction with biodiesel–diesel–ethanol blend fuels have shown a significant improvement in overall engine performance along with reduced emissions of particulates and noxious gases. In fact, bio-based additives improve overall engine efficiency, especially nano-biochar (i.e., sugarcane bagasse nano-biochar), which can be used as an adequate substitute for metal-based fuel additives (Ampah et al. 2022).

7.6 Applications of Biochar-Based Nanocomposites in Energy Storage

The need for infrastructure for energy storage has been a critical phenomenon during the commercialization of fuel cells. This may involve storing the energy cost-effectively and with little or no energy loss. Besides all conventional energy storage methods, especially hydrogen, the physical adsorption method is more suitable for reversibility, faster kinematics, and storage capacity (Zhou et al. 2021). Furthermore, porous carbon-based materials have significantly improved energy-storage properties. Figure 2 shows the different energy applications of biochar-based nanocomposites, including energy storage. Moreover, the nature of

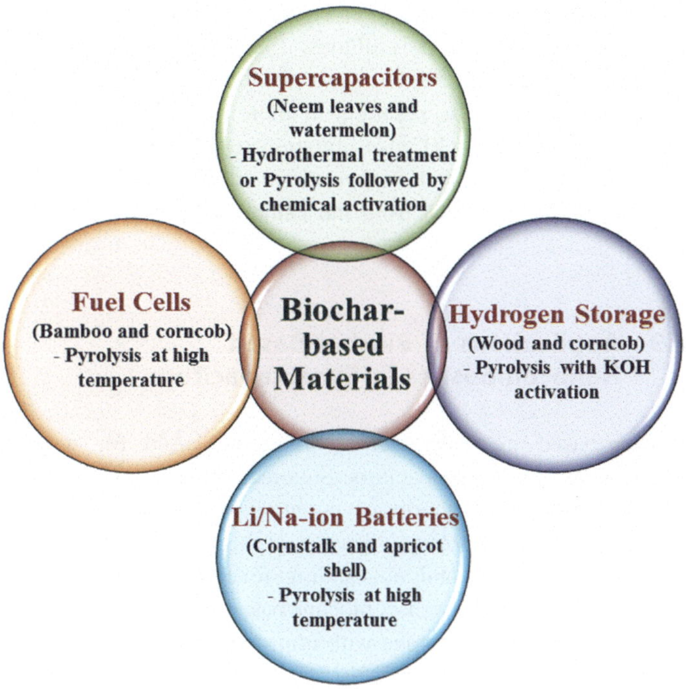

Fig. 2 Applications of biochar-based nanocomposites in different energy applications

carbon-based storage materials has a large surface area, higher pore volume, higher chemical stability, etc. In addition, biochar-based carbon materials for energy storage have been considered to show significant improvement due to their flexible nature of surface and pore structures. Biochar-based materials have more micropores, contributing significantly due to capillary forces during adsorptions (Liu et al. 2019).

7.7 Applications of Biochar-Based Nanocomposites in Fuel Cells

As for fuel cell energy generation and storage, biochar-based materials have been used as catalysts for evolution reactions and cathodic oxygen reduction (Liu et al. 2019). Due to the nature of biochar, it has been used as direct carbon-based fuel cells (Jafri et al. 2021). The diffusivity of the biochar for the reaction has been improved due to the smooth interaction of the molten carbon electrolytes (Rui et al. 2020). The oxidation of carbon to oxides in the biochar has resulted in the initiation of electron transport that can enhance electricity generation (Kumar et al. 2020). Also, it has been more suitable for fuel cell applications due to its enhanced nature of electrical conductivity. The presence of micropores and a larger surface area can lead to an improved reaction that causes better fuel cell performance. Higher ash content leads to reduced energy output due to reduced ion flow between the electrodes and the fuel. To increase the efficiency of fuel cells, it is required to optimize the content of ash and inorganic materials in biochar (Palniandy et al. 2019).

7.8 Applications of Biochar-Based Nanocomposites in Batteries

Biochar-based nanocomposite materials have significant features such as large surface functional groups with a reasonable surface area containing an abundance of mesopores. It has the potential to be efficiently employed for ion diffusion and interface for electrolytes/electrodes in commercially available batteries, such as sodium-ion and lithium-ion batteries, to enhance electrochemical reactions (Senthil and Lee 2021). Subsequently, the capacity of the sodium-ion and lithium-ion-based battery can be increased with the improved pore and surface area by fine-tuning the respective electrodes' conductivity. Furthermore, doping nitrogen with biochar can significantly improve the storage space or the ability to store energy in commercially available batteries, especially lithium-ion and sodium-ion batteries (Bhattarai et al. 2022). The process involves the establishment of defects through active chemical reactions, which leads to greater availability of space and more excellent conductivity. Nitrogen doping also leads to a higher electronegativity of the carbon in the biochar, significantly increasing the space to store the ions with stable electrochemical reactions and electron conduction. It can also result in the development of interfaces during charge transfer with reduced internal resistance. Due to the favorable surface properties of biochar, such as higher specific capacity and higher conductivity, the metal oxide can be efficiently doped with biochar (Zou et al. 2021). In commercial batteries, biochar combined with metal oxides (i.e., cobalt, manganese, and iron oxide) can be used as the anode and electrodes. In addition, it can combine and contain polysulfides, the sulfur present in lithium-ion batteries, which leads to the suppression of active material growth, resulting in more excellent battery stability (Yang et al. 2022). Limitation to the above combination has been noticed, such as the inability to restrict the diffusion of polysulfides from the cathodes (Shebert et al. 2020).

7.9 Applications of Biochar-Based Nanocomposites in Supercapacitors

Biochar-based materials can be efficiently used for the development of supercapacitors. Supercapacitors have been established to have enormous potential regarding energy storage, possessing energy density at higher altitudes, a complete life cycle, and outstanding reversibility (Solyali et al. 2022). So, in this sense, biochar-based materials can be fine-tuned for energy storage with improved performance of supercapacitors (Liu et al. 2015). The improvement of supercapacitors' performance has been achieved by changing the surface area and conductivity; the porosity and the surface of the functional groups are enriched with oxygen. Some of the significant parameters for supercapacitors' enhanced electrical conductivity nature depend on the excess surface area, with increased porosity in addition to an oxygen-enriched surface group and an enriched energy density (Zhao et al. 2022). These disadvantages can be eliminated by improving the micropore and surface area by activating phosphoric acid, potassium hydroxide, or zinc chloride in the biochar-based material. Therefore, it can lead to the development of a more significant number of mesopores, enhancing the diffusion reaction and leading to effective ion transfer resulting in improved capacitance (Cuong et al. 2021). In addition, some of the heteroatoms doping (such as nitrogen and phosphorus) in biochar can improve electron transport and the performance of capacitors (Ghosh et al. 2020). Subsequently, the doped heteroatoms resulted in higher cyclic stability and higher conductivity. Similarly, doping of biochar with metal oxides leads to pseudo-capacitance (Husain et al. 2022).

7.10 Biochar-Based Nanocomposites in Other Applications

In addition to the abovementioned application, biochar and nano-biochar can be used as sensing materials in various electrochemical devices in electrochemistry. To detect microcystin-LR (MCLR) toxin in water, a biochar-based immunosensor paper or immunosensor system has been developed as coating materials, which serve as highly conductive and dispersible biochar-based nanocomposites particles with anti-MCLR antibodies in filter paper by the dipping-drying method (Yao et al. 2021). Recently, it has been observed that the waste lignin was converted into nano-biochar, which can be used as a renewable substitute for carbon black in the reinforcement of styrene-butadiene rubber in the rubber industry. Overall, biochar-based nanocomposite and nano-biochar have future applications in various fields, and there is a great need for further research.

8 Conclusion and Future Prospects

Biochar and biochar-based nanocomposites have been considered as a material that has various feedstock options and can be used for different applications (such as adsorbent, sensor, capacitor, energy storage, photocatalytic activity, and soil amendment) due to their larger surface area, high porosity, presence of appropriate functional groups, etc. This chapter extensively has deliberated the preparation, characteristics, modification or activation, and application of biochar and biochar-based nanocomposites (particularly for energy, environmental, and agricultural applications). Biochar physical activation methods are one of the promising approaches to address current environmental concerns as a powerful remedial agent. Carbon-rich material helps to mitigate carbon dioxide while maintaining a sustainable environment. The development of biochar-based sensors is still in progress, and a suitable relationship between biochar and electrochemical sensing applications has yet to be established. In the preparation of biochar, various aspects, such as the characteristics of the feedstock, the production process, the temperature, the residence time, the yield of the produced char, the elemental composition, the surface area, the porosity, have been profoundly explored. And, the preparation/conversion of biochar/nano-biochar has been involved in different mechanical processes that are discussed with examples. Apart from these, biochar alteration has become a trend to improve physicochemical and mechanical properties. The use of chemical or physical modification techniques has notably increased. Techniques such as traditional gas modifications (using steam, CO_2, ozone, and air), thermal modifications (conventional and microwave), and other recent methods, including ultrasonic modification, plasma modification, and electrochemical modification, have been described to improve the prospective application, as well as the opportunity to explore the material with intense analysis. These modifications have represented the improvement in the size and volume of the pores, the surface area, etc., of the synthesized nano-biochar. All the essential segments of nano-biochar have been comprehensively discussed with suitable examples and their crucial viewpoints on the topic.

Acknowledgements The authors wish to express their gratitude to the Department of Green Energy Technology and Department of Ecology and Environmental Sciences, Pondicherry University, for the facilities and support.

References

Ahmad M, Lee SS, Dou X, Mohan D, Sung JK, Yang JE, Ok YS (2012) Effects of pyrolysis temperature on soybean stover-and peanut shell-derived biochar properties and TCE adsorption in water. Biores Technol 118:536–544

Alam MS, Gorman-Lewis D, Chen N, Safari S, Baek K, Konhauser KO, Alessi DS (2018) Mechanisms of the removal of U (VI) from aqueous solution using biochar: a combined spectroscopic and modeling approach. Environ Sci Technol 52(22):13057–13067

Alvarez J, Lopez G, Amutio M, Bilbao J, Olazar M (2015) Physical activation of rice husk pyrolysis char for the production of high surface area activated carbons. Ind Eng Chem Res 54(29):7241–7250

Al-Wabel MI, Al-Omran A, El-Naggar AH, Nadeem M, Usman AR (2013) Pyrolysis temperature induced changes in characteristics and chemical composition of biochar produced from conocarpus wastes. Biores Technol 131:374–379

Ampah JD, Yusuf AA, Agyekum EB, Afrane S, Jin C, Liu H, Kamel S (2022) Progress and recent trends in the application of nanoparticles as low carbon fuel additives—a state of the art review. Nanomaterials 12(9):1515

Antón-Herrero R, García-Delgado C, Alonso-Izquierdo M, García-Rodríguez G, Cuevas J, Eymar E (2018) Comparative adsorption of tetracyclines on biochars and stevensite: looking for the most effective adsorbent. Appl Clay Sci 160:162–172

Ardebili SMS, Taghipoor A, Solmaz H, Mostafaei M (2020) The effect of nano-biochar on the performance and emissions of a diesel engine fueled with fusel oil-diesel fuel. Fuel 268:117356

Bakraoui M, Karouach F, Ouhammou B, Aggour M, Essamri A, El Bari H (2020) Biogas production from recycled paper mill wastewater by UASB digester: optimal and mesophilic conditions. Biotechnol Rep 25:e00402

Bartoli M, Arrigo R, Malucelli G, Tagliaferro A, Duraccio D (2022) Recent advances in biochar polymer composites. Polymers 14 (12):2506

Bezerra CW, Zhang L, Liu H, Lee K, Marques AL, Marques EP, Zhang J (2007) A review of heat-treatment effects on activity and stability of PEM fuel cell catalysts for oxygen reduction reaction. J Power Sources 173(2):891–908

Bhattarai RM, Chhetri K, Natarajan S, Saud S, Kim SJ, Mok YS (2022) Activated carbon derived from cherry flower biowaste with a self-doped heteroatom and large specific surface area for supercapacitor and sodium-ion battery applications. Chemosphere 135290

Bian R, Chen D, Liu X, Cui L, Li L, Pan G, Chang A (2013) Biochar soil amendment as a solution to prevent Cd-tainted rice from China: results from a cross-site field experiment. Ecol Eng 58:378–383

Chausali N, Saxena J, Prasad R (2021) Nanobiochar and biochar based nanocomposites: advances and applications. J Agric Food Res 5:100191

Chen W, Meng J, Han X, Lan Y, Zhang W (2019) Past, present, and future of biochar. Biochar 1(1):75–87

Chen WH, Hsu HC, Lu KM, Lee WJ, Lin TC (2011) Thermal pretreatment of wood (Lauan) block by torrefaction and its influence on the properties of the biomass. Energy 36(5):3012–3021

Chen Z, Rao Y, Usman M, Chen H, Białowiec A, Zhang S, Luo G (2021) Anaerobic fermentation of hydrothermal liquefaction wastewater of dewatered sewage sludge for volatile fatty acids production with focuses on the degradation of organic components and microbial community compositions. Sci Total Environ 777:146077

Chi NTL, Anto S, Ahamed TS, Kumar SS, Shanmugam S, Samuel MS, Pugazhendhi A (2021) A review on biochar production techniques and biochar based catalyst for biofuel production from algae. Fuel 287:119411

Chiappero M, Berruti F, Mašek O, Fiore S (2021) Analysis of the influence of activated biochar properties on methane production from anaerobic digestion of waste activated sludge. Biomass Bioenerg 150:106129

Chintala R, Mollinedo J, Schumacher TE, Malo DD, Julson JL (2014) Effect of biochar on chemical properties of acidic soil. Arch Agron Soil Sci 60(3):393–404

Chu G, Zhao J, Chen F, Dong X, Zhou D, Liang N, Steinberg CE (2017) Physi-chemical and sorption properties of biochars prepared from peanut shell using thermal pyrolysis and microwave irradiation. Environ Pollut 227:372–379

Cuong DV, Matsagar BM, Lee M, Hossain MSA, Yamauchi Y, Vithanage M, Hou CH (2021) A critical review on biochar-based engineered hierarchical porous carbon for capacitive charge storage. Renew Sustain Energy Rev 145:111029

Dahal RK, Acharya B, Saha G, Bissessur R, Dutta A, Farooque A (2019) Biochar as a filler in glassfiber reinforced composites: experimental study of thermal and mechanical properties. Compos B Eng 175:107169

Das S, Mohanty S, Sahu G, Rana M, Pilli K (2021) Biochar: a sustainable approach for improving soil health and environment. In: Soil erosion-current challenges and future perspectives in a changing world, pp 121–138

Dey S, Mehta NS (2020) Automobile pollution control using catalysis. Resour Environ Sustain 2:100006

Ding Y, Liu YX, Wu WX, Shi DZ, Yang M, Zhong ZK (2010) Evaluation of biochar effects on nitrogen retention and leaching in multi-layered soil columns. Water Air Soil Pollut 213(1):47–55

Do Minh T, Song J, Deb A, Cha L, Srivastava V, Sillanpää M (2020) Biochar based catalysts for the abatement of emerging pollutants: a review. Chem Eng J 394:124856

Dutta S, Banerjee I, Jaiswal KK (2020) Graphene and graphene-based nanomaterials for biological and environmental applications for sustainability. Octa J Biosci 8(2):106–112

Fan M, Marshall W, Daugaard D, Brown RC (2004) Steam activation of chars produced from oat hulls and corn stover. Biores Technol 93 (1):103–107

Feng D, Zhao Y, Zhang Y, Sun S, Meng S, Guo Y, Huang Y (2016) Effects of K and Ca on reforming of model tar compounds with pyrolysis biochars under H_2O or CO_2. Chem Eng J 306:422–432

Fidel RB, Laird DA, Thompson ML, Lawrinenko M (2017) Characterization and quantification of biochar alkalinity. Chemosphere 167:367–373

Funke A, Ziegler F (2010) Hydrothermal carbonization of biomass: a summary and discussion of chemical mechanisms for process engineering. Biofuels, Bioprod Biorefin 4(2):160–177

Gao Y, Yue Q, Gao B, Li A (2020) Insight into activated carbon from different kinds of chemical activating agents: a review. Sci Total Environ 746:141094

Ghosh S, Barg S, Jeong SM, Ostrikov K (2020) Heteroatom-doped and oxygen-functionalized nanocarbons for high-performance supercapacitors. Adv Energy Mater 10(32):2001239

Gümüş D, Akbal F (2017) A comparative study of ozonation, iron coated zeolite catalyzed ozonation and granular activated carbon catalyzed ozonation of humic acid. Chemosphere 174:218–231

Gupta RK, Dubey M, Kharel P, Gu Z, Fan QH (2015) Biochar activated by oxygen plasma for supercapacitors. J Power Sources 274:1300–1305

Gururani P, Bhatnagar P, Bisht B, Jaiswal KK, Kumar V, Kumar S, Rindin KG (2022) Recent advances and viability in sustainable thermochemical conversion of sludge to bio-fuel production. Fuel 316:123351

Hassan M, Naidu R, Du J, Liu Y, Qi F (2020) Critical review of magnetic biosorbents: their preparation, application, and regeneration for wastewater treatment. Sci Total Environ 702:134893

Huang H, Wang YX, Tang JC, Zhu WY (2014) Properties of maize stalk biochar produced under different pyrolysis temperatures and its sorption capability to naphthalene. Huan Jing ke Xue= Huanjing Kexue 35(5):1884–1890

Huang J, Zimmerman AR, Chen H, Gao B (2020) Ball milled biochar effectively removes sulfamethoxazole and sulfapyridine antibiotics from water and wastewater. Environ Pollut 258:113809

Hue N (2020) Biochar for maintaining soil health. In: Soil health. Springer, Cham, pp 21–46

Husain Z, Raheman AS, Ansari KB, Pandit AB, Khan MS, Qyyum MA, Lam SS (2022) Nano-sized mesoporous biochar derived from biomass pyrolysis as electrochemical energy storage supercapacitor. Mater Sci Energy Technol 5:99–109

Jafri N, Wong WY, Yoon LW, Cheah KH (2021) Pretreated mesocarp fibre biochars as carbon fuel for direct carbon fuel cells. Int J Hydrogen Energy 46(31):16762–16775

Jain A, Balasubramanian R, Srinivasan MP (2016) Hydrothermal conversion of biomass waste to activated carbon with high porosity: a review. Chem Eng J 283:789–805

Jain MS, Paul S, Kalamdhad AS (2019) Utilization of Biochar as an amendment during lignocellulose waste composting: impact on composting physics and realization (probability) amongst physical properties. Process Saf Environ Prot 121:229–238

Jaiswal KK, Banerjee I, Mayookha VP (2021a) Recent trends in the development and diversification of sericulture natural products for innovative and sustainable applications. Bioresour Technol Rep 13:100614

Jaiswal KK, Chowdhury CR, Yadav D, Verma R, Dutta S, Jaiswal KS, SelvaKumar KK (2022) Renewable and sustainable clean energy development and impact on social, economic, and environmental health. Energy Nexus 7:100118

Jaiswal KK, Dutta S, Banerjee I, Pohrmen CB, Kumar V (2021b) Photosynthetic microalgae–based carbon sequestration and generation of biomass in biorefinery approach for renewable biofuels for a cleaner environment. Biomass Convers Biorefinery 1–19. https://doi.org/10.1007/s13399-021-01504-y

Jaiswal KK, Kumar V, Verma R, Verma M, Kumar A, Vlaskin MS, Kim H (2021c) Graphitic bio-char and bio-oil synthesis via hydrothermal carbonization-co-liquefaction of microalgae biomass (oiled/de-oiled) and multiple heavy metals remediations. J Hazard Mater 409:124987

Jaiswal KK, Kumar V, Vlaskin MS, Nanda M, Verma M, Ahmad W, Kim H (2021d) Hydropyrolysis of freshwater macroalgal bloom for bio-oil and biochar production: kinetics and isotherm for removal of multiple heavy metals. Environ Technol Innov 22:101440

Jaiswal KK, Sudhakar S, Ramaswamy AP (2018) 'Graphene'–world's thinnest material for revolutionizing applications. Everyman's Sci 53:219–223

Jimenez-Cordero D, Heras F, Alonso-Morales N, Gilarranz MA, Rodriguez JJ (2015) Ozone as oxidation agent in cyclic activation of biochar. Fuel Process Technol 139:42–48

Jung KW, Hwang MJ, Jeong TU, Ahn KH (2015) A novel approach for preparation of modified-biochar derived from marine macroalgae: dual purpose electro-modification for improvement of surface area and metal impregnation. Biores Technol 191:342–345

Kalargaris I, Tian G, Gu S (2017) The utilisation of oils produced from plastic waste at different pyrolysis temperatures in a DI diesel engine. Energy 131:179–185

Kinney TJ, Masiello CA, Dugan B, Hockaday WC, Dean MR, Zygourakis K, Barnes RT (2012) Hydrologic properties of biochars produced at different temperatures. Biomass Bioenerg 41:34–43

Kleber M, Hockaday W, Nico PS (2015) Characteristics of biochar: macro-molecular properties. In: Biochar for environmental management. Routledge, pp 143–170

Kovalskii AM, Volkov IN, Evdokimenko ND, Tkachenko OP, Leybo DV, Chepkasov IV, Shtansky DV (2022) Hexagonal BN-and BNO-supported Au and Pt nanocatalysts in carbon monoxide oxidation and carbon dioxide hydrogenation reactions. Appl Catal B 303:120891

Kumar A, Bhattacharya T, Hasnain SM, Nayak AK, Hasnain MS (2020) Applications of biomass-derived materials for energy production, conversion, and storage. Mater Sci Energy Technol 3:905–920

Kumar V, Jaiswal KK, Vlaskin MS, Nanda M, Tripathi MK, Gururani P, Joshi HC (2022) Hydrothermal liquefaction of municipal wastewater sludge and nutrient recovery from the aqueous phase. Biofuels 13(5):657–662

Kuppusamy S, Thavamani P, Megharaj M, Venkateswarlu K, Naidu R (2016) Agronomic and remedial benefits and risks of applying biochar to soil: current knowledge and future research directions. Environ Int 87:1–12

Lee JW (ed) (2012) Advanced biofuels and bioproducts. Springer Science & Business Media

Lehmann J (2007) Bio-energy in the black. Front Ecol Environ 5 (7):381–387

Leng L, Huang H, Li H, Li J, Zhou W (2019) Biochar stability assessment methods: a review. Sci Total Environ 647:210–222

Li R, Zhang Y, Deng H, Zhang Z, Wang JJ, Shaheen SM, Du J (2020) Removing tetracycline and Hg(II) with ball-milled magnetic nanobiochar and its potential on polluted irrigation water reclamation. J Hazard Mater 384:121095

Lian F, Huang F, Chen W, Xing B, Zhu L (2011) Sorption of apolar and polar organic contaminants by waste tire rubber and its chars in single-and bi-solute systems. Environ Pollut 159(4):850–857

Liu G, Zheng H, Jiang Z, Zhao J, Wang Z, Pan B, Xing B (2018) Formation and physicochemical characteristics of nano biochar: insight into chemical and colloidal stability. Environ Sci Technol 52 (18):10369–10379

Liu M, Almatrafi E, Zhang Y, Xu P, Song B, Zhou C, Zhu Y (2022) A critical review of biochar-based materials for the remediation of heavy metal contaminated environment: applications and practical evaluations. Sci Total Environ 806:150531

Liu WJ, Jiang H, Yu HQ (2015) Development of biochar-based functional materials: toward a sustainable platform carbon material. Chem Rev 115(22):12251–12285

Liu WJ, Jiang H, Yu HQ (2019) Emerging applications of biochar-based materials for energy storage and conversion. Energy Environ Sci 12(6):1751–1779

Lupul I, Yperman J, Carleer R, Gryglewicz G (2015) Tailoring of porous texture of hemp stem-based activated carbon produced by phosphoric acid activation in steam atmosphere. J Porous Mater 22 (1):283–289

Lyu H, Gao B, He F, Zimmerman AR, Ding C, Tang J, Crittenden JC (2018) Experimental and modeling investigations of ball-milled biochar for the removal of aqueous methylene blue. Chem Eng J 335:110–119

Mandal S, Pu S, Adhikari S, Ma H, Kim DH, Bai Y, Hou D (2021) Progress and future prospects in biochar composites: application and reflection in the soil environment. Crit Rev Environ Sci Technol 51(3):219–271

Mariana M, Abdul Khalil HPS, Mistar EM, Yahya EB, Alfatah T, Danish M, Amayreh M (2021) Recent advances in activated carbon modification techniques for enhanced heavy metal adsorption. J Water Process Eng 43:102221

Mayakaduwa SS, Kumarathilaka P, Herath I, Ahmad M, Al-Wabel M, Ok YS, Vithanage M (2016) Equilibrium and kinetic mechanisms of woody biochar on aqueous glyphosate removal. Chemosphere 144:2516–2521

Mohan D, Pittman CU Jr, Steele PH (2006) Pyrolysis of wood/biomass for bio-oil: a critical review. Energy Fuels 20(3):848–889

Naghdi M, Taheran M, Pulicharla R, Rouissi T, Brar SK, Verma M, Surampalli RY (2019) Pine-wood derived nanobiochar for removal of carbamazepine from aqueous media: adsorption behavior and influential parameters. Arab J Chem 12(8):5292–5301

Narzari R, Bordoloi N, Sarma B, Gogoi L, Gogoi N, Borkotoki B, Kataki R (2017) Fabrication of biochars obtained from valorization of biowaste and evaluation of its physicochemical properties. Biores Technol 242:324–328

Nielsen TD, Hasselbalch J, Holmberg K, Stripple J (2020) Politics and the plastic crisis: a review throughout the plastic life cycle. Wiley Interdiscip Rev Energy Environ 9(1):e360

Niu Q, Luo J, Xia Y, Sun S, Chen Q (2017) Surface modification of bio-char by dielectric barrier discharge plasma for Hg^0 removal. Fuel Process Technol 156:310–316

Oleszczuk P, Ćwikła-Bundyra W, Bogusz A, Skwarek E, Ok YS (2016) Characterization of nanoparticles of biochars from different biomass. J Anal Appl Pyrol 121:165–172

Palniandy LK, Yoon LW, Wong WY, Yong ST, Pang MM (2019) Application of biochar derived from different types of biomass and treatment methods as a fuel source for direct carbon fuel cells. Energies 12(13):2477

Panahi HKS, Dehhaghi M, Ok YS, Nizami AS, Khoshnevisan B, Mussatto SI, Lam SS (2020) A comprehensive review of engineered biochar: production, characteristics, and environmental applications. J Clean Prod 270:122462

Panwar NL, Pawar A (2022) Influence of activation conditions on the physicochemical properties of activated biochar: a review. Biomass Convers Biorefinery 12:925–947

Py X, Guillot A, Cagnon B (2003) Activated carbon porosity tailoring by cyclic sorption/decomposition of molecular oxygen. Carbon 41 (8):1533–1543

Qian L, Zhang W, Yan J, Han L, Gao W, Liu R, Chen M (2016) Effective removal of heavy metal by biochar colloids under different pyrolysis temperatures. Biores Technol 206:217–224

Qin Y, Li G, Gao Y, Zhang L, Ok YS, An T (2018) Persistent free radicals in carbon-based materials on transformation of refractory organic contaminants (ROCs) in water: a critical review. Water Res 137:130–143

Ramanayaka S, Vithanage M, Alessi DS, Liu WJ, Jayasundera AC, Ok YS (2020) Nanobiochar: production, properties, and multifunctional applications. Environ Sci Nano 7(11):3279–3302

Rodriguez-Narvaez OM, Peralta-Hernandez JM, Goonetilleke A, Bandala ER (2019) Biochar-supported nanomaterials for environmental applications. J Ind Eng Chem 78:21–33

Rui B, Yang M, Zhang L, Jia Y, Shi Y, Histed R, Fan L (2020) Reduced graphene oxide-modified biochar electrodes via electrophoretic deposition with high rate capability for supercapacitors. J Appl Electrochem 50(4):407–420

Saha D, Sinha A, Roy B (2020) A critical review of emission and performance characteristics of CI engine using bio-additives. Clean Technol Environ Policy 22(8):1613–1638

Sajjadi B, Chen WY, Egiebor NO (2019) A comprehensive review on physical activation of biochar for energy and environmental applications. Rev Chem Eng 35(6):735–776

Sangani MF, Abrishamkesh S, Owens G (2020) Physicochemical characteristics of biochars can be beneficially manipulated using post-pyrolyzed particle size modification. Bioresour Technol 306:123157

Saxena M, Maity S, Sarkar S (2014) Carbon nanoparticles in 'biochar' boost wheat (Triticum aestivum) plant growth. RSC Adv 4 (75):39948–39954

Senthil C, Lee CW (2021) Biomass-derived biochar materials as sustainable energy sources for electrochemical energy storage devices. Renew Sustain Energy Rev 137:110464

Seow YX, Tan YH, Mubarak NM, Kansedo J, Khalid M, Ibrahim ML, Ghasemi M (2021) A review on biochar production from different biomass wastes by recent carbonization technologies and its sustainable applications. J Environ Chem Eng 10:107017

Sharma N, Jaiswal KK, Kumar V, Vlaskin MS, Nanda M, Rautela I, Ahmad W (2021) Effect of catalyst and temperature on the quality and productivity of HTL bio-oil from microalgae: a review. Renew Energy 174:810–822

Shebert GL, Zamani S, Yi C, Joo YL (2020) Polysulfide entrapment and retardation in gel electrolyte Li–S batteries: experiments and modeling. J Mater Chem A 8(8):4341–4353

Siddiqui MTH, Baloch HA, Nizamuddin S, Haris M, Mubarak NM, Khalid M, Srinivasan M (2020) Synthesis of novel magnetic carbon nano-composite from waste biomass: a comparative study of industrially adoptable hydro/solvothermal co-precipitation route. J Environ Chem Eng 8(2):103519

Silber A, Levkovitch I, Graber ER (2010) pH-dependent mineral release and surface properties of corn straw biochar: agronomic implications. Environ Sci Technol 44(24):9318–9323

Singh E, Kumar A, Mishra R, You S, Singh L, Kumar S, Kumar R (2021) Pyrolysis of waste biomass and plastics for production of biochar and its use for removal of heavy metals from aqueous solution. Bioresour Technol 320:124278

Smith M (2011) Evolution of acidic functional groups on biochars by ozone oxidation to improve performance as a soil amendment. Doctoral dissertation, Washington State University

Sohi SP (2012) Carbon storage with benefits. Science 338(6110):1034–1035

Solyali D, Safaei B, Zargar O, Aytac G (2022) A comprehensive state-of-the-art review of electrochemical battery storage systems for power grids. Int J Energy Res 46:17786-17812

Song B, Chen M, Zhao L, Qiu H, Cao X (2019) Physicochemical property and colloidal stability of micron-and nano-particle biochar derived from a variety of feedstock sources. Sci Total Environ 661:685–695

Sun J, He F, Pan Y, Zhang Z (2017) Effects of pyrolysis temperature and residence time on physicochemical properties of different biochar types. Acta Agric Scand Sect B—Soil Plant Sci 67(1):12–22

Tan XF, Liu YG, Gu YL, Xu Y, Zeng GM, Hu XJ, Li J (2016) Biochar-based nano-composites for the decontamination of wastewater: a review. Bioresour Technol 212:318–333

Tan Z, Wang Y, Zhang L, Huang Q (2017) Study of the mechanism of remediation of Cd-contaminated soil by novel biochars. Environ Sci Pollut Res 24(32):24844–24855

Tomczyk A, Sokołowska Z, Boguta P (2020) Biomass type effect on biochar surface characteristic and adsorption capacity relative to silver and copper. Fuel 278:118168

Uzoma KC, Inoue M, Andry H, Fujimaki H, Zahoor A, Nishihara E (2011) Effect of cow manure biochar on maize productivity under sandy soil condition. Soil Use Manag 27(2):205–212

Vaccari FP, Baronti S, Lugato E, Genesio L, Castaldi S, Fornasier F, Miglietta F (2011) Biochar as a strategy to sequester carbon and increase yield in durum wheat. Eur J Agron 34(4):231–238

Venegas A, Rigol A, Vidal M (2016) Changes in heavy metal extractability from contaminated soils remediated with organic waste or biochar. Geoderma 279:132–140

Vijayaraghavan K (2019) Recent advancements in biochar preparation, feedstocks, modification, characterization and future applications. Environ Technol Rev 8(1):47–64

Wallace CA, Afzal MT, Saha GC (2019) Effect of feedstock and microwave pyrolysis temperature on physio-chemical and nano-scale mechanical properties of biochar. Bioresour Bioprocess 6(1):1–11

Wang B, Gao B, Fang J (2017) Recent advances in engineered biochar productions and applications. Crit Rev Environ Sci Technol 47 (22):2158–2207

Wang H, Yan S, Ren T, Yuan Y, Kuang G, Wang B, Liu G (2020) Novel environmental factors affecting microbial responses and physicochemical properties by sequentially applied biochar in black soil. Environ Sci Pollut Res 27(30):37432–37443

Wang T, Li G, Yang K, Zhang X, Wang K, Cai J, Zheng J (2021) Enhanced ammonium removal on biochar from a new forestry waste by ultrasonic activation: characteristics, mechanisms and evaluation. Sci Total Environ 778:146295

Weidong Z, Jiangwei Y, Xiaoyin Z, Xiaolong Q (2020) Review on the progress of the first-generation biodiesel hydrogenation and upgrading. Energy Sour Part A: Recovery, Util Environ Effects 42(21):2704–2714

Yaashikaa PR, Kumar PS, Varjani S, Saravanan A (2020) A critical review on the biochar production techniques, characterization, stability and applications for circular bioeconomy. Biotechnol Rep 28:e00570

Yang Q, Deng N, Zhao Y, Gao L, Cheng B, Kang W (2022) A review on 1D materials for all-solid-state lithium-ion batteries and all-solid-state lithium-sulfur batteries. Chem Eng J 451:138532

Yao L, He L, Yang Y, Zhang Y, Liu Z, Liang L, Piao Y (2021) Nanobiochar paper based electrochemical immunosensor for fast and ultrasensitive detection of microcystin-LR. Sci Total Environ 750:141692

Yi P, Pignatello JJ, Uchimiya M, White JC (2015) Heteroaggregation of cerium oxide nanoparticles and nanoparticles of pyrolyzed biomass. Environ Sci Technol 49(22):13294–13303

Yuan JH, Xu RK, Zhang H (2011) The forms of alkalis in the biochar produced from crop residues at different temperatures. Bioresour Technol 102(3):3488–3497

Zailan Z, Tahir M, Jusoh M, Zakaria ZY (2021) A review of sulfonic group bearing porous carbon catalyst for biodiesel production. Renew Energy 175:430–452

Zeghioud H, Fryda L, Djelal H, Assadi A, Kane A (2022) A comprehensive review of biochar in removal of organic pollutants from wastewater: characterization, toxicity, activation/functionalization and influencing treatment factors. J Water Process Eng 47:102801

Zhang M, Gao B, Yao Y, Xue Y, Inyang M (2012) Synthesis, characterization, and environmental implications of graphene-coated biochar. Sci Total Environ 435:567–572

Zhang Q, Li Q, Zhang L, Yu Z, Jing X, Wang Z, Huang W (2017) Experimental study on co-pyrolysis and gasification of biomass with deoiled asphalt. Energy 134:301–310

Zhang YP, Adi VSK, Huang HL, Lin HP, Huang ZH (2019) Adsorption of metal ions with biochars derived from biomass wastes in a fixed column: adsorption isotherm and process simulation. J Ind Eng Chem 76:240–244

Zhao X, Zhou Y, Xu Y, Huang C, Shen Y, Zhang Y, Chen X (2022) Customizing oxygen–containing functional groups for reduced graphene oxide film supercapacitor with high volumetric performance. J Energy Storage 52:104642

Zhao Y, Feng D, Zhang Y, Huang Y, Sun S (2016) Effect of pyrolysis temperature on char structure and chemical speciation of alkali and alkaline earth metallic species in biochar. Fuel Process Technol 141:54–60

Zhou Y, Yan T, Pan W (2021) Study on the sorption mechanism of middle-low temperature sorption thermal storage materials from the microscale simulation: a review. Int J Energy Res 45(7):9719–9752

Zou H, Zhao J, He F, Zhong Z, Huang J, Zheng Y, Gao B (2021) Ball milling biochar iron oxide composites for the removal of chromium (Cr (VI)) from water: Performance and mechanisms. J Hazard Mater 413:125252

Synthesis of Biochar-Based Heterostructures/ Composites and Their Characteristics

Physical Activation and Nanoscale Transformation of Biochar Using Different Mechanochemical Techniques

Mridusmita Sarmah, Arup Borgohain, Jiban Saikia, Diganta Deka, Harisadhan Malakar, Puja Khare, and Tanmoy Karak

Abstract

Biochar is a solid carbonaceous pyrolyzed product prepared from various lignocellulosic biomass (agriculture and crop residues), sewage sludge, waste products obtained from different sources, litters, animal manures, etc., under limited oxygen conditions. In recent times, biochar applications in soil remediation, removal of hazardous organic and inorganic contaminants, wastewater treatment, electrode material, catalysis, etc., have drawn worldwide curiosity because of its unique properties. Instead of having unique properties, biochar application range is still limited due to its heterogeneous surface functionality, low porosity, and less adsorption capacity. The physicochemical properties of carbonaceous materials can be improved through various physical and chemical techniques, and biochar may be a better carbonaceous material due to its inexpensive and easy production process. Among physical modification techniques, steam activation is an environmentally and economically sustainable easy modification technique free from the use of any chemicals. With the advancement of nanotechnology, nanoscale engineering of biochar has captured care from scientific societies due to its nano-sized fabrication with excellent physical, chemical, and surface properties. Various mechanochemical techniques such as ball milling, microwave irradiation, and magnetic modification are vastly used for the nanoscale fabrication of biochar materials. Nanoscale engineering further increases the stability and efficiency of raw biochar material over physical modification techniques. This chapter includes steam activation used for physical activation of biochar as well as three selected mechanochemical techniques, viz. ball milling, microwave irradiation, and magnetic modification used for nanoscale modification of biochar along with their application in various fields. Biochar modification and its fabrication into nanoscale form is not only a novel approach but also provides a more advanced and efficient waste management technique.

Keywords

Ball milling • Magnetic modifications • Microwave irradiation • Modification of biochar • Nano-biochar materials • Steam activation

Abbreviations

AFM	Atomic force microscope
BCMW-β-CD	β-Cyclodextrin polymer coupled rice husk biochar
BMHC	Hickory chip ball-milled biochar
BMHC-H_2O_2	Hickory chip ball-milled biochar modified with 10% H_2O_2 solution
BM-PBC	Ball-milled pristine biochar
CB	Corn straw biochar
CCC	Cation coagulation concentration
CNT	Carbon nanotubes
FB	Fruit tree branch biochar
GLMB	Graphene-like magnetic biochar
MBC	Micro-nano-ball-milled bone biochar

M. Sarmah (✉) · A. Borgohain · D. Deka · T. Karak
Upper Assam Advisory Centre, Tea Research Association, Dikom, 786101, Assam, India
e-mail: mridus.chem@gmail.com

T. Karak
e-mail: tanmay.karak@gmail.com

M. Sarmah · A. Borgohain · J. Saikia
Department of Chemistry, Dibrugarh University, Dibrugarh, 786004, Assam, India

H. Malakar
Tocklai Tea Research Institute, Tea Research Association, Cinnamara, Jorhat, 785008, Assam, India

P. Khare
Crop Production and Protection Division, CSIR-Central, Institute of Medicinal and Aromatic Plants, P.O.-CIMAP, Near Kukrail Picnic Spot, Lucknow, 226015, India

D. Mishra et al. (eds.), *Biochar-Based Nanocomposites for Contaminant Management*, Advances in Science, Technology & Innovation, https://doi.org/10.1007/978-3-031-28873-9_2

MBCNP	Magnetic functionalized bagasse biochar nanoparticles
MBM-HCBC	Magnetic ball-milled hickory chips biochar
MCBC	Magnetic corn biochar
MC-PDA-Ag	Ag nanoparticle loaded on polydopamine-coated magnetic biochar catalyst
PBCNZVI	Magnetic modification of palm BC with nano-zero-valent iron
SMT	Sulfamethazine
β-CD	β-Cyclodextrin polymer

1 Introduction

The growing population and pollution have become global issues and demand for novel, sustainable, and eco-friendly solutions (Luo et al. 2020). During the past several years, biochar has been considered as an eco-friendly and economic remediator in support of various environmental and agricultural issues due to its low cost, porous structure, and shallow large-scale fabrication (Akhil et al. 2021; Boraah et al. 2022; Borgohain et al. 2022; Sarmah et al. 2023). Biochar application has covered multidisciplinary areas such as agricultural and environmental protection (Sarmah et al. 2023), enzyme immobilization, catalysis, energy, sensing, and product development (Akhil et al. 2021; Borgohain et al. 2020). Because of nutrient enrichment, biochar has been immensely utilized to improve soil quality and crop yield (Nigam et al. 2019; Borgohain et al. 2020; Nigam et al. 2021; Vijay et al. 2021). The unique properties of biochar, such as easy availability of feedstock, large surface area, high porosity, low ash content, outstanding sorption capacities, and high amount of surface activity, make it an emerging tool to mitigate the growing organic and inorganic contaminants produced from both natural and anthropogenic activities (Balajii and Niju 2019). Biochar production provides a good engineering technique for properly managing solid waste biomass to use in a productive way (Ahmad et al. 2014).

Biochar (BC) is a porous carbonaceous product prepared from readily available solid organic waste biomass such as crop residues from agriculture, forestry, municipal solid waste, sewage sludge, food processing waste, paper mill waste, litters, and animal manures through pyrolysis with little or unavailable air (Sohi 2012; Varjani et al. 2019; Joseph et al. 2021). Various thermochemical methods, such as fast pyrolysis, slow pyrolysis, hydrothermal carbonization, gasification, and torrefaction, are used to convert unutilized nutrient-rich solid organic biomass to biochar through biomass decomposition (Li et al. 2016). However, the thermochemical conversion of biomass into BC is accompanied by several physicochemical and morphological changes depending on pyrolysis temperature and associated conditions (Kumar et al. 2020). Instead of having unique properties, the use of biochar is inhibited due to its heterogeneous surface functionality. With the upgraded technology of nanoscience, carbon-based nanomaterials are immensely used in agricultural and environmental protection, catalysis, energy, and biomedical engineering (Amusat et al. 2021; Chausali et al. 2021). In recent times, synthesis and application of nano-based carbonaceous materials have attracted both science and industrial sector due to their nano-sized, larger specific surface area, micro-porosity, surface hydrophobicity, more negative zeta potentials, smaller hydrodynamic radius, and presence of a large number of oxygen-containing functional groups that are responsible for their mechanical and thermal stability. The mostly used nano-based carbonaceous materials till date include carbon nanotubes, fullerenes, carbon nanofibers, carbon black, and carbon onions. Nowadays, continuous research effort has been carried out to make an environmentally and economically sustainable nano-based carbonaceous product, and with this concern, biochar is an excellent base material for such production and nanoscale modification (Zaytseva and Neumann 2016). Carbon-rich biochar is an excellent selection for developing nano-based carbonaceous material in such a situation.

In recent times, nanotechnology has been applied to bring some improvement in biochar properties that constructs an attractive tool due to its nano-enabled properties like nanoparticle size, better sorption tendency, higher specific surface area, higher porosity, the higher surface-to-volume ratio, etc. (Ramanayaka et al. 2020; Anupama and Khare 2021; Rajput et al. 2022). Currently, nano-biochar materials are exclusively utilized in soil amendment, enzyme immobilization, catalysis, energy production, wastewater treatment, and environmental remediation due to their significantly high adsorption capacity, high mechanical and thermal stability, more negative zeta potentials, smaller hydrodynamic radius, the higher surface-to-volume ratio, and an additional number of oxygen-containing functional groups (Ramanayaka et al. 2020; Rajput et al. 2022). In sustainable agriculture, the use of nano-biochar material as a soil conditioner not only gives it an ultra-protection from various contaminants but also improves soil health by changing its various physiochemical qualities and microbial activities, which in turn increases plant growth (Awad et al. 2018; Liu et al. 2018; Xiao et al. 2018). Therefore, biochar's nanoscale engineering is essential to remove the problems arising from small size and rich surface functional groups (Lyu et al. 2018b; Li et al. 2020).

Presently, biochar-based nanocomposites have been synthesized by introducing nano-formulation into the biochar matrix to increase its working potential (Shukla et al. 2021). Integrating biochar with nanomaterials (oxides, hydroxides, and magnetic substances) forms biochar-based

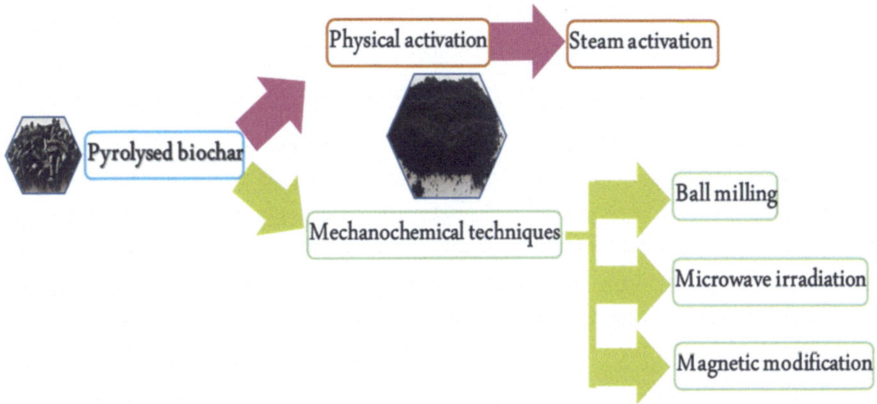

Fig. 1 Modification of pyrolyzed biochar through physical activation and mechanochemical techniques

nanocomposites with more efficient characteristics (Anupama and Khare 2021; Chausali et al. 2021). The significant limitations during biochar-based nanocomposites preparation are as toxicity arising from used chemicals, high production cost, and reduction of surface area due to pore filling of biochar with metal-oxide particles (Anupama and Khare 2021). So recent studies are interested in raw biochar activation and biochar engineering through various mechano chemical techniques such as ball milling, microwave irradiation, and magnetic modifications (Fig. 1) because of their easy and cheap production cost, low risk of chemical toxicity, and environment susceptibility (Amusat et al. 2021). During this processing, biochar size is reduced from micrometer (μm) to nanometer (nm) (Rajput et al. 2022). In this chapter, we have summarized the physical activation of biochar, biochar nanoscale modification using different mechanochemical techniques, their applications in various fields, and the future aspects of this study. Therefore, this chapter will provide information about different mechano chemical techniques most abundantly used for the synthesis of nano-biochar materials and their application efficiency in various fields.

2 Physical Activation of Biochar

In physical activation, biochar is thermally oxidized in the presence of an oxidizing medium such as steam, carbon dioxide, ozone, or air. This activation process increases the surface area by expanding the pore volume. This section will discuss the steam activation technique used to activate the biochar surface physically.

2.1 Steam Activation

Steam activation (Fig. 2) is a suitable physical activation method to improve the surface physicochemical properties of biochar (Amusat et al. 2021). During steam activation, pyrolyzed biochar is activated by mild oxidation caused by hot steam (700–800 °C) almost for 30 min to 3 h and removes unwanted impurities from the biochar surface (Shim et al. 2015; Rajapaksha et al. 2016; Wang et al. 2020). During steam activation, the reaction between water and carbon is thermally feasible due to the small size of a water molecule (Sajjadi et al. 2019). It is an environmentally and economically sustainable activation process due to no use of chemicals, low processing cost, easy operating procedure, and easy removal of volatile matter like bio-oil and tar during activation (Wang et al. 2016a). Steam activation of biochar is a complex reaction. Initially, oxides ions are produced from steam, chemisorbed on active surface carbon of biochar, and oxidized to CO_2 and CO (Pandey et al. 2020; Amusat et al. 2021). Then, the gasification rate is increased either by CO or CO_2 evolved from surface oxides. At the same time, water vapors undergo a water gas shift reaction to produce high CO_2 and H_2, which further activates the biochar surface by enhancing the carbon gasification rate (Lussier et al. 1998; Sattar et al. 2014; Ferreira et al. 2017). The active carbon sites are partly deactivated by produced

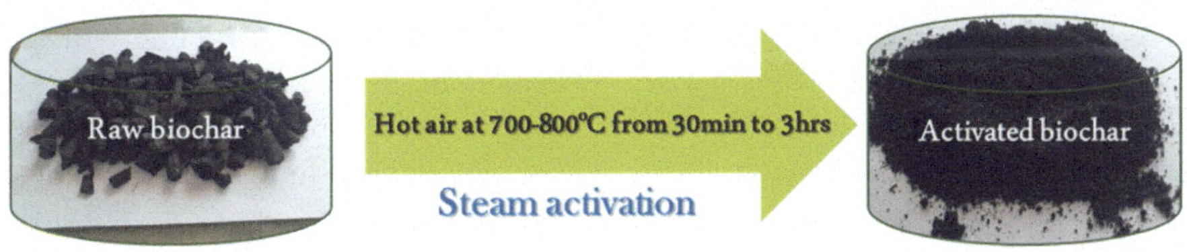

Fig. 2 Schematical illustration for the modification of pyrolyzed biochar by steam activation

H_2 either through dissociative hydrogen adsorption, reverse oxygen exchange, or deactivation of surface oxides (Hermann and Huttinger 1986; Akhil et al. 2021). After activation for several times, surface area of biochar will be increased (Kwak et al. 2019; Amusat et al. 2021). Kwak et al. (2019) explained the increased surface area of wheat straw biochar from 5.8 to 310.2 $m^2\ g^{-1}$, through steam activation due to exposure to a new surface caused by the expansion of pore volume.

Such activation is mainly affected by activation temperature, time, gaseous flow rate, and mass ratio of steam to biochar (Kołtowski et al. 2017). The activation rate was found to be highest between the temperature range 700 and 800 °C, accompanied by modification in the aromatic and surface structure of biochar that further enhanced its cation exchange capacity (Feng et al. 2017). So, the reason for maximum activation at the beginning of the heating process can be explained due to expansion of pores coupled with the increased rate of formation of surface functional groups catalyzed by the steam-alkali metal catalyst, and this further increases the surface area of activated biochar (Ferreira et al. 2017; Wang et al. 2020). But above 800 °C, low molar mass, aromatic compounds, and easily exchangeable metal species are fixed in a well-ordered arrangement that reduces biochar activity (Kołtowski et al. 2016). During steam activation, no significant change in hydrocarbon content is observed due to low residence time. The steam mainly activates the surface carbon content site of biochar due to its low flow rate (Sattar et al. 2014).

The carbon conversion rate increases with temperature, as suggested by Zhai et al. (2015) in their study on the characterization of steam gasification of rice husk biochar. At lower temperatures, activation is almost independent of flow rate and period. Low-temperature steam activation activates the biochar surface by increasing the number of acidic functional groups such as carboxylic, phenolic, and lactonic groups with a subsequent increase in biochar acidity. But at high temperatures, both polarity and acidity of biochar decreased due to the degradation of acidic functional groups as volatile matter and increasing ash content (Shim et al. 2015; Herath et al. 2016; Mayakaduwa et al. 2016; Feng et al. 2017; Bardestani and Kaliaguine 2018). Steam activation of biochar at higher temperatures (between 700 and 800 °C) can be explained by the dilution effect. Initially, steam fill in both the external and internal surface of biochar molecules but the steam at the internal surface gets diluted by carbon gasification (by CO and H_2), leaving behind the pure and concentrated steam on the external surface that causes surface activation (Kwak et al. 2019; Amusat et al. 2021). The long activation period is negatively correlated with biochar sorption activity because pore walls get deformed but positively related to % ash content and surface area (Feng et al. 2017).

Bardestani and Kaliaguine (2018) observed the pH raised to 10.3 during the steam activation of biochar at 900 °C for 60 min. Degradation of functional groups at higher temperatures results in the lowering of carbon, hydrogen, and oxygen content in steam-activated biochar compared to that prepared at lower temperatures (300–500 °C) (Arriagada et al. 1997; Rajapaksha et al. 2016; Kim et al. 2016; Park et al. 2016). Bardestani and Kaliaguine (2018) observed higher fixed carbon and ash content in steam-activated biochar prepared from biomass at 900 °C for 60 min compared to original biochar due to the degradation of volatile components. Biochars produced from cucumber plant and tea waste showed a higher sorption rate (k_2 ~ 1.11–1.57 mg $g^{-1}\ min^{-1}$) toward Sulfamethazine (SMT) after steam activation with a flow rate of 5 mL min^{-1} for 45 min because of higher porosity and adjustable pH of activated biochar as reported by Rajapaksha et al. (2016). Therefore, steam activation of biochar is an environmentally sustainable, cost favorable, and easy activation process to produce a carbon-rich activated substance using cheap biochar materials obtained from easily available environmental waste.

3 Preparation and Application of Nano-Biochar Materials

3.1 Ball Milling

Ball milling (Fig. 3a) is a cost-effective, efficient, and environmentally friendly mechanochemical technique to introduce nanoscale improvement in carbon-based materials like biochar where pyrolyzed biochar is grounded using both agate and stainless-steel balls in a cylinder (Lyu et al. 2017; Amusat et al. 2021). It is applied for reducing the particle size and for modification of the surface structure. During the milling process, biochar microparticles are converted to fine or nanoscale particles under non-equilibrium conditions (Kumar et al. 2020). The ball's weight varies depending on its diameter and the material used. During this milling process, the biochar surface is homogeneously fabricated and the particle size of biochar is reduced by high energy shear forces and rapid extrusion through solid–solid interaction (Zhang et al. 2019a; Li et al. 2020). Ball milling is a sequential mechanochemical process including diffusion, strain generation, and interaction among milling substrate and milling balls to form kinetic energy to produce ultrafine biochar particles by disrupting both physical and chemical interactions between large molecules of milling substrate (Hick et al. 2010). The energy released during the above distortion process forms some new functional groups on the material surface (Kumar et al. 2020). Hence, the ball-milling process results in a solid phase reaction on the biochar

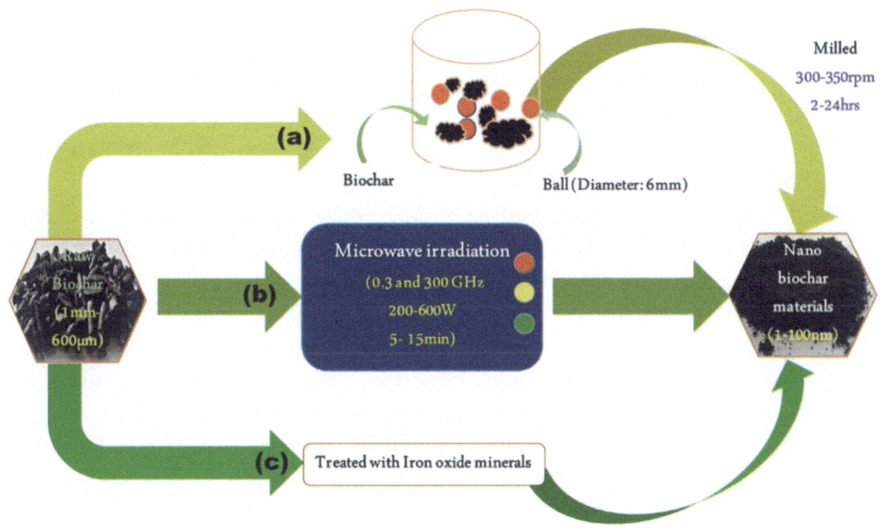

Fig. 3 Schematical illustration for the preparation of nano-biochar materials through various mechanochemical techniques; **a** Ball milling; **b** Microwave irradiation; **c** Magnetic modification

surface using heat and mechanical energy generated during the milling process and introduces –OH bond, C=O bond, phenolic-OH bond, and aromatic C=C bond which confirms the decreases of zeta potential of biochar (Rojas-Chávez et al. 2009; Lyu et al. 2018a, b; Li et al. 2020). Biochar treated with ball milling was found to have higher adsorption capability, increased oxygen-containing functional groups, and improved efficiency in environmental applications compared to their un-milled part (Naghdi et al. 2017; Wang et al. 2019a, b; Kumar et al. 2020). Recently, this technology has been widely used to produce novel engineered nanomaterials by reducing the particle size to the nanoscale (Fan et al. 2016; Wang et al. 2016b; Lyu et al. 2018a, b). The main influencing factors of this technique are milling period, ball size, rotation speed, the mass ratio of the ball, and biochar (Wang et al. 2016b; Lyu et al. 2017; Lyu et al. 2018a, b; Li et al. 2020). Ball milling can be carried out either in the dry phase or in the wet phase (Yuan et al. 2019). Yuan et al. (2019) performed a comparative study of dry milling and wet milling technique in sawdust biochar and milled the biochar for two different periods (2 and 12 h) following a 30-min rest time. Irrespective of milling type, specific surface area of ball-milled biochar increased by 200% compared to un-milled biochar; but, Fourier-transform infrared (FTIR) spectroscopy, scanning electron microscope (SEM) analysis revealed increased functional groups and smaller and better-distributed particles in long-term wet-ball milling (12 h) than that in dry milled biochar.

Recently, the use of ball-milled engineering techniques to improve biochar's physiochemical and sorption performance is a most focused research topic. Lyu et al. (2018a, b) investigated the synthesis of ball-milled pristine biochar (BM-PBC) to reduce the Ni(II) contamination from an aqueous medium via electrostatic interaction, cation-π interaction, and surface complexation. They milled the pyrolyzed biochar with balls (diameter 6 mm) at 300 rpm for 3, 6, 12, and 24 h, respectively, where the biochar to ball-milled ratio was 1:100. Comparatively high adsorption capacity (230–650 mmol kg^{-1}) of ball-milled biochar than that of original pristine biochar (26–110 mmol kg^{-1}) was due to increase in both external and internal surface areas, larger pore volume, and addition of new acidic functional groups on biochar's surface as confirmed by the results of alkaline pH (1.6–6.0), zeta potential, stability, and Boehm titration explained by Lyu et al. (2018a, b). The SEM of BM-PBC analysis showed granular ultrafine particles with a diameter of around 100 nm.

A preparation study of biochar nanoparticles from fruit tree branch biochar, corn straw biochar, and peanut straw biochar using ball milling followed by sonication and centrifugation extraction was carried out by Xu et al. (2020). For the preparation of biochar nanoparticles, they processed the milling operation with the ball: biochar mass ratio 1:20 at a speed of 350 rpm for 2 h. The highest yield and nanoparticle size of peanut straw biochar was found due to less crystallization with the increase in temperature. That showed the high dependency of yield and particle size of nano-biochar on feedstock type and pyrolyzed temperature. The measured Brunauer–Emmett–Teller (BET) surface area of fruit tree branch biochar, corn straw biochar, and peanut straw biochar nanoparticles ranged between 5 and 70 m^2 g^{-1}, and these data agreed with Oleszczuk et al. (2016) who prepared biochar nanoparticles from miscanthus, willow, and wheat straw. The cation coagulation concentration of biochar nanoparticles increases with the alkaline pH of the solution which in turn increases the suspension stability of biochar in Na^+ and Ca^{2+} solution (Xu et al. 2020).

A ball-milled cow bone meal biochar showed excellent adsorption efficiency to reduce Cd(II), Cu(II), and Pb(II) contamination from an aqueous environment (Xiao et al. 2020). Three pyrolyzed cow bone meal biochars obtained at temperatures 300, 450, and 600 °C were ball-milled at a speed of 300 rpm for 12 h with bone biochar to agate spheres ratio of 1:100 when rotation direction was changed after every 3 h and then they were centrifuged at 9000 rpm for 5 min to produce micro-nano-ball-milled bone biochar (MBC). The milling process increased the specific surface area of biochar from 2.76–52.78 $m^2 g^{-1}$ (un-milled) to 35.49–313.09 $m^2 g^{-1}$ (milled) with an average pore diameter of 6.46–14.48 nm. The Cd(II), Cu(II), and Pb(II) adsorption capacities of MBC-600 in the aquatic environment were 165.77, 287.58, and 558.88 mg g^{-1}, respectively (T 298 K, pH 5.0) which were 93.91%, 75.56%, and 64.61% higher compared to un-milled bone BC. The MBC showed long-term ($\sim$48 h) stability in aqueous solution due to restricted accumulation of the submicron MBC particles caused by strong repulsive electrical double layer forces inserted by acidic functional groups on the biochar surface (Lyu et al. 2018a; Zhang et al. 2019a). To protect the environment and living beings from adverse effects of methylene blue (organic dye), Zhang et al. (2021) prepared a hickory chip ball-milled biochars (BMHC) with biochar: balls mass ratio 1:100 by operating for 12 h at 300 rpm which was then again modified with 10% H_2O_2 solution (BMHC–H_2O_2). The increase in acidic functional groups on the BMHC–H_2O_2 surface made it an outstanding adsorbent for cationic methylene in aqueous solutions with 1.9–18.1 $m^2 g^{-1}$ specific surface area and 310 mg g^{-1} adsorption efficiency. The adsorption capacity of BMHC–H_2O_2 (310 mg g^{-1}) was 1.7 times and 45.7 times higher than ball-milled hickory chip biochars and simple hickory chip biochars, respectively.

Another ball-milled biochar nanocomposite was prepared with copper oxide nanoparticles and found to be potentially more effective in the removal of anionic dye (reactive red 120) whose large molar mass and complex aromatic structure caused its less decomposition (Zhang et al. 2019a). Ball-milled biochar modified with 10% CuO photocatalyst showed 58% higher adsorption capacity (57.59 mg g^{-1}) than that of simple ball-milled biochar (24.37 mg g^{-1}) due to uniform distribution of CuO nanoparticles in biochar matrix but, decrease in adsorption capacity was monitored with an increase in CuO percentage because of its accumulation in the composites. Zhang et al. (2022) fabricated phosphorus (5% red P)-loaded biochar using ball milling at 700 °C for 12 h and that novel ball-milled biochar was found abundantly applicable in refining soil physiochemical properties, maintaining neutral soil pH, promoting alkaline phosphatase, catalase, and urease activities, decreasing leaching, and reducing Cd and Pb stress on maize plants by adsorbing plant-available Cd and Pb fractions. Ball-milling approach has great potential to structure the biochar material on a nanoscale without using any solvent, and hence, it is a most promising green technique to develop environmentally friendly nano-biochar materials with improved physicochemical and sorptive properties.

3.2 Microwave Irradiation

A microwave is an electromagnetic wave with a frequency between 0.3 and 300 GHz higher than radio waves. In the microwave technique (Fig. 3b), materials are irradiated at a certain frequency when only the polar molecules rotate by absorbing microwave radiation and this further generates heat due to collisions and frictions among rotating polar molecules, ions, and non-polar molecules (Li et al. 2016; Singla et al. 2017; Ahmed et al. 2021). Microwave radiation is uniformly dispersed within the oven chamber when the wave gets reflected, transmitted through, or absorbed (Kumar et al. 2014; Moravec and Mares 2017). During collision, kinetic energy along with vibrational, rotational, and translational energy of valence electron is converted to heat energy as stated by Li et al. (2016). Hence, in microwave irradiation materials are heated through dielectric heating without allowing any direct contact between the hot heating source and the treated materials, and this protects the materials from thermal damage (Huang et al. 2016; Arafat et al. 2017). Several factors affecting microwave techniques are the size of the load, its direction concerning waves, the dielectrics, and the thermal characteristics of the material (Kumar et al. 2014; Ahmed et al. 2021). It is a fast-heating process as compared to other conventional heating techniques and microwave processing temperature is inversely correlated with product flow rate through the oven chamber (Li et al. 2016). Several advantages of microwave irradiation are energy and cost efficiency, uniform, non-contact and control heating, high yield, rapid reaction time, easy and fast startup and shutdown, inert gas consumption, etc. (Kumar et al. 2014; Li et al. 2016; Ahmed et al. 2021). Microwave irradiation followed by the oscillation of molecules produces a highly porous structure.

In recent times, advanced microwave irradiation is predominantly used in biochar and its composite preparation coupled with its modification through the increased surface area, pore volume, and uniform distribution of micropore (Arafat et al. 2017; Chu et al. 2017; Gabhane et al. 2020). Wang et al. (2018) explained the application of the microwave irradiation technique in the modification of nitrogen-doped *Phragmites australis* biochars using ammonium chloride and ammonium acetate mixed phosphoric acid as activating agents. After microwave irradiation (at 700 W for 15 min), pore size ranged between 1.5 and 3 nm and surface area increased from the original 27–1074 $m^2 g^{-1}$. At the same time, basicity of biochar increased due to the formation of nitrogen-containing functional groups. After modification, these nitrogen-doped *P. australis* biochars had been proven as a very good adsorbent for Acid Red 18 caused by pore-filling mechanisms, π–π interactions, Lewis's acid–base interaction, electrostatic attraction, and hydrogen bonding (Wang et al. 2018). β-cyclodextrin polymer (β-CD) coupled with rice husk biochar (BCMW-β-CD) was modified in microwave irradiation at 350 W for 15 min to remove estrogen as well as heavy metal impurities from wastewater (Qu et al. 2020). The adsorption capacity of BCMW-β-CD increased from 142.4 to 240.13 mg g^{-1} for plumbum (Pb^{2+}) while that for bisphenol A (estrogen) increased from 72.1 to 209.20 mg g^{-1} because uniform dielectric heating in microwave irradiation built up a highly effective biochar surface. BCMW-β-CD produced using low pyrolyzed biochar showed the highest adsorption for both bisphenol A and Pb^{2+} due to the highest availability of β-CD bound to oxygen-containing surface functional groups because bisphenol A adsorption was due to supramolecular interactions with β-CD accompanied by π–π interaction while that of Pb^{2+} due to its complexation and electrostatic attraction with anionic oxygen-containing functional groups at pH range between 2.3 and 4.0. According to thermodynamic analysis, adsorption of Pb^{2+} and bisphenol A on BCMW-β-CD was feasible with positive and negative values of ΔH^{o}, respectively (Qu et al. 2020).

In another study, lignocellulosic waste (wheat straw, oat hulls, rapeseed cake, and hazelnut hulls) biochars were modified with ferrocene catalyst and microwave irradiation at 80 °C at a power of 200 W for 5 min to form carbon nanotubes (CNT) (Hildago-Oporto et al. 2019). The higher CNT concentration and smaller hydrodynamic diameter in modified biochar were reported in hazelnut hull and wheat straw-derived biochar pyrolyzed at 600 °C because of lower full-width half maximum ($FWHM_G$) of G-band, lower intensity of I_D/I_G caused by non-uniform shapes, and diameter of CNTs. Wu et al. (2021) explained that biochar-hydrogel polymer composite is an excellent and cheap soil amendment for the improvement of soil fertility and water holding capacity and microwave irradiation further enhanced its thermal stability through the crosslinking between cellulose and gelatin also due to the uniform distribution of biochar in a polymer matrix (Sethi et al. 2019). In the study by Bai et al. (2022), microwave-pyrolyzed sugarcane bagasse biochar/ferrites nanocomposite was described as a superb adsorbent in wastewater with an adsorption capacity of 12.8 mg g^{-1}. The reason for high adsorption capacity was explained due to synergistic effects including electrostatic interaction, redox reaction, and complexation. Hence, microwave irradiation is a novel, fast, and highly efficient irradiation technique used abundantly to bring about nanoscale modification in biochar and its composites.

3.3 Magnetic Modifications

Presently magnetic modification of carbon-rich and porous biochar structures is the most attractive and innovative research topic. The negative surface charges of biochar are extremely helpful for the removal of toxic organic contaminants and heavy metals from contaminant water. Wastewater treatment and magnetic modification of biochar (Fig. 3c) with iron-based oxides is a promising way to further enhance its sorption capacity and more selectively for negatively charged oxoanions of heavy metals (As, Cr, etc.) (Zhang et al. 2007; Wang et al. 2015). Magnetic biochar was prepared by pyrolyzing a treated mixture of hematite mineral and biomass when hematite minerals get decompose to give iron oxide nanoparticles and finally biochar in combination with nano-iron oxide particles gives biochar nanocomposites (Zhang et al. 2019b). It is a simple and cost-effective method to modify biochar with iron oxide minerals. Magnetic modification of biochar can be carried out by three processes, viz. by direct pyrolysis of waste biomass and iron oxides mixture, by the distribution of iron oxides particles on biochar surface through precipitation, and final one, by heating treatment of biomass and iron oxides mixture (Thines et al. 2017). To carry out the uniform, fast, and controlled pyrolysis microwave heating technique is most preferably used (Wang et al. 2015). Such modification effectively increases specific surface area, sorption capacity, porosity, pore volume, etc. (Wang et al. 2015).

Coupling of biochar with magnetic nanoparticles is not only an excellent way to improve its sorption capacity but also resolves the difficulties that arise during biochar separation after adsorption (Reddy and Lee 2013). Khan et al. (2020) treated corn waste straw biochar with ferric nitrate followed by sonication and calcination to modify its adsorption capacity for Cd(II). Magnetic corn BC (MCBC) possessed a higher specific surface area (225.90–313.88 m^2 g^{-1}) compared to raw BC (93.76–97.18 m^2 g^{-1}) due to larger pore volume. Iron oxides present in MCBC stimulate the Cd(II) adsorption affinity with an increase in pH (around 6–8) through complexation, precipitation, electrostatic interaction, and ion exchange with anionic functional groups on the MCBC surface but at acidic pH, MCBC showed lower adsorption due to simultaneous competition between hydronium ions and Cd(II) ions for surface adsorption (Zhou et al. 2018). After magnetic modification, adsorption capacity of MCBC increased from 28.2 to 46.90 mg g^{-1} due to an increased surface area, and an increase in the number of O-containing functional groups which accelerated the inner sphere complexation with iron particles (Khan et al. 2020).

Magnetic modification of palm biochar with nano-zerovalent iron (PBCNZVI) not only improves biochar physiochemical properties but also stabilizes NZVI by limiting its self-accumulation (Jiang et al. 2018). PBCNZVI showed lower specific surface area but higher absorbability for glyphosate due to uniform distribution of magnetically active nano-iron oxide particles and also an increase in oxygen-containing functional groups on PBCNZVI surface as compared to original biochar. Almost two times higher adsorption capacity (80.00 mg g^{-1}) of biochar-NZVI for glyphosate at pH 4.0 was influenced by various processes such as electrostatic interactions, hydrogen bonding, complexation, and pore-filing (Jiang et al. 2018). Zubrika et al. (2018) found much higher adsorption capacity [25.6 mg g^{-1} for As(V) and 144.9 mg g^{-1}] and a more positive zeta potential value of magnetic carbon biochar prepared from wheat straw and ferrofluid. They prepared magnetic wheat straw biochar either by microwave irradiation at 900 W for 30 min or by conventional heating. The maximum percentage of maghemite (γ-Fe_2O_3) nanoparticles in wheat straw and ferrofluid biochar was 17.4% and during microwave pyrolysis maghemite nanoparticles changed chemically to form α-Fe, γ-Fe, and Fe_3C. The high sorption capacity of the above microwave pyrolyzed magnetic carbon biochar was explained due to the presence of several iron-bearing composites such as superparamagnetic Fe_2O_3, α-Fe, and non-stoichiometric Fe_3C, γ-Fe_2O_3, and γ-Fe formed from nano-maghemite presented in the ferrofluid. The adsorption of As(V) by magnetic wheat straw biochar was negatively correlated with pH and that for methylene blue was correlated positively.

No doubt that ball-milled biochar has higher adsorption capacity and surface enriched with functional groups still, there are some problems in adsorbent recovery after contaminant removal. But magnetic ball-milled biochar not only possessed comparatively higher sorption capacity but also has easy adsorbent recovery benefits due to its in-built magnetic property (Thines et al. 2017; Lyu et al. 2018a, b; Li et al. 2019a). A ball-milled hickory chip biochar was magnetically modified with Fe_3O_4 with a mass ratio of 1:3 using the ball-milling technique in the air with a speed of 500 rpm for 12 h as explained by Li et al. (2020). Magnetic ball-milled hickory chips biochar (MBM-HCBC) showed 27 times higher adsorption capacity (500.5 mg g^{-1} at alkaline pH) of cationic methylene blue as compared to raw biochar as well as ball-milled biochar due to an increase in an oxygen-containing functional group, aromaticity, and large pore volume. At lower pH, methylene blue removal was restricted due to electrostatic repulsion between cationic dye and surface H^+ ions (Li et al. 2019a). A comparable Langmuir adsorption capacity of MBM–HCBC with graphene, C nanotubes, coal, and zeolite was due to its increased surface area, opened pore volume, oxygen-containing functional groups, and aromatic C=C bonds that insisted π–π and electrostatic interactions. According to Liu et al. (2020), 17β-estradiol adsorption onto graphene-like magnetic biochar (GLMB) in presence of Al_2O_3 nanoparticles and microplastics was influenced by several factors like π–π interactions, micropore filling effects, electrostatic interaction, etc. Lotus seedpod biochar was magnetically modified by using a potassium ferrate activator to produce graphene-like magnetic biochar (GLMB), and its maximum surface roughness of 55.8 nm was observed in atomic force microscope (AFM) analysis. The high specific surface area (828.37 m^2 g^{-1}) and microporous structure of GLMB attractively improved its adsorption capacity for 17β-estradiol. Adsorption kinetics of GLMB was investigated in both the homo-adsorption system and hetero-adsorption system. The GLMB adsorption capacity increased from 48.71 mg g^{-1} (in the homo-adsorption system) to a maximum of 61.60 mg g^{-1} (in the Al_2O_3/GLMB system).

In another study, the use of magnetic functionalized bagasse biochar nanoparticles (MBCNP) for remediation of tyrosinase and then their reuse for bisphenol detection by constructing electrochemical biosensors was explained by He et al. (2019). First bagasse biochar nanoparticles were shaken (with 200 rpm for 6 h) with H_2SO_4 (18 M) and HNO_3 (12 M) mixture to get functionalized biochar nanoparticles which were further converted to MBCNP by treating with $FeCl_3 \cdot 6H_2O$ and $FeSO_4 \cdot 7H_2O$. The BET surface area and pore volume of MBCNP were 526.49 m^2 g^{-1} and 0.259 cm^3 g^{-1}, respectively. The uniform distribution of nano-sized magnetic particles (10–20 nm) over the MBCNP surface was confirmed from SEM analysis. The pollution and toxicity exerted by the bioaccumulation of mercury is a growing environmental problem (Li et al. 2019a; Wang et al. 2019a). Modification of microwave-active cotton straw biochar with mixed Mn–Fe oxides followed by ultrasound treatment to eliminate Hg(0) from flue gas was studied by Shan et al. (2019). Integration of Mn-oxide with Fe-oxide was proved profitable not only for Hg(0) adsorption due to the formation of non-stoichiometric Mn–Fe spinel which enhanced Hg(0) oxidation on Mn^{4+} active sites but also for easy separation of adsorbent from flue gas (Yang et al. 2017). Initially, microwave treatment increased BET surface area from 57.047 to 189.804 m^2 g^{-1} due to increased pore structure formed from the removal of ash and volatile matter (Liu et al. 2018). Then, the magnetic modification followed by ultrasound treatment of microwave cotton straw biochar further enhanced the BET surface area (451.569 m^2 g^{-1}) and pore volume (0.3011 cm^3 g^{-1}) due to the formation of new pores by removing blockages, smoothness of cotton straw BC, and better distribution of Mn and Fe oxides along with optimization of surface structure (Yang et al. 2017; Xu et al. 2018). The optimum Mn/Fe ratio in magnetic cotton straw BC was 3/10 with the highest Hg(0) removing efficiency (87.1%) caused by the presence of active MnO_2 but its excess amount of Mn showed a reverse effect by decreasing the redox reaction rate (Yang et al. 2017). The best optimization of cotton straw biochar with Fe–Mn-oxide was observed at 700 °C calcination temperature and 120 °C reaction temperature.

Furthermore, an Ag nanoparticle loaded on polydopamine-coated magnetic biochar (MC–PDA–Ag) catalyst showed excellent adsorption efficiency in organic dye (methylene blue) contaminations in an aqueous environment (Li et al. 2019b). Here, biochar acted as excellent support for Ag nanoparticles to increase its stability and catalytic activity while a green polymer Polydopamine coated the magnetic biochar catalyst with surface available amine groups and hydroxyl groups and thus acted as a non-toxic reducing agent for Ag^+. Catalytic reduction of methylene blue by MC–PDA–Ag in the presence of $NaBH_4$ achieved around 99.09% caused by π–π interaction and electrostatic interaction. According to Li et al. (2019b), initially, electron donor species (BH_4^-) and electron acceptor species (methylene blue) were absorbed on the catalyst surface followed by the reduction of methylene blue molecules by accepting electrons from the catalytic surface and finally get desorbed. Therefore, green and solvent-free magnetic modification of biochar not only provides an alluring mechanochemical technique to improve biochar physiochemical properties and activities but is also helpful in adsorbent regeneration by fixing magnetic properties, and this further predicts an efficient large-scale application of magnetic nano-biochar materials. Thus, there are immense possibilities and applicability of nano-biochar materials in the present time for the sustainable and eco-friendly management of environmental contaminants.

4 Conclusion and Future Perspectives

Biochar is a carbonaceous nutrient source and its diverse applications have achieved optimum attention because of its unique properties such as large surface area, high porosity, low ash content, and high amount of surface activity, as well as for sustainable management of agricultural waste, animal litter, and sewage sludge in low cost. Biochar has the potential to eliminate many hazardous organic and inorganic contaminants like heavy metals, dyes, drugs, etc., due to its sorption capacity even higher than that of activated carbon. Biochar activities are mainly influenced by feedstock type and pyrolysis conditions but, there are some harmful impacts such as a drop of soil water content, promote soil erosion and salinity, and reduction of soil porosity by enhancing its density from the long-time presence of biochar. Its effectiveness and sustainability can be further improved to maintain global sustainability in terms of the carbon cycle. In this chapter, we have discussed the physical activation of biochar through steam activation as well as nano-biochar fabrication through three selected mechanochemical techniques, viz., ball milling, microwave irradiation, and magnetic modification. These engineered nano-biochar materials have a comparatively higher surface area, pore volume, adsorption capacity, stability, and nanoparticle size. We have elucidated the synthesis of steam-activated BC along with the synthesis of various nano-biochar materials using ball milling, microwave irradiation, and magnetic modification methods for enhancing the properties of biochar and analyzed in detail how every property of biochar changes as compared to unmodified biochar. Ball milling, microwave irradiation, and magnetic modification of biochar are green and cost-effective techniques to fabricate nano-biochar materials with improved qualities. Depending on the method type and application purpose, the increase in specific surface area, pore size, pore fraction, pore volume, functional group formation, adsorption capacity, and catalytic activity of modified biochar will be different. The strong adsorption capacity along with high activity and high transport potential of nano-biochar materials may be a reason for environmental risk to soil organisms and human health rendered by nano-size of biochar particles but due to the easy migration of nutrients, nano-biochar materials can be applied as an organic substituent for inorganic fertilizers, pesticides, and herbicides. Some of the future perspectives of the present area are given below:

- Because of unique properties, nano-biochar nanomaterials demonstrated higher potential as compared to other nanomaterials used for similar applications. Therefore, in such high demanding situation of nano-biochar nanomaterials, it requires urgent field study to quantify the possible environment as well as health-related risk results from the use of these materials.
- At the same time, accurate specification and optimization of synthesis methods relying on various feedstock materials is utmost essential for proper management of both time and material stocks.
- The major drawback found during the production of biochar-based nanomaterials was its low yield. So, future research should be set up for more optimized and efficient technique for the large-scale production of biochar-based nano-material along with the possible environmental risk factors during their application.
- Further optimization of synthesis conditions of nano-BC materials may support the advancement of BC-based nanotechnology.

References

Ahmad M, Rajapaksha AU, Lim JE, Zhang M, Bolan N, Mohan D, Vithanage M, Lee SS, Ok YS (2014) Biochar as a sorbent for contaminant management in soil and water: a review. Chemosphere 99:19–33

Ahmed EEA, Salih MM, Babker AM, Elmahi OM, Eltayeb LB, Waggiallah HA (2021) Microwave irradiation exposure enhancing melanin staining quality and act as time reducing factor in the Masson-Fontana method. J Radiat Res Appl Sci 14(1):105–110

Akhil DK, Lakshmi D, Kartik AK, Vo DN, Jayaseelan A, Panachamoorthy GK (2021) Production, characterization, activation and environmental applications of engineered biochar: a review. Environ Chem Lett 19(6):2261–2297

Amusat SO, Kebede TG, Dube S, Nindi MM (2021) Ball-milling synthesis of biochar and biochar–based nanocomposites and prospects for removal of emerging contaminants: a review. J Water Process Eng 41:101993

Anupama, Khare P (2021) A comprehensive evaluation of inherent properties and applications of nano-biochar prepared from different methods and feedstocks. J Clean Prod 320:28759

Arafat HM, Ganesan P, Jewaratnam J, Chinna K (2017) Optimization of process parameters for microwave pyrolysis of oil palm fiber (OPF) for hydrogen and biochar production. Energy Convers Manag 133:349–362. https://doi.org/10.1016/j.enconman.2016.10.046

Arriagada R, Garcia R, Molina-Sabio M, Rodriguez-Reinoso F (1997) Effect of steam activation on the porosity and chemical nature of activated carbons from Eucalyptus globulus and peach stones. Microporous Mater 8:123–130

Awad YM, Lee SS, Kim KH, Ok YS, Kuzyakov Y (2018) Carbon and nitrogen mineralization and enzyme activities in soil aggregate-size classes: effects of biochar, oyster shells, and polymers. Chemosphere 198:40–48. https://doi.org/10.1016/j.chemosphere.2018.01.034

Bai L, Huang A, Feng J, SuX MW, Ma S, Lin H (2022) One-step microwave pyrolysis synthesis of bagasse biochar/ ferrites nanocomposite and synergistic effect on As(V) adsorption in water. Mater Chem Phys 283:12603

Balajii M, Niju S (2019) Biochar-derived heterogeneous catalysts for biodiesel production. Environ Chem Lett 17:1447–1469

Bardestani R, Kaliaguine S (2018) Steam activation and mild air oxidation of vacuum pyrolysis biochar. Biomass Bioenergy 108:101–112

Boraah N, Chakma S, Kaushal P (2022) Attributes of wood biochar as an efficient adsorbent for remediating heavy metals and emerging contaminants from water: a critical review and bibliometric analysis. J Environ Chem Eng 10(3):107825

Borgohain A, Sarmah M, Konwar MK, Gogoi R, Gogoi BB, Khare P, Paul RK, Handique JG, Malakar H, Deka D, Saikia J, Karak T (2022) Tea pruning litter biochar amendment in soil reduces arsenic, cadmium, and chromium in made tea (Camellia sinensis L.) and tea infusion: a safe drink for tea consumers. Food Chem X 13:100255. https://doi.org/10.1016/j.fochx.2022.100255

Borgohain A, Konwar K, Buragohain D, Varghese S, Dutta AK, Paul RK, Khare P, Karak T (2020) Temperature effect on biochar produced from tea (Camellia sinensis L.) pruning litters: A comprehensive treatise on physico-chemical and statistical approaches. Bioresour Technol 318:124023

Chausali N, Saxena J, Prasad R (2021) Nanobiochar and biochar based nanocomposites: advances and applications. J Agric Food Res 5:100191

Chu G, Zhao J, Chen F, Dong X, Zhou D, Liang N, Wu M, Pan B, Steinberg CEW (2017) Physi-chemical and sorption properties of biochars prepared from peanut shell using thermal pyrolysis and microwave irradiation. Environ Pollut 227:372–379

Fan X, Chang DW, Chen X, BaekJB DL (2016) Functionalized graphene nanoplatelets from ball milling for energy applications. Curr Opin Chem Eng 11:52–58

Feng D, Zhao Y, Zhang Y, Gao J, Sun S (2017) Changes of biochar physiochemical structures during tar H_2O and CO_2 heterogeneous reforming with biochar. Fuel Process Technol 165:72–79

Ferreira SD, Lazzarotto IP, Junges J, Manera C, Godinho M, Osorio E (2017) Steam gasification of biochar derived from elephant grass pyrolysis in a screw reactor. Energy Convers Manag 153:163–174

Gabhane JW, Bhange VP, Patil PD, Bankar ST, Kumar S (2020) Recent trends in biochar production methods and its application as a soil health conditioner: a review. SN Appl Sci 2:1307

He L, Yang Y, Kim J, Yao L, Dong X, Li T, Piao Y (2019) Multi-layered enzyme coating on highly conductive magnetic biochar nanoparticles for bisphenol A sensing in water. Chem Eng J 384:123276

Hick SM, Griebel C, Restrepo DT, Truitt JH, Buker EJ, Bylda C, Blair RG (2010) Mechanocatalysis for biomass-derived chemicals and fuel. Green Chem 12 (3):468–474

Hildago-Oporto P, Navia R, Hunter R, Coronado G, Gonzalez ME (2019) Synthesis of carbon nanotubes using biochar as precursor material under microwave irradiation. J Environ Manage. 244:83–91

Herath I, Kumarathilaka P, Al-Wabel MI, Abduljabbar A, Ahmad M, Usman ARA, Vithanage M (2016) Mechanistic modeling of glyphosate interaction with rice husk derived engineered biochar. MicroporMesopor Mater 225:280–288

Hermann G, Huttinger KJ (1986) Mechanism of water vapour gasification of carbon—a new model. Carbon 24:705–713

Huang YF, Chiueh P-T, Kuan WH, Lo SL (2016) Microwave pyrolysis of lignocellulosic biomass: heating performance and reaction kinetics. Energy 100:137–144

Jiang X, Ouyang Z, Zhang Z, Yang C, Li X, Dang Z, Wu P (2018) Mechanism of glyphosate removal by biochar supported nano-zero-valent iron in aqueous solutions. Colloids Surf A 547:64–72

Joseph S, Cowie AL, Zwieten LV, Bolan N, BudaiA BW, Cayuela ML, Graber ER, Ippolito JA, Kuzyakov Y, Luo Y, Ok YS, Palansooriya KN, Shepherd J, Stephens S, Weng Z, Lehmann J (2021) How biochar works, and when it doesn't: a review of mechanisms controlling soil and plant responses to biochar. GCB Bioenergy 13:1731–1764

Khan ZH, Gao M, Qiu W, Islam MS, Song Z (2020) Mechanisms for cadmium adsorption by magnetic biochar composites in an aqueous solution. Chemosphere 246:125701

Kim BS, Lee HW, Park SH, Baek K (2016) Removal of Cu^{2+} by biochars derived from green macroalgae. Environ Sci Pollu Res Int 23:985–994

Kołtowski M, Hilber I, Bucheli TD, Oleszczuk P (2016) Effect of steam activated biochar application to industrially contaminated soils on bioavailability of polycyclic aromatic hydrocarbons and ecotoxicity of soils. Sci Total Environ 566:1023–1031

Kołtowski M, Charmas B, Skubiszewska-Zięba J, Oleszczuk P (2017) Effect of biochar activation by different methods on toxicity of soil contaminated by industrial activity. Ecotoxicol Environ Saf 136:119–125

Kumar H, Kalkal P, Buch A, Chandanwale SS, Bamanikar S, Jain A (2014) Role of microwaves in rapid processing of tissue for histopathology. Med J Dr D Y Patil Vidyapeeth 7(4):458–462

Kumar M, Xiong X, Wan Z, Sun Y, Tsang DCW, Gupta J, Gao B, Cao X, Tang J, Ok YS (2020) Ball milling as a mechanochemical technology for fabrication of novel biochar nanomaterials. Bioresour Technol 312:123613

Kwak JH, Islam MS, Wang S (2019) Biochar properties and lead(II) adsorption capacity depend on feedstock type, pyrolysis temperature, and steam activation. Chemosphere 231:393–404

Li J, Dai J, Liu G, Zhang H, Gao Z, Fu J, He Y, Huang Y (2016) Biochar from microwave pyrolysis of biomass: a review. Biomass Bioenergy 94:228–244

Li R, Zhang Y, Deng H, Zhang Z, Wang JJ, Shaheen SM, Du J (2019a) Removing tetracycline and Hg(II) with ball-milled magnetic nanobiochar and its potential on polluted irrigation water reclamation. J Hazard Mater 384:121095

Li H, Jiang D, Huang Z, He K, Zeng G, Chen A, Yuan L, Peng M, Huang T, Chen G (2019b) Preparation of silver-nanoparticle-loaded magnetic biochar/poly (dopamine) composite as catalyst for reduction of organic dyes. J Colloid Interface Sci 555:460–469

Li Y, Zimmermanc AR, Hed F, Chene J, Hana L, Chen H, Hug X, Bin G (2020) Solvent-free synthesis of magnetic biochar and activated carbon through ball-mill extrusion with Fe_3O_4 nanoparticles for enhancing adsorption of methylene blue. Sci Total Environ 722:137972

Liu G, Zheng H, Jiang Z (2018) Formation and physicochemical characteristics of nano biochar: insight into chemical and colloidal stability. Environ Sci Technol 52:10369–10379

Liu N, Liu Y, Tan X, Li M, Liu S, Hu X, Zhang P, Dai M, Xu W, Wen J (2020) Synthesis a graphene-like magnetic biochar by potassium ferrate for 17β-estradiol removal: effects of Al_2O_3 nanoparticles and microplastics. Sci Total Environ 715:136723

Luo M, Lin H, He Y, Zhang Y (2020) The influence of corncob-based biochar on remediation of arsenic and cadmium in yellow soil and cinnamon soil. Sci Total Environ 717:13701

Lussier MG, Zhang Z, Miller DJ (1998) Characterizing rate inhibition in steam/hydrogen gasification via analysis of adsorbed hydrogen. Carbon 36(9):1361–1369

Lyu H, Gao B, He F, Ding C, Tang J, Crittenden JC (2017) Ball-milled carbon nanomaterials for energy and environmental applications. ACS Sustain Chem Eng 5:9568–9585

Lyu HH, Gao B, He F, Zimmerman AR, Ding C, Huang H, Tang JC (2018a) Effects of ball milling on the physicochemical and sorptive properties of biochar: experimental observations and governing mechanisms. Environ Pollut 233:54–63

Lyu H, Gao B, He F, Zimmerman AR, Ding C, Tang J, Crittenden JC (2018b) Experimentaland modeling investigations of ball-milled biochar for the removal of aqueous methylene blue. Chem Eng J 335:110–119

Mayakaduwa SS, Herath I, Ok YS, Mohan D, Vithanage M (2016) Insights into aqueous carbofuran removal by modified and non-modified rice husk biochars. Environ Sci Pollu Res 24:22755–22763

Moravec J, Mares J (2017) A simple, time-saving, microwave-assisted periodic acid-Schiff's staining of glycoproteins on 1D electrophoretic gels. Electrophoresis 38(24):3100–3103

Naghdi M, Taheran M, Brar SK, Rouissi T, Verma M, Surampalli RY, Valero JR (2017) A green method for production of nanobiochar by ball milling-optimization and characterization. J Clean Prod 164:1394–1405

Nigam N, Yadav V, Mishra D, Karak T, Khare P (2019) Biochar amendment alters the relation between the Pb distribution and biological activities in soil. Int J Environ Sci Technol 16:8595–8606

Nigam N, Khare P, Ahsan M, Yadav V, Shanker K, Singh RP, Panday V, Das P, Singh A, Yadav R, Tripathi P, Sinam G, Shukla AK, Karak T (2021) Biochar amendment reduced the risk associated with metal uptake and improved metabolite content in medicinally important herbs. Physiol Plant 173(1):321–339

Oleszczuk P, Ćwikła-Bundyra W, Bogusz A, Skwarek E, Ok YS (2016) Characterization of nanoparticles of biochars from different biomass. J Anal Appl Pyrol 121:165–172

Pandey D, Daverey A, Arunachalam K (2020) Biochar: production, properties and emerging role as a support for enzyme immobilization. J Clean Prod 255:120267

Park SH, Cho HJ, Ryu C, Park YK (2016) Removal of copper (II) in aqueous solution using pyrolytic biochars derived from red macroalga Porphyratenera. J Ind Eng Chem 36:314–319

Qu J, Dong M, Wei S, Meng Q, Hu L, Hu Q, Wang L, Han W, Zhang Y (2020) Microwave-assisted one pot synthesis of β-cyclodextrin modified biochar for concurrent removal of Pb(II) and bisphenol a in water. Carbohydr Polym 250:117003

Rajapaksha AU, VithanageM LeeSS, SeoDC TDCW, Ok YS (2016) Steam activation of biochars facilitates kinetics and pH-resilience of sulfamethazine sorption. J Soils Sediments 16:889–895

Rajput VD, Minkina T, Ahmed B, Singh VK, Mandzhieva S, Sushkova S, Bauer T, Verma KK, Shan S, van Hullebusch ED, Wang B (2022) Nano-biochar: a novel solution for sustainable agriculture and environmental remediation. Environ Res 210:112891

Ramanayaka S, Vithanage M, Alessi D, Liu WJ, Jayasundera ACA, Ok YS (2020) Nanobiochar: production, properties, and multifunctional applications. Environ Sci Nano 7:3279–3302

Rojas-Chávez H, Díaz-de la Torre S, Jaramillo-Vigueras D, Plascencia G (2009) PbTemechanosynthesis from PbO and Te. J Alloys Compd 483(1–2):275–278

Reddy DHK, Lee SM (2013) Three-dimensional porous spinel ferrite as an adsorbent for Pb(II) removal from aqueous solutions. Ind Eng Chem Res 52:15789–15800

Sajjadi B, Chen WY, Egiebor NO (2019) A comprehensive review on physical activation of biochar for energy and environmental applications. Rev Chem Eng 35:735–776

Sarmah M, Borgohain A, Gogoi BB, Yeasin Md, Paul RK, Malakar H, Handique JG, Saikia J, Deka D, Khare P, Karak T (2023) Insights into the effects of tea pruning litter biochar on major micronutrients (Cu, Mn, and Zn) pathway from soil to tea plant: an environmental armour. J Hazard Mater 442:129970. https://doi.org/10.1016/j.jhazmat.2022.129970

Sattar A, Leeke GA, Hornung A, Wood J (2014) Steam gasification of rapeseed, wood, sewage sludge and miscanthus biochars for the production of a hydrogen-rich syngas. Biomass Bioenergy 69:276–286

Sethi S, Kaith BS, Kumar SV (2019) Fabrication and characterization of microwave assisted carboxymethyl cellulose-gelatin silver nanoparticles imbibed hydrogel: its evaluation as dye degradation. React Funct Polym 142:134–146

Shan Y, Yang W, Li Y, Liu Y, Pan J (2019) Preparation of microwave-activated magnetic bio-char adsorbent and study on removal of elemental mercury from flue gas. Sci Total Environ 697:134049

Shim T, Yoo J, Ryu C, Park Y, Jung J (2015) Effect of steam activation of biochar produced from a giant miscanthus on copper sorption and toxicity. Bioresour Technol 197:85–90

Shukla P, Giri BS, Mishra RK, Pandey A, Chaturvedi P (2021) Lignocellulosic biomass-based engineered biochar composites: a facile strategy for abatement of emerging pollutants and utilization in industrial applications. Renew Sustain Energy Rev 152:111643

Singla K, Sandhu MSV, Pal RG, Bansal H, Bhullar RK, Preetinder Kaur P (2017) Comparative evaluation of different histoprocessing methods. Int J Health Sci 11(2):28–34.

Sohi SP (2012) Carbon storage with benefits. Science 338:1034–1035

Thines KR, Abdullah EC, Mubarak NM, Ruthiraan M (2017) Synthesis of magnetic biochar from agricultural waste biomass to enhancing route for waste water and polymer application: a review. Renew Sustain Energy Rev 67:257

Varjani S, Kumar G, Rene ER (2019) Developments in biochar application for pesticide remediation: current knowledge and future research directions. J Environ Manag 232:505–513
Vijay V, Shreedhar S, Adlak K, PayyanadS, Sreedharan V, Gopi G, Voort TS van der Malarvizhi P, Yi S, Gebert J, Aravind PV (2021) Review of large-scale biochar field-trials for soil amendment and the observed influences on crop yield variations. Front Energy Res 9:710766. https://doi.org/10.3389/fenrg.2021.710766
Wang S, Gao B, Zimmerman AR, Li Y, Ma L, Harris WG, Migliaccio KW (2015) Removal of arsenic by magnetic biochar prepared from pinewood and natural hematite. Bioresour Technol 175:391-395
Wang F, Zeng X, Wang Y, Yu J, Xu G (2016a) Characterization of coal char gasification with steam in a micro-fluidized bed reaction analyzer. Fuel Process Technol 141:2–8
Wang Z, Wang Y, Liao J, Yang Y, Liu N, Tang J (2016b) Improving the adsorption ability of graphene sheets to uranium through chemical oxidation, electrolysis and ball-milling. J Radioanal Nucl Chem 308:1095–1102
Wang L, Yan W, He C, Wen H, Cai Z, Wang Z, Chen Z, Liu W (2018) Microwave-assisted preparation of nitrogen-doped biochars by ammonium acetate activation for adsorption of acid red 18. Appl Surf Sci 433:222–231
Wang L, Chen L, Tsang DCW, Kua HW, Yang J, Ok YS, Ding S, Hou D, Poon CS (2019a) The roles of biochar as green admixture for sediment-based construction products. Cement Concrete Comp 104:103348
Wang Y, Xu H, Liu Y (2019b) Oxidative absorption of elemental mercury from flue gas using a modified Fenton-like wet scrubbing system. Energy Fuel 33(4):3028–3033
Wang RZ, Huang DL, Liu YG, Zhang C, Lai C, Wang X, Zeng GM, Zhang Q, Gong XM, Xu P (2020) Synergistic removal of copper and tetracycline from aqueous solution by steam-activated bamboo-derived biochar. J Hazard Mater 384:121470
Wu Y, Brickler C, Li S, Chen G (2021) Synthesis of microwave-mediated biochar-hydrogel composites for enhanced water absorbency and nitrogen release. Polym Test 93:106996
Xiao X, Chen B, Chen Z (2018) Insight into multiple and multilevel structures of biochars and their potential environmental applications: a critical review. Environ Sci Technol 52:5027–5047
Xiao J, Hu R, Chen G (2020) Micro-nano-engineered nitrogenous bone biochar developed with a ball-milling technique for high-efficiency removal of aquatic Cd(II), Cu(II) and Pb(II). J Hazard Mater 387:121980
Xu W, Adewuyi YG, Liu Y, Wang Y (2018) Removal of elemental mercury from fluegas using CuOx and CeO_2 modified rice strawchars enhanced by ultrasound. Fuel Process Technol 170:21–31
Xu CY, Li QR, Geng ZC, Hu FN, Zhao SW (2020) Surface properties and suspension stability of low-temperature pyrolyzed biochar nanoparticles: effects of solution chemistry and feedstock sources. Chemosphere 259:127510
Yang W, Liu Y, Wang Q, Pan J (2017) Removal of elemental mercury from flue gas using wheat straw chars modified by Mn-Ce mixed oxides with ultrasonic-assisted impregnation. Chem Eng J 326:169–181
Yuan Y, Zhang N, Hu X (2019) Effects of wet and dry ball milling on the physicochemical properties of sawdust derived-biochar. Instrum Sci Technol 48:287–300
Zaytseva O and Neumann G (2016) Carbon nanomaterials: production, impact on plant development, agricultural and environmental applications. Chem Biol Technol Agric 3:17
Zhai M, Zhang Y, Dong P, Liu P (2015) Characteristics of rice husk char gasification with steam. Fuel 158:42–49
Zhang G, Qu J, Liu H, Cooper AT, Wu R (2007) $CuFe_2O_4$/activated carbon composite: a novel magnetic adsorbent for the removal of acid orange II and catalytic regeneration. Chemosphere 68: 1058–1066
Zhang Q, Wang J, Lyu H, Zhao Q, Jiang L, Liu L (2019a) Ball-milled biochar for galaxolide removal: sorption performance and governing mechanisms. Sci Total Environ 659:1537–1545
Zhang Y, Joel MZ, He Y, Weathersby D, Han F, Rimal G, Tang J, Dai Q (2019b) Synthesis of Fe_2O_3/biochar nanocomposites by microwave method for magnetic energy-storage concentration cells. Mater Lett X(3):100020
Zhang P, Xue B, Jiao L, Meng X, Zhang L, Li B, Sun H (2022) Preparation of ball-milled phosphorus-loaded biochar and its highly effective remediation for Cd- and Pb-contaminated alkaline soil. Sci Total Environ 813:152648
Zhang Y, Zheng Y, Yang Y, Huang J, Zimmerman AR, Chen H, Hu X, Gao B (2021) Mechanisms and adsorption capacities of hydrogen peroxide modified ball milled biochar for the removal of methylene blue from aqueous solutions. Bioresource Technol 337:125432
Zhou Q, Liao B, Lin L, Qiu W, Song Z (2018) Adsorption of Cu(II) and Cd(II) from aqueous solutions by ferromanganese binary oxide–biochar composites. Sci Total Environ 615:115–122
Zubrik A, Matik M, Lovás M, Štefušová K, Danková Z, Hredzák S, Václavíková M, Bendek F, Briančin J, Machala L, Pechoušek J (2018) One-step microwave synthesis of magnetic biochars with sorption properties. Carbon Lett 26:31–42

Biochar-Based Hydrogel Nanocomposites: An Innovative Technique for Contaminant-Free Environment

Arup Borgohain, Madhusmita Baruah, Mridusmita Sarmah, Jiban Saikia, Diganta Deka, Harisadhan Malakar, Puja Khare, and Tanmoy Karak

Abstract

Growing demand for foods for billions of hungry mouths is imitating high-yielding crop production which needs the application of an enormous amount of chemicals. Several human activities in recent years for societal development are also on the front foot. These agricultural activities coupled with non-systematic societal development create serious environmental pollution. Many technologies have been developed by impregnating biochar with hydrogel composites, as a sorbent to remove pollutants from water and soil. Biochar is a stable carbon-dominant material with porous structure, high surface area, developed pore structure, and low cost, which ensure its applicability as a porous absorbent. Agricultural waste biomass can be successfully converted into biochar by pyrolysis or thermochemical conversion technique. On the other hand, hydrogels are three-dimensional hydrophilic polymer molecules that can accommodate a significant fraction of fluids in their swelling form. Due to its unique bio-degradability, biocompatibility, and water retention properties, hydrogels are used as a carrier of biochar-based compounds to remediate environmental pollution. The formation of the biochar hydrogel composite systems could synergistically improve the inherent property of the sorbent system for the removal of contaminants. The most beneficial use of hydrogel-biochar nanocomposites is to remove the organic contaminant and toxic elements including chromium, arsenic, mercury, and lead in the aquatic environment via simultaneous adsorption and degradation. Most importantly, these composites contribute to waste management, climate change mitigation, and improving soil fertility. Water contaminants like phenol, dyes, and pharmaceutical hazardous compounds are effectively degraded by biochar-supported photocatalysts. In this chapter, an overview of current progress in the preparation of biochar-loaded hydrogel composites system is provided along with the potential application for removing the unwanted environmental threat. Herein, the chapter begins with a fundamental environmental contaminant and their effect on social ecosystem. We provide the basic information regarding practical information of biochar-based nanocomposites in the treatment of environmental contaminants.

Keywords

Biodegradability • Bio-compatibility • Environmental contaminant • Heavy metal • Porous structure • Surface area • Water treatment

A. Borgohain (✉) · M. Sarmah · D. Deka · T. Karak
Upper Assam Advisory Centre, Tea Research Association, Dikom, Dibrugarh, 786101, Assam, India
e-mail: arupgohain02@gmail.com

T. Karak
e-mail: tanmay.karak@gmail.com

A. Borgohain · M. Baruah · M. Sarmah · J. Saikia
Department of Chemistry, Dibrugarh University, Dibrugarh, 786004, India

H. Malakar
Tocklai Tea Research Institute, Tea Research Association, Cinnamara, Jorhat, 785008, Assam, India

P. Khare
Crop Production and Protection Division, CSIR-Central, Institute of Medicinal and Aromatic Plants, P.O.-CIMAP, Near Kukrail Picnic Spot, Lucknow, 226015, India

1 Introduction

Nowadays, environmental pollution is one of the most concerning foci for civilization. Due to numerous human activities, the cultivated soil, surface water, and groundwater are continuously polluted by metals and metalloids. The pollutants released from several sources applied adverse effects on the ecosystem as well as the living body (Islam et al. 2018; Mishra et al. 2019). Different organic and inorganic pollutants such as heavy metals, persistent organic

D. Mishra et al. (eds.), *Biochar-Based Nanocomposites for Contaminant Management*, Advances in Science, Technology & Innovation, https://doi.org/10.1007/978-3-031-28873-9_3

pollutants, synthetic dyes, trace elements (e.g., As, Cd, Cr), and some other heavy metals can enter the food stream and show a harmful effect on the living body. Usually, water-soluble toxic elements run through rainwater to the surface water such as sea, river, and lake water and also polluted the groundwater through leaching (Nieder et al. 2018; Hojjati-Najafabadi et al. 2022). Untreated wastewater released from several industrial or domestic activities carried a high concentration of inorganic pollutants to the surface water and the soil, and ultimately to the groundwater. The bio-accumulated toxic element shows different adverse effects on human health and the ecosystem. Plants also accumulate heavy metals in their whole parts in different concentrations, and they can also enter the food chain of the animal as well as humans. To minimize the hazardous effect of toxic trace elements, several techniques have been developed by researchers. Borgohain et al. (2022) used tea pruning litter biochar in tea-growing soil to reduce the concentration of As, Cd, and Cr in successive tea infusions. Bhatnagar and Sillanpää (2009) mentioned the application of chitin and chitosan derivatives as effective biosorbents with high content of amino and hydroxyl functional groups that resulted in significant adsorption potential for the removal of different aquatic pollutants. To remove methyl orange from aqueous solutions, Yao et al. (2017) applied magnesium/aluminum-layered double hydroxide (Mg/Al-LDH) modified by glycerin and its calcined product at 500 °C. Moreover, clay mineral materials, carbon-based nanomaterials, metal–organic framework materials, zero-valent iron materials, etc., can also be applied as adsorbents to remove various types of pollutants from aqueous environment (Zhang et al. 2020a, b). Among all applications of biochar-based hydrogel nanocomposite (BHNCs), it can be considered one of the best adsorbers due to its large surface area, functional groups, high micro-porosity, active sites of surface, catalytic degradation, and abundance in the environment, as well as outstanding adsorption capability of a wide variety of contaminants (Tan et al. 2016). The efficiency of adsorption and mechanical characteristics such as thermal stability, anti-swelling property, and effective adsorption of the BHNCs can be enhanced by the amendment of biochar in the hydrogel (Liu et al. 2014; Elsaeed et al. 2021).

Biochar is a super adsorber that adsorbs different pollutants due to its unique physical and chemical properties. This carbon-rich compound is prepared from different biomass through the pyrolysis process at high temperatures in absence of air or the presence of a trace amount of air. Feedstocks for the preparation of biochar can be used, like chips and pellets of wood, tree cuttings, bagasse, rice husks, crop residues, sewage sludge, poultry or cattle manure, and bones from slaughterhouses and the waste produced during the production of the distillery, edible oil, canning industries, etc. (Borgohain et al. 2020). Biochar has several uses in agriculture, industrials, and water purification. The by-products produced during the production of biochar are biogas and bio-oil is also a good source of energy since these can be used for different purposes (Reza et al. 2019). Due to its excellent porous structure, the researcher has been showing interest in different studies such as water, soil, and air purification studies, and nutrient slow-release studies. Wu et al. (2022) developed tough porous hydrogel nanocomposite by interacting biochar nanoparticle with polyacrylamide hydrogel matrix. The tensile strength of BHNC was 128 kPa, with a stretch ability of 5.9 and a toughness of 538 J m^{-2} that was applied to adsorbed methylene blue. To remove ciprofloxacin from aqueous environment, hydrogel beads were prepared by introducing chitosan with biochar particles. These hydrogel beads were become very efficient to remove ciprofloxacin (Afzal et al. 2018). To remove phenol from aqueous environments, Karakoyun et al. (2011) prepared different nanocomposites with p(acrylamide)-chicken biochar, p(acrylamide)-wood biochar, and p (acrylamide)-tire biochar with common cross-linker and initiator, i.e., N,N′-methylene bisacrylamide and ammonium persulfate, respectively. The composites were effective adsorber of phenol from aqueous environments. Ohemeng Boahen and co-workers (2019) prepared an alginate-kelp biochar composite hydrogel bead by physical cross-linking with Ca^{2+}, and it was applied for removal of dye from aquatic environment. Biochar prepared from rice husk was introduced into poly(acrylamide) with a cross-linker N,N′-methylene bisacrylamide. These beads were utilized for the removal of As from polluted water (Sanyang et al. 2014). To achieve contaminant-free water as well as soil environment by application of BHNC, several studies have been going on. In this chapter, we are discussing the application of biochar and preparation of nanocomposite from biochar and hydrogel for a contaminant-free environment.

Hydrogels are hydrophilic polymeric compounds that constitute a group of monomers linked with three-dimensional cross-linked networks. Due to the hydrophilic nature of hydrogel, it can hold large amounts of water in its three-dimensional structure and make the hydrogel high serviceability (Ahmed 2015; Sinha and Chakma 2019; Zainal et al. 2020). Recently, hydrogels get huge importance in different research areas due to their special characteristics such as flexibility based on swelling kinetics, changeable surface area, functionality, fast dispersal in liquid solutions, modifiable porous system, catalysis amiability, thermal stability, interesting range of pH, etc. (Sinha and Chakma 2019). The hydrophilic 3D network of hydrogel is formed by cross-linking either physically, chemically, or via polymerization in definite reaction conditions. This cross-linking between the monomers enhanced molecular weight and improved mechanical strength and stability, as well as polymer elasticity, viscosity, and polymer insolubility of the synthesized hydrogel, and it may depend upon the functional group and density of the cross-linking network present on it (Utech and Boccaccini 2015). Due to the capillary and surface tension, a significant amount of water gets absorbed by the hydrogel and gets slow release from its 3D networks. The starting chemical compound for the preparation of a hydrogel can be used gum arabic, polyethylene glycol, hydroxyethylcellulose, carboxymethyl cellulose, hydroxypropyl methylcellulose, hydroxypropyl cellulose, starch, polyvinyl alcohol, sterculia gum, polyacrylamide, and chitosan (Zainal et al. 2020). Hydrogel has the water absorption property along with aquatic contaminants due to its three-dimensional cross-link polymer structure. Likewise, it had been used in agriculture for its excellent water and nutrient-holding capacity (Nassaj-Bokharaei et al. 2021; Wu et al. 2021). Along with other improvements in the surface property as mentioned previously, cross-linking hydrogel (peel-hydrogel) with cation (such as Ca^{2+} ion) and fused with Fe-tartaric acid also provides mechanical strength to the nanocomposite as well as water swelling capacity up to 1500%, maximum adsorption capacities of Cr (VI) ions (825.97 mg g^{-1}) and Pb(II) ions (913.88 mg g^{-1}) (Mahmoud and Mohamed 2020). Therefore, incorporating nanoparticles in a polymeric network of hydrogel can be considered an advanced technique for purification of the environment, and as a nutrient supplier in agriculture (Nassaj-Bokharaei et al. 2021).

A novel nanocomposite was prepared from sodium alginate poly(acrylic acid) and ZnO for the kinetic, isotherm, and thermodynamic study to evaluate the sequestration of methylthioninium chloride dye, and the conspicuous outcome of the study proved that it was an excellent absorber of methylthioninium chloride dye (Makhado et al. 2020). By incorporating magnetic Fe_3O_4 nanoparticles and coating with poly(ethylene glycol) in aminated graphene oxide, He et al. (2020) prepared a novel hydrogel-based nanocomposite. The composite was applied for the removal of Ca^{2+} and Mg^{2+} from oilfield wastewater and the adsorption percentage of the respective cation resulted in 69.8% and 61.1% at 10 min. Moharrami and Motamedi (2020) prepared a hydrogel-based nanocomposite using starch-grafted copolymers of 2-acrylamide-2methyl propane sulfonate and acrylic acid hydrogel with magnetite-functionalized cellulose nanocrystals. The hydrogel-based nanocomposite was applied to remove cationic dyes crystal violet and methylene blue from the aquatic environment. As a result, it was found that the adsorption capacities for crystal violet and methylene blue were 2500.0 mg g^{-1} and 1428.6 mg g^{-1}, respectively from the aquatic environment. Nowadays, purification of contaminated water is becoming one of the major issues for research globally. Therefore, researchers are searching, designing, and developing advanced and particularly low-cost adsorption polymeric matrices with efficient treatment materials for the removal of toxic organic and inorganic pollutants, especially from wastewater. Better performance of BHNC gained attention in recent research field due to their broad-spectrum applications in the field of agriculture, environment, catalysis, energy, and equally important in the field of biomedical. Moreover, a number of BHNCs have been reported as an absorbent of inorganic and organic pollutants for the purification of water. Therefore, several types of research have been done on BHNC for its efficient absorption properties. In this chapter, we are trying to evaluate the utility of BHNC. The main objectives of this book chapter are:

(1) To provide accumulated information for future research on biochar-based nanocomposite,
(2) How does biochar-based hydrogel nanocomposites were being applied for removing different hazardous materials from nature?
(3) To establish the advantages of biochar-based nanocomposites in different fields of research and their application, and
(4) Discussion of some preparation techniques of hydrogels and biochar-based nanocomposites.

2 Types of Contaminants

There are three types of contaminants present in the environment, i.e., biological, physical, and chemical. In this chapter, we are focusing on chemical contaminants with sub-branching to organic, inorganic, and radioactive. These major sources of chemical contamination originate from the anthropogenic as well as natural activities.

2.1 Organic Contaminants

Synthetic and volatile chemically active organic compounds that release from different chemical activities are termed organic contaminants; they can be by-products of industrial processes and petroleum production and laboratory synthesis. Moreover, it may originate from gas stations, urban stormwater runoff, and septic systems. Common organic contaminants are benzene, toluene, xylene, polycyclic aromatics, and some other petroleum hydrocarbon-based fuel used for different purposes (Fleeger et al. 2003; Brusseau and Artiola 2019). The different chlorinated solvents used in the industrial and chemical laboratory act as organic contaminants, e.g., trichloroethene, tetrachloroethene, trichloroethane, carbon tetrachloride, etc. One of the major sources of organic contaminants is DDT (dichloro-diphenyl-trichloro-ethane), 2,4-D (2,4-Dichlorophenoxyacetic acid), and atrazine, and these toxic organic compounds can apply a negative impact in soil along with surface and groundwater. The polychlorinated biphenyls are highly carcinogenic organic contaminants, used in industrial and different products such as insulating fluids, plasticizers, and pigments which cause pollution of soil and aquatic environments. Some other organic contaminants are polycyclic aromatic compounds that originate from coal tar and creosote which is considered as environmental pollutant. The organic compounds used in pharmaceuticals production, food additives, and cosmetics are also considered as organic contaminants for the environments, and these can be applied negative impact to the environment (Brusseau and Artiola 2019). Phthalate is also organic contaminant, which is commonly used in plastics, synthetic coatings, and paints to improve their flexibility, mechanical performance, and resistance to temperature variations and in cosmetics (Chatonnet et al. 2014).

2.2 Inorganic Contaminants (ICs)

Based on the source, inorganic contaminants can be heavy metals, trace elements, excess nutrients used in agriculture, and several salts. These inorganic contaminants get dissolved in a surrounding solvent and ionized to anions and cations with high mobility. Heavy toxic elements can remain in the environment for a long time, some of Cd, Pb, As, Mn, Ni, Cr, etc. (Krstic et al. 2021), and some other inorganic contaminants remain in the environment for a very short time. For example, in a short time nitrate (NO_3^-) or ammonia (NH_4^+) can be chemically transformed into other compounds and remain as N_2 or N_2O gas in the environment. The salts and metals inorganic contaminants may occur naturally or may result from urban stormwater runoff. Most of these originated from industrial or domestic discharges and oil–gas production and may be from mining or farming. Again application of the excess amount of inorganic fertilizers can cause inorganic contaminants since it has leaching capacity to groundwater and is also easily mobilized in soil and aqueous environment after application of it in the agricultural field. Since plant absorbs a definite amount of nutrient-based on their needs, therefore use of the excess amount of N, P, and K also can cause environmental contamination by exceeding the permissible limit. For example, excess application of nitrogen fertilizers can cause leaching of nitrate to the groundwater and run off to the surface water; as a result, it harms the ecosystem and imbalances the soil–water diversity (Brusseau and Artiola 2019).

2.3 Radioactive Contaminants (RCs)

Radioactive contaminants are one of the major pollutants that harm the environment. Due to the harmfulness of RCs to human beings and the ecosystem, the presence of radioactive elements in the environment is considered radioactive contamination irrespective of its source. Radioactive elements are hazardous because of their radioactive decaying property and possess a harmful effect through emitting alpha, beta, and gamma particles in the form of radiation (Rich and Crosby 2013). The common radioactive contaminations are radium, selenium, thallium, and uranium (Lytle et al. 2014), cesium, strontium, plutonium, technetium, and strontium (Dahlgaard et al. 2004). The major sources of radioactive contaminants are anthropogenic oriented, such as the production, development, and testing of a nuclear weapon for defense, from commercial power reactors, through reprocessing of nuclear fuel and disposal of radioactive waste, coal, and rare-earth mining and most importantly energy production through the nuclear reactor (Deng et al. 2020). Moreover, it can also be originated from oil and gas production and mining activities.

3 Impact of Contaminants on Social Life and the Ecosystem

There are several ways of entering contaminants into the ecosystem as well as into the human body, and its way of affecting them is also different. These contaminants can cause several hazardous impacts on the living body and the environment; however, only some of the impacts are presented in this chapter. From a study, it has been proved that intake of toxic heavy metals such as lead (Pb), mercury (Hg), and cadmium (Cd) damages the immune, nervous, and reproductive systems of humans (Krstic et al. 2021), and some other animals. In addition, the long-time consumption of As, Cd, and Cr can cause cancer, cardiovascular disease, dermal changes, and respiratory, pulmonary, gastrointestinal, hematological, hepatic, renal, neurological, developmental, immunologic, genotoxic, and mutagenic effects on humans. It was reported that even a small concentration of Cd and Pb in the food can cause Alzheimer's disease, alarming poisoning, enzyme dysfunction, mental retardation, even prostate, and most importantly breast cancers (Borgohain et al. 2022). Furthermore, inorganic and organic toxic compounds can show harmful effects on human health by causing cancer, immune suppression, or genotoxic effects, and even at low exposure, these can cause endocrine-disrupting effects in both animals and humans (Rodríguez-Hernández et al. 2016).

The organic contaminants released into the aquatic environment directly affect the ecosystem, e.g., anti-inflammatory drugs sodium diclofenac, and ibuprofen showed applied negative effect on freshwater fish by damaging their tissues of gills and kidneys, as well as reducing the population of the freshwater fish for a long term. Moreover, a mixture of organic compounds atenolol, bezafibrate, carbamazepine, cyclophosphamide, ciprofloxacin, furosemide, hydrochlorothiazide, ibuprofen, lincomycin, ofloxacin, ranitidine, salbutamol, and sulfamethoxazole prohibit the rapid increase in growth of human embryonic kidney cells, i.e., HEK293. Furthermore, discharge of antibiotic chloramphenicol, thiamphenicol, veterinary medicine florfenicol, and antibiotic thiamphenicol to the aquatic environment prohibited the growth of freshwater green alga *Chlorella pyrenoidosa* (Pal et al. 2010). Therefore, such types of organic compounds can damage the ecosystem including humans.

4 Hydrogel

Biochar-based hydrogel nanocomposite system draws much attention to the removal of environmental contaminants because hydrogel improves both the mechanical and thermal stability of biochar. Hydrogels are either physically or chemically cross-linked polymeric network structure materials that can imbibe a large amount of fluid into their three-dimensional network structures. Water retention properties of hydrogel arise due to the presence of hydrophilic groups such as carboxylic, amino sulphonic, and hydroxyl (Lim et al. 2017; Mahmud et al. 2018). Based on cross-linker density, hydrogels possess different swelling capacity and degradability behavior at physiological temperature and pH. Swell hydrogels sunk again when removed from the swelling medium. This flexible property of hydrogels opens a new window for its application in cosmetics, water purification, drug delivery, agrochemical delivery (fertilizer, herbicides, pesticides) wound dressing, and environment-friendly adsorbents (Sirajuddin et al. 2014; Zainal et al. 2020).

4.1 Significance of Cross-Linking

Cross-linking is the most crucial step for hydrogel preparation. The physical properties of the hydrogel have changed by changing the cross-linking concentration and crystalline nature of polymer molecules. Cross-linking results in the insolubility of polymer molecules in an aqueous medium and increases the strength and toughness of hydrogels (Bharskar 2020). Vulcanization of rubber is also an example of a chemical cross-linking process; in this process of vulcanization, sulfur atoms are used as a cross-linker between chains of polyisoprene units. For the preparation of environment-friendly hydrogels, cellulose-based materials such as carboxy methyl cellulose (CMC), hydroxyl ethyl cellulose, hydroxyl propyl methyl cellulose, sodium alginate, and hydroxyl propyl cellulose are used as a starting material (Zainal et al. 2020). There are three types of cross-linking methods—physical, chemical, and polymerizations to produce hydrogels from cellulose-based materials. In physical cross-linking, the polymer chains are tied together through non-covalent interactions such as hydrogen bonding, ion–ion interactions, ion–dipole interactions, and Van der Waal interactions. Ionic cross-linking has been accelerated by the addition of gelling solutions (Singh et al. 2010). Gelling agents have oppositely charged multivalent ions such as Ca^{2+}, Fe^{3+}, Ni^{2+}, and Zn^{2+}. Akalin and Pulat (2018) prepared a series of CMC/ferric chloride hydrogel beads via ionic cross-linking reaction by adding drop-wise CMC solution into ferric chloride solutions. Freeze-thawing, photo-initiator, and radiation-induced techniques are the commonly used techniques for the preparation of physically cross-linked hydrogel. In the freezing process, low molecular mass polymers crystallize as the intermolecular distance between polymer chains decreases and connect the polymer network with either hydrogen bond or covalent interaction (Zhao et al. 2016).

Chemical cross-linking is a highly versatile method for the preparation of both mechanically and thermally stable hydrogel. Chemical cross-linker includes citric acid, glutaraldehyde, epichlorohydrin, fumaric acid, and maleic acid (Hennink and Nostrum 2012). Citric acid is used as an effective cross-linker between CMC and hydroxyl ethyl cellulose. In chemical cross-linking; the cross-linking agent connects the polymer backboned through direct bonding interactions. The interaction can occur through the reaction of functional groups such as –COOH, $-NH_2$, and –OH with the cross-linker. The natural polymer containing amine groups are cross-linked chemically with glutaraldehyde and poly(ethylene glycol) through the mechanism of Schiff base reactions. Polymer chains containing hydroxyl and amino groups react readily with episodes, such as epichlorohydrin, ethylene glycol diglycosidic ether, and butanedioldiglycidyl. Episodes readily undergo reactions at room temperature due to large ring strain (Wang et al. 2016; Nagireddi et al. 2017). In photo-initiator techniques, the cross-linking process is initiated by irradiation such as ultra-violet and ruthenium-based light. Qi et al. (2018) prepared sericin hydrogel through photo-cross-linking techniques for biomedical application. In radiation induce techniques, radiations like gamma and electron beams have been used to initiate the polymerization reaction without any catalyst or chemicals (Gonzalez-Torres et al. 2018).

4.2 Classification of Hydrogels

There are numerous ways to classify hydrogel. Figure 1 depicts the most prominent parameters to classify hydrogel. The behavior of the hydrogel may change by changing any of the features listed in Fig. 1. Permanent gels are formed by covalently cross-linked with polymer chains leading to stable gel products. In physically cross-linked gels, polymer chains have bound together by molecular entanglements, which

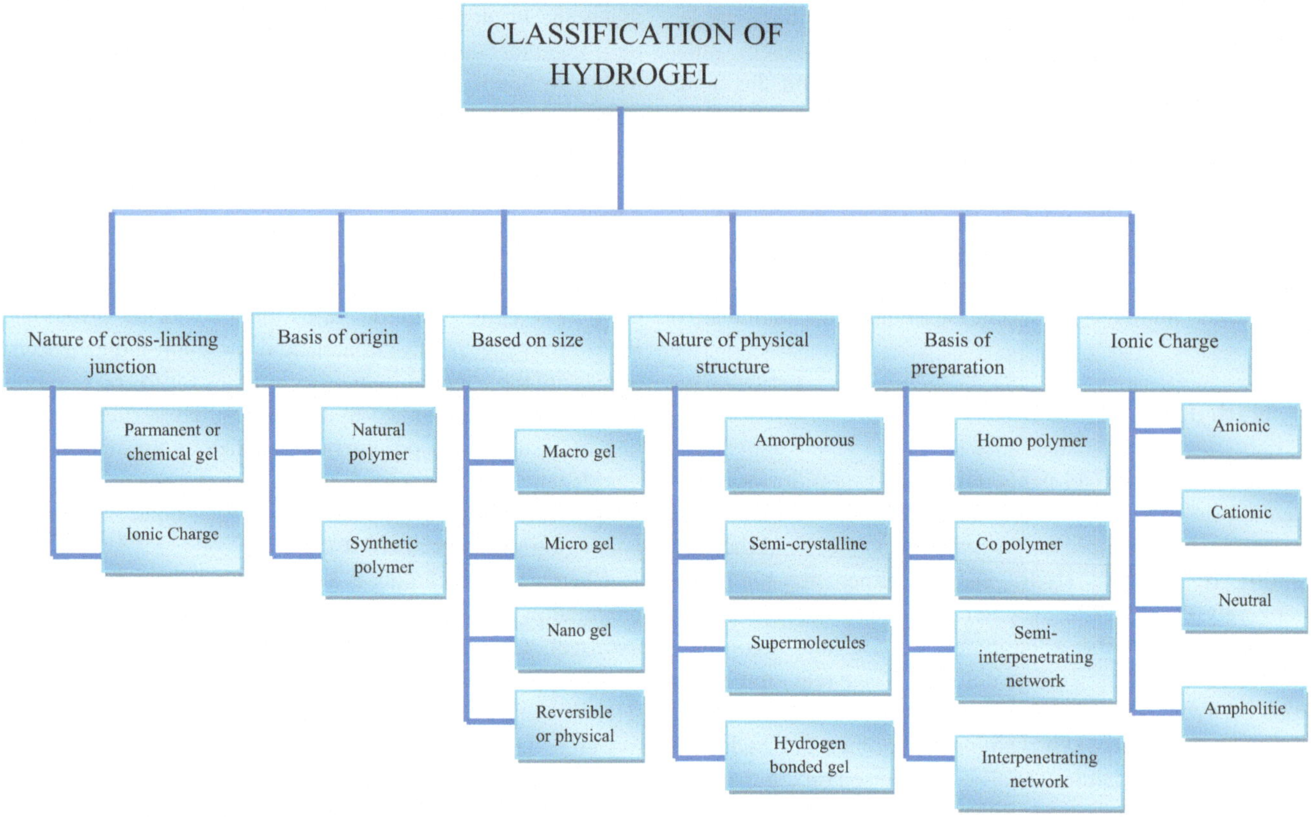

Fig. 1 Classification of hydrogel based on different parameters

prevent the dissolution of gels in the corresponding medium. Nature-origin polymers have advantages of bio-degradability, biocompatibility, renewability, and good mechanical strength properties (Bharskar 2020). Cellulose derivatives have become the center of attention of many researchers as it is the most abundant natural material found in trees, vegetables, and waste biomass products. As cellulose is an insoluble biomaterial, researchers have developed different dissolution mediums for the homogeneous distribution of bio-polymer. Such dissolution solvent system includes NaOH/urea, NaOH/thiourea, and ionic liquids. The bioactive properties of polymers have enhanced synthetic polymeric materials. Homopolymers are derived from a single monomer species with numerous applications in drug delivery and for making contact lenses. In a co-polymeric network, two or more species combine to form a hydrogel chain with at least one monomer having hydrophilic groups and following any configuration to form the three-dimensional network structures (Bharskar 2020; Zainal et al. 2020).

4.3 Adsorption of Contaminants by Hydrogel

Hydrogel has been applied in the field of pollutant adsorbent, due to its high water retention and adsorption capacity. The adsorption capacity of hydrogel is influenced by various factors including hydrogel structure, porosity, nature of hydrogel material, the concentration of pollutant, temperature, ionic strength, and pH of the pollutant medium (Abd El-Mohdy et al. 2013). Hydrogel adsorbs heavy metal either through bonding or non-bonding interactions. Bonding interactions include the formation of the covalent bond and non-bonding interactions include electrostatic, ion–ion, ion–dipole interactions, and hydrogen bonding interactions between hydrogel networks with the adsorbent (Khan and Lo 2016). Ngwabebhoh et al. (2016) used chitosan-montmorillonite hydrogel for adsorption of Cu^{2+} and anionic nitrazine yellow dye, where the functional $-NH_2$ group is responsible for adsorption of Cu^{2+} and corresponding dye. When pH is less than 5.6, the $-NH_2$ group is protonated to NH_3^+; this positive charge on the hydrogel surface attracts negatively charged dye through electrostatic interactions. Moreover, with increasing pH, the lone pair of the amino group is shared with the electron acceptor of the dye, and the bond is formed. Thakur et al. (2016) established the electrostatic attraction force for adsorption of methylene blue by sodium alginate-cl-poly (acrylic acid)/TiO_2-hydrogel. Zheng et al. (2011) reported chitosan/poly-acrylic acid hydrogel as adsorption of Ni^{2+} ions and found that adsorption is highly controlled by the pH of the medium. Chitosan

Table 1 Different starting materials used for the preparation of hydrogel, and its application for removal of the contaminants

Hydrogel	Removal of contaminant	References
Acrylamide/acrylic acid copolymers chitosan hydrogel	Water/oil separation	Liu et al. (2018)
Cellulose hydrogel with magnetic nanoparticles	Methylene blue	Chen et al. (2017)
Cellulose-lignin composite hydrogel	Pb(II)	Yin et al. (2016)
Chelator-mimetic multi-functionalized hydrogel	p-Nitrophenol and p-Aminophenol	Shan et al. (2018)
Chitosan-based multicomponent functional gel	Cr(VI)	Kim et al. (2015b)
Chitosan/Sodium alginate double network hydrogel	Pb^{2+}, Cd^{2+}	Tang et al. (2020)
Chitosan hydrogel	Cu(II), Cd(II)	Liu et al. (2012) Paulino et al. (2011)
Lignocellulose-based composites	Cu(II)	Zhang et al. (2018)
Lignosulphonate-based graphene hydrogel	Pb(II)	Thakur et al. (2017)
Magnetic chitosan hydrogel	Cr(VI)	Mirabedini et al. (2016)
Methacrylic acid hydrogel	Methylene blue, Rhodamine-6G	Shafiq et al. (2019)
Modified polyacrylamide hydrogel	Cu^{2+}, Pb^{2+}	Kasgoz et al. (2003)

contains both carboxylic and amino groups; at lower pH, carboxylic acid groups are not ionized but amino groups are protonated, which in turn causes electrostatic repulsion between Ni^{2+} ions and hydrogel. With the increase in pH, carboxylic groups lose their proton and become negatively charged. They concluded that more Ni^{2+} ions were adsorbed by hydrogel at pH range 3–7; for pH greater than 7, metal ions are precipitated. Recently, hydrogel-based nanocomposite was prepared by encapsulating nano-magnetite into the starch hydrogel and doped it in porous metal–organic frameworks to reduce lovastatin antibiotics from water. The organic framework was prepared by doping magnetic iron oxide nanoparticles (NFe_3O_4) into the pores of Zn-(glutaric acid) through a microwave irradiation process. From the study, it was found that the maximum swelling ratio was 1000% at pH 10 (25 °C), surface area 528.39 $m^2\ g^{-1}$, and pore size 2.90 nm with a highly crystalline structure. As a result, the nanocomposite adsorption of Fluvastatin, a statin drug, provided a maximum equilibrium adsorption capacity which was 782.05 mg g^{-1} (Mohamed and Mahmoud 2020a) (Table 1).

5 What is Biochar?

Biochar is a carbon-rich compound that resulted from the thermal decomposition of biomass during pyrolysis. In pyrolysis of feedstocks in presence of trace amount of O_2 at a temperature range 350–900 °C, the compounds such as cellulose, hemicellulose, and lignin present in it get converted to solid (biochar), liquid (bio-oil), and gaseous (e.g., H_2, CO, and CO_2) (Borgohain et al. 2020). The physicochemical properties of biochar depend upon the pyrolysis temperature, heating rate, and residence time. During pyrolysis, with increasing the temperature the acidic functional groups (e.g., –COOH) are reduced into a basic functional group (–OH) on the surface of the product, i.e., biochar and achieved a more stable carbon structure. Borgohain et al. (2020) reported that the empirical formula of biochar is $CH_{0.75}N_{0.04}O_{0.67}$ and the major components of biochar are carbon (C), hydrogen (H), oxygen (O), ash, trace amounts of nitrogen (N) and sulfur (S), etc. The advantages of application of biochar in cultivated soil are presented in Fig. 2.

5.1 Interaction of Biochar with Heavy Metals

Biochar has immobilized metal in contaminated soil through several interactions with heavy metals, either by direct interactions or by indirect interactions with metals. Agricultural manure or plant waste-based biochar can interact with heavy metals through non-meditative mechanisms, which include electrostatic interactions, ion exchange, precipitation, and complexation. Electrostatic interactions take place between oppositely charged ions. The surface charge of biochar is measured through zeta potential, which in turn determines the electrostatic interaction tendency of biochar. Pyrolysis at high temperature, the pH, surface area, and negatively charged surface functional groups of biochar increase; therefore, the biochar gets more electro-negativity properties and attracts more positively charged ions present in contaminated soil as well as in water. Tong et al. (2011) showed that Cu(II) adsorption increases with an increase in pH. Biochar-containing carboxylic groups (–COOH) and

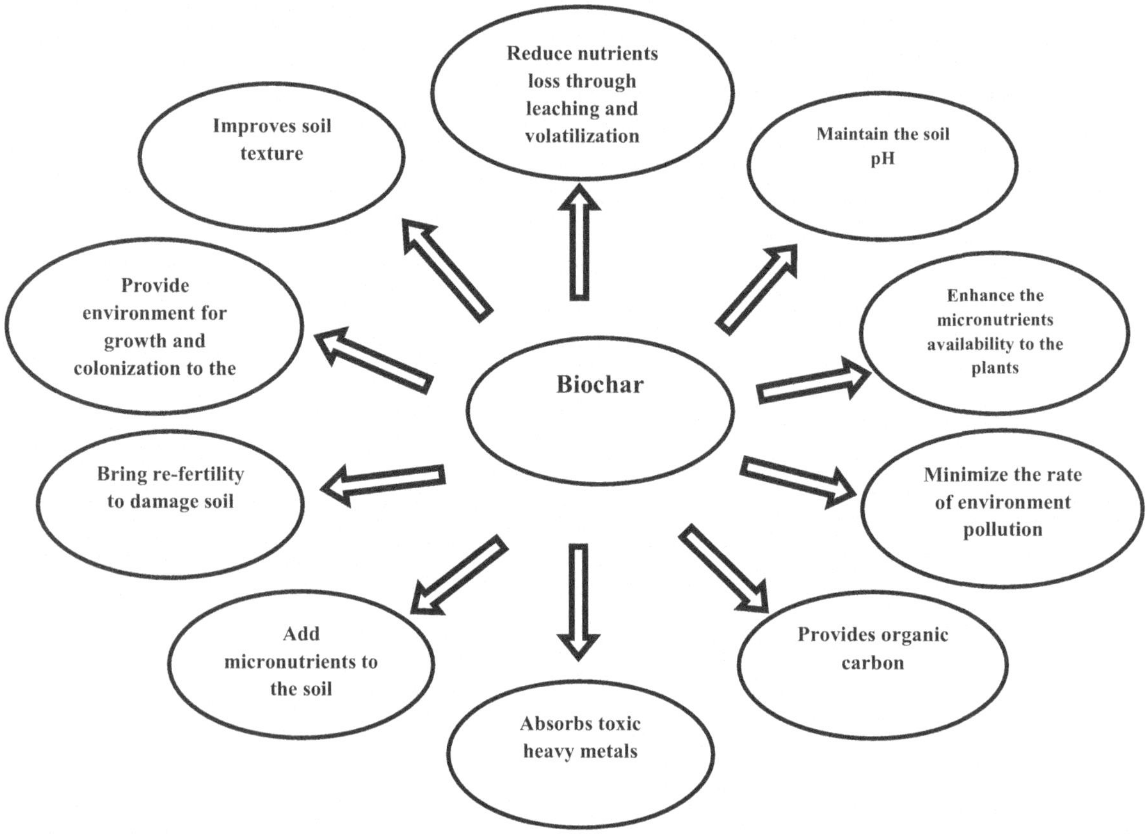

Fig. 2 Beneficial role of applied biochar in soils

other oxygen-containing groups remove metal ions through the cation-exchange process, where cations like Ca^{2+} and Mg^{2+} are replaced by heavy metals at the surface of biochar. Biochar has a high cation-exchange capacity and adsorbed more heavy metals (Tang et al. 2013). Plant-derived biochar contains less concentration of calcium ions than animal-derived biochar, and immobilization of Cu(II) and Cd(II) is more dominant for animal-derived biochar. Biochar is made up of crop residue, adsorbed contaminant through the formation of surface complexation. The functional groups such as carboxylic (–COOH), hydroxyl (–OH), and carbonyl (C=O) present in biochar formed a complex with heavy metals. For example, the intensity of the Fourier-transform infrared spectroscopy (FTIR) peak is changed from 2349 to 2360 cm^{-1}, when biochar is applied to Cd(II)-rich soil (He et al. 2019). Many minerals present in biochar removed environmental contaminants, forming an insoluble precipitate. Based on different mineral compositions, Pb formed precipitated as its oxide, chloride, and sulfate form (Liu et al. 2014). The indirect effect of biochar has changed the metal mobility and bioavailability in soil (Mansoor et al. 2021).

5.2 Microbial Activity on the Application of Biochar

As already discussed, the application of biochar in the soil can alter the pH; therefore, it may cause an impact on microbial activity in the soil. As different microorganisms are alive at different pH, biochar either reduces or increases the colonization of microorganism. Soil microorganism plays a crucial role in the nutrient cycle, for example, rhizobium fixed the atmospheric nitrogen into plant-available nitrate form, and *Pseudomonas fluorescence* converts the inorganic form of phosphate into organic form. Some biochar acts as an inorganic nutrient for soil microorganisms and protects microbes from predators. Rutigliano et al. (2014) established the effect of soil amendment of wood-derived biochar in wheat and also studied the soil pH, total organic carbon, extractable carbon, and microbial biomass carbon. They showed that the application of biochar has not changed microbial activity and its diversity. Lehmann et al. (2011) forwarded the idea that biochar produced micro-location in soil and optimized the microbial growth by co-location of microorganisms on the biochar surface.

Steinbeiss et al. (2009) investigated the behavior of ^{13}C glucose (0%N) and yeast (5%)-based biochar to promote carbon content in the soil. Yeast-derived biochar increases fungi in the soil, while glucose-derived biochar provides a carbon source for the production of biomass.

5.3 Adsorption of Contaminants by Biochar

During the last decades, a huge amount of toxic elements had released to the environment in the form of heavy metals and metalloids, and their rehabilitation is most important for healthy food production. Heavy metals rapidly contribute to environmental pollution and finally participate in the food chain with an extremely harmful effect on flora and fauna. Plants absorb various micronutrients in the form of metal ions for their growth and food production. An excessive amount of heavy metal in the soil environment can cause toxicity to plants either directly or indirectly by tuning the availability of other primary macronutrients. Thus, it is most important to decrease heavy metals accessibility and provide non-contaminated soil to plants (Rees et al. 2016). A range of laboratory studies has been done to evaluate the effectiveness of biochar in retaining both organic and inorganic contaminants. The rehabilitation of metal contaminants depends on remediation techniques, mainly immobilization and phytoremediation to remediate metal contaminants in soil (Al-Wabel et al. 2015). Abdelhafez et al. (2014) prepared bagasse and orange peel-based biochars for rebooting the initial condition of a contaminant soil for agricultural uses and to determine its application for stabilizing Pb and As in soil. They found that the application of biochar, water retention capacity, cation-exchange capacity, and concentration of organic matter in soil increase considerably. Lucchini et al. (2014) investigated the effect of wood-derived BC on the varying heavy metal dynamics in soil and plants. They observed that repeated application of BC into soil did not accelerate the increase of metal concentration. Bakshi et al. (2018) synthesized a zero-valent Fe-biochar for arsenic removal from drinking water via an oxidation–reduction mechanism. Here, arsenic has been reduced and iron has been oxidized to form a complex of Fe (As)OOH and FeOOH on the biochar surface. Song et al. (2014) applied sewage sludge BC products as a soil amendment to investigate the accumulation of As, Zn, Pb, Ni, Cd, Cr, and Cu on the garlic plant *(Allium sativum* L.). They showed that metal accumulation by the plant was controlled by changing pyrolysis temperature. Biochar produced at 450 °C has more micro-pores with highly stable functional groups, resulting in the lowest accumulation of heavy metals by plants (Song et al. 2014). Rice hull-derived biochar was found to immobilize Cd, Pb, and Zn in soil but its high dose application led to a decrease in phyto-available soil nitrogen, reducing the plant growth (Kim et al. 2015a).

6 Current Status of Application of Biochar-Based Nanocomposites

From the study, it has been proved that the nano-sized biochar can show better performance than its regular size biochar particles. Since the nano-sized biochar particles have a larger surface area, therefore the application of nano-sized particles has more advantages for the adsorption of various organic, inorganic, and heavy metal contaminants, vitamins, and several drugs from aqueous solutions. Nano-biochar has prominent physical, chemical, and surface properties such as micro-porosity and large surface area. Therefore, nano-biochar can enhance the soil quality as well as plant growth, plant-disease management, bioremediation of contaminants and pesticides, and water purification, as a support material for enzymes (Chausali et al. 2021), and microbes (Borgohain et al. 2020). These excellent properties of nano-biochar become more advanced with converting it to nanocomposite with organic polymer resulting in a more surface functional group, higher surface area, and better catalytic activity, become better supporting material for the enzyme (Chausali et al. 2021), and facilitate the colonization of beneficial microbes in the soil. Several researches have been done on the preparation of biochar-based nanocomposite and its application in the purification of different contaminants from an aqueous environment. Moreover, it has been used for water storage as well as a slow nutrient release source in the agricultural field.

6.1 Application of Biochar-Based Hydrogel Nanocomposite in Agriculture

Since biochar is an excellent solution for agricultural challenges such as leaching, volatilization, denitrification, and surface runoff of nutrients; therefore, biochar-based fertilizer can be used as a slow-release fertilizer. The application of biochar-based fertilizer can improve soil fertility and other essential properties, most importantly, the controlled release of nutrients (Sim et al. 2021). Elsaeed et al. (2021) recently prepared a biochar-based hydrogel nanocomposite (BHNC) by microwave and conventional techniques through grafting of biochar in hydrogel and applied it as a super absorbent of water. The BC was prepared by pyrolysis of orange peel, and the hydrogel was prepared by grafting of biochar based on copolymerization of acrylic acid, acrylamide, and acrylamide-co-2-acrylamide-2-methylpropane sulfonic acid monomers along the chains of carboxymethyl. As a result, they have concluded that the application of this biochar nanocomposite in the soil can improve the percentage of water-holding capacity of the soil and also decrease several contaminants in soil (Elsaeed et al. 2021). Therefore, BHNC can be utilized in sustainable agricultural irrigation.

In a recent study, Shan et al. (2022) loaded $MgAl_2O_4$ in biochar prepared from rice husks. The modified biochar was embedded with carboxymethyl cellulose-lanthanum hydrogel through co-precipitation followed by calcination. This biochar-based hydrogel composite was used to capture phosphate from an aquatic environment. From the study, it was found that the composite can absorb a maximum of 89.65 mg P g^{-1} at pH 3. In another study, Zhang et al. (2020b) prepared a composite by grafting poly(acrylic acid) in chitosan and BC composite. The composite was tested for the absorption and desorption mechanism of ammonium (NH_4^+–N) from the soil environment. The absorption efficiency was too fast that within 20 min it absorbed >90% off NH_4^+–N from the soil. As a result, the composite absorbed a maximum amount of 149.25 mg g^{-1} of ammonium (NH_4^+–N) at 25 °C, and it was observed that the slow-release tendency was reduced by 24.18–31.77% at a 1.57–2.02% material dosage in 10 weeks of study (Zhang et al. 2020b). Moreover, a cost-effective, eco-friendly, and multifunctional porous protein-based hydrogel was prepared from waste leather treated with KOH. The prepared waste leather-based hydrogel was found to release micronutrient nitrogen and potassium very slowly for more than 40 days; therefore, it can be used as a slow-release bio-fertilizer in the field of agriculture. Moreover, the prepared hydrogel showed high water absorption capacity, i.e., 2208 g g^{-1}, and a good absorber of Cr(III), i.e., 149.3 g g^{-1}. This hydrogel is highly decomposable in soil and contributes carbon for supporting the growth of *Ensifer* sp. *Y1* (Hu et al. 2021).

6.2 Application of Biochar-Based Hydrogel Nanocomposite in Treatment of the Polluted Environment

Water can be purified in different ways such as filtration, osmosis, adsorption, sedimentation, and disinfection; however, adsorption is the most advantageous method over other methods (Kemp et al. 2013). Sanyang et al. (2014) used a novel hydrogel-biochar composite for the removal of arsenic from waste water. Acrylamide-based hydrogel was prepared by using N, N′-methylenebisacrylamide as a cross-linker. The cross-linking reaction was initiated by ammonium persulfate as the initiator. They synthesized the hydrogel-biochar composite system by embedding rice husk biochar into a hydrogel network. The results of these tests exhibited arsenic adsorption of 28.32 mg g^{-1}, which were followed by pseudo-first-order kinetics. The experimental results also demonstrated that the adsorption capacity of composite is highly dependent on solution pH, sorption dosages, and initial concentration of contaminant (Sanyang et al. 2014). Jung et al. (2016) reported that phosphate removal with maximum efficiency of 97.02% can be achieved by using biochar-fabricated alginate hydrogel beads. Biochar was prepared from three different marine macroalgae *Undaria pinnatifida*, *Laminaria japonica*, and *Porphyra tenera,* and hydrogel was prepared by an ionic cross-linking method having calcium chloride as a cross-linker. They reported that the adsorption process was endothermic in nature and adsorption was best fitted by the Langmuir–Freundlich model. The prepared material was employed effectively for the adsorption of phosphate with an adsorption capacity of 157.7 mg g^{-1} at 30 °C (Jung et al. 2016).

The absorption of contaminants by a biochar-based nanocomposite can be enhanced by the increase in the ratio of biochar. In excellent research, Simeng and Chen (2018) applied the different ratios of biochar encapsulated with acrylamide. They also achieved a maximum swelling ratio of 688% greater than the untreated polyacrylamide hydrogel. From the isotherm experiment result, they have found that the adsorption was maximum for the composite where the biochar ratio was maximum, i.e., 63.58 mg g^{-1}. Zhang et al. (2020a) reported a new method for efficient removal of Cr(IV) from aqueous solution, using highly selective chitosan-modified magnetic bamboo biochar. In comparison with ordinary biochar, the adsorption capacity of chitosan-modified biochar was increased from 75.8 to 127 mg g^{-1} at 25 °C. They concluded that adsorption takes place through several interactions such as electrostatic interaction, redox interaction, ion exchanges, and chelation. The result showed that after the fifth adsorption–desorption cycle, the adsorption efficiency of chitosan-modified biochar remains above 90%.

Mohamed and Mahmoud (2020a) recently prepared an advanced nanocomposite through grafting starch-gelatin mixed hydrogels with super magnetic $CoFe_2O_4$, waste mallow stems biochar, and MoO_3. They found that the swelling efficiency of the prepared nanocomposite was quite higher, i.e., 438%, which was an excellent material for the adsorption of different contaminants. Therefore, they applied the nanocomposite to adsorb Pb(II) and a satisfactory result was provided by the composite with maximum adsorption of 5000 μmol g^{-1} of 0.15 mol L^{-1} Pb(II) ions in 40-min reaction time. Nassaj-Bokharaei et al. (2021) prepared a BHNC by reinforced starch-based hydrogel in biochar nanoparticles and studied the nutritional and morphological responses of tomato plant (*Lycopersicon esculentum* Mill.) and mentioned that under the non-stress condition, it shows maximum water-holding capacity, i.e., 85%, with serving water by the BHNC, also increased the P, Fe, and Zn up to 16–29%, and enhanced the microbial population size up to 89% (Nassaj-Bokharaei et al. 2021). A novel three-dimensional ZnO-loaded biochar-based hydrogel nanocomposite was synthesized for removing U(VI) from contaminated aqueous solution. As a result, the hydrogel nanocomposite shows very effectively due to its well mechanical characteristics, thermal stability, anti-swelling property, and effective adsorption of U(VI) than the pure form of a hydrogel. From the Langmuir model, it was found that the maximum adsorption of U(VI) was 239.21 mg g^{-1} (Guo et al. 2022). A similar type of study was conducted by Wu et al. (2019) where they had undertaken a rapid redox reaction between $KMnO_4$ and Mn(II) acetate in presence of biochar, and this modified biochar was incorporated into a polyacrylamide gel network through a rapid and facile free-radical polymerization. They reported that the BHNC can hold 92.1% Cd(II) and 80.5% Pb(II) of its adsorbed elements during its several adsorption–desorption cycles. Moreover, the maximum adsorption capacity of the BHNC from the Langmuir adsorption capacity for Cd(II) and Pb(II) was 84.76 and 70.90 mg g^{-1}, respectively (Wu et al. 2019). A nanocomposite was prepared by Mohamed and Mahmoud (2020b) through encapsulating *Pisum sativum* pods based on nano-scale biochar in the starch hydrogel. They found that the swelling capacity of the nanocomposite was 500% (25 °C) with an excellent surface area of 226.94 m^2 g^{-1} and pore volume of 9.88 cm^3 g^{-1}. Therefore, from the study, it has been clear that the BHNC has huge potential for purifying contaminated water.

7 Future Perspectives and Conclusion

From several studies, it can be concluded that different types of BHNCs systems have several applications for a contaminant-free environment. Due to its chemically stable and biocompatible properties, with highly porous structures that resulted in high adsorption capacity, rapid diffusion, and excellent recyclability, BHNCs provide a contaminant-free environment. It shows a very high affinity toward adsorption in an aqueous medium and indeed be considered for the removal of dye from wastewater. It is a uranium absorber in nuclear waste management. On the other hand, in future BHNCs can be applied in agriculture as a nutrient supplier and also as an adsorber of toxic metals and metalloids from the soil environment. Biochar hydrogel nanocomposite-like materials have great potential for the preparation of several metal nano-catalysts and can be used in reduction/oxidation and degradation processes. It has a huge opportunity to use as a slow-release nutrient releaser as well as a carbon source for the soil. For future studies, further research is necessary to explore the long-term applicability of BHNCs and some other environmental contaminant detectors. A detailed investigation should be carried out to determine the interaction among pollutants for the treatment of heavy metal contamination in soil.

References

Abd El-Mohdy HL, Hegazy EA, El-Nesr EM, El-Wahab MA (2013) Metal sorption behavior of poly(N-vinyl-2-pyrrolidone)/(acrylic acid-co-styrene) hydrogels synthesized by gamma radiation. J Environ Chem Eng 1:328–338. https://doi.org/10.1016/j.jece.2013.05.013

Abdelhafez AA, Li J, Abbas MHH (2014) Feasibility of biochar manufactured from organic wastes on the stabilization of heavy metals in a metal smelter contaminated soil. Chemosphere 117:66–71. https://doi.org/10.1016/j.chemosphere.2014.05

Afzal MZ, Sun X-F, Liu J, Song C, Wang S-G, Javed A (2018) Enhancement of ciprofloxacin sorption on chitosan/biochar hydrogel beads. Sci Total Environ 639:560–569. https://doi.org/10.1016/j.scitotenv.2018.05.129

Ahmed EM (2015) Hydrogels: preparation, characterization, and applications: a review. J Adv Res 6:105–121. https://doi.org/10.1016/j.jare.2013.07.006

Akalin GO, Pulat M (2018) Preparation and characterization of nanoporous sodium carboxymethyl cellulose hydrogel beads. J Nanomater 2018:1–12. https://doi.org/10.1155/2018/9676949

Al-Wabel MI, Usman ARA, El-Naggar AH, Aly AA, Ibrahim HM, Elmaghraby S, Al-Omran A (2015) Conocarpusbiochar as a soil amendment for reducing heavy metal availability and uptake by maize plants. Saudi J Biol Sci 22:503–511. https://doi.org/10.1016/j.sjbs.2014.12.003

Bakshi S, Banik C, Rathke SJ, Laird DA (2018) Arsenic sorption on zero-valent iron-biochar complexes. Water Res 137:153–163. https://doi.org/10.1016/j.watres.2018.03.021

Bharskar RG (2020) A review on hydrogel. World J Pharm Pharmace Sci 7:1288–1298

Bhatnagar A, Sillanpää M (2009) Applications of chitin and chitosan-derivatives for the detoxification of water and wastewater—a short review. Adv Colloid Interface Sci 152:26–38. https://doi.org/10.1016/j.cis.2009.09.003

Borgohain A, Konwar K, Buragohain D, Varghese S, Dutta AK, Paul RK, Karak T (2020) Temperature effect on biochar produced from tea (*Camellia sinensis* L.) pruning litters: a comprehensive treatise on physico-chemical and statistical approaches. Bioresour Technol 318:124023. https://doi.org/10.1016/j.biortech.2020.12402

Borgohain A, Sarmah M, Konwar K, Gogoi R, Gogoi BB, Khar P, Paul RK, Handique JG, Malakar H, Deka D, Saikia J, Karak T (2022) Tea pruning litter biochar amendment in soil reduces arsenic, cadmium, and chromium in made tea (*Camellia sinensis*L.) and tea infusion: a safe drink for tea consumers. Food Chem X 13:100255

Brusseau ML, Artiola JF (2019) Chemical contaminants. Envir Pollut Sci 12:175–190. https://doi.org/10.1016/b978-0-12-814719-1.00012-4

Chatonnet P, Boutou S, Plana A (2014) Contamination of wines and spirits by phthalates: types of contaminants present, contamination sources and means of prevention. Food Add Cont: Part A 31:1605–1615. https://doi.org/10.1080/19440049.2014.941947

Chausali N, Saxena J, Prasad R (2021) Nanobiochar and biochar based nanocomposites: advances and applications. J Agri Food Res 5:100191. https://doi.org/10.1016/j.jafr.2021.100191

Chen P, Liu X, Jin R, Nie W, Zhou Y (2017) Dye adsorption and photo-induced recycling of hydroxypropyl cellulose/molybdenum

disulfide composite hydrogels. Carbohydr Polym 167:36–43. https://doi.org/10.1016/j.carbpol.2017.02.094
Dahlgaard H, Eriksson M, Nielsen S, Joensen H (2004) Levels and trends of radioactive contaminants in the greenland environment. Sci Total Envir 331:53–67. https://doi.org/10.1016/J.Scitotenv.2004.03.0
Deng D, Zhang L, Dong M, Samuel R, Ofori-Boadu A, Lamssali M (2020) Radioactive waste: a review. Water Environ Res 92:1818–1825. https://doi.org/10.1002/wer.1442
Elsaeed SM, Zaki EG, Ibrahim TM, Talha, NI, Saad HA, Gobouri AA, Mohamed el-kousy S (2021) Biochar grafted on carboxy methyl cellulose-terpolymer by green microwave route for sustainable agriculture. Agriculture 11(4):350. https://doi.org/10.3390/agriculture11040350
Fleeger JW, Carman KR, Nisbet RM (2003) Indirect effects of contaminants in aquatic ecosystems. Sci Total Environ 317:207–233. https://doi.org/10.1016/s0048-9697(03)00141-4
Gonzalez-Torres M, Leyva-Gómez G, Rivera M, Krötzsch E, Rodríguez-Talavera R, Rivera AL, Cabrera-Wrooman A (2018) Biological activity of radiation-induced collagen-polyvinyl pyrrolidone-PEG hydrogels. Mat Let 214:224–227. https://doi.org/10.1016/j.matlet.2017.12.006
Guo Y, Liu X, Xie S, Liu S, Wang C, Wang L (2022) 3D ZnO modified biochar-based hydrogels for removing U(VI) in aqueous solution. Colloids Surf a: Physicochem Eng Asp 642:128606. https://doi.org/10.1016/j.colsurfa.2022.128606
He L, Yang L, Zhang L, Wang Z, Cheng H, Wang X, Lv C, Zhang J, Mo H, Shen J (2020) Removal of Ca^{2+} and Mg^{2+} from oil field waste water using re-usable PEG/Fe_3O_4/GO–NH_2 nano adsorbents and its efficiency for oil recovery. J Environ Chem Eng 9:104653. https://doi.org/10.1016/j.jece.2020.104653
He L, Zhong H, Liu G, Dai Z, Brookes PC, Xu J (2019) Remediation of heavy metal contaminated soils by biochar: mechanisms, potential risks and applications in China. Environ Pollut 252:846–855. https://doi.org/10.1016/j.envpol.2019.05.151
Hennink WE, van Nostrum CF (2012) Novel crosslinking methods to design hydrogels. Adv Drug Deliv Rev 64:223–236. https://doi.org/10.1016/j.addr.2012.09.009
Hojjati-Najafabadi A, Mansoorianfar M, Liang T, Shahin K, Karimi-Maleh H (2022) A review on magnetic sensors for monitoring of hazardous pollutants in water resources. Sci Total Environ 824:153844. https://doi.org/10.1016/j.scitotenv.2022.153844
Hu Z-Y, Chen G, Yi S-H, Wang Y, Liu Q, Wang R (2021) Multifunctional porous hydrogel with nutrient controlled-release and excellent biodegradation. J Environ Chem Eng 9:106146. https://doi.org/10.1016/j.jece.2021.106146
Islam R, Kumar S, Karmoker J, Kamruzzaman M, Rahman MM, Rahman MA, Rahman MM (2018) Pathways of persistent organic pollutants (POPs) bioaccumulation and its adverse effects on ecosystems and human exposure: a review study on Bangladesh perspectives. Environ Technol Innov 12:115–131. https://doi.org/10.1016/j.eti.2018.08.002
Jung K-W, Jeong T-U, Kang H-J, Ahn K-H (2016) Characteristics of biochar derived from marine macroalgae and fabrication of granular biochar by entrapment in calcium-alginate beads for phosphate removal from aqueous solution. Bioresour Technol 211:108–116. https://doi.org/10.1016/j.biortech.2016.03.06
Karakoyun N, Kubilay S, Aktas N, Turhan O, Kasimoglu M, Yilmaz S, Sahiner N (2011) Hydrogel-biochar composites for effective organic contaminant removal from aqueous media. Desalination 280:319–325. https://doi.org/10.1016/j.desal.2011.07.014
Kasgoz H, Ozgumuş S, Orbay M (2003) Modified polyacrylamide hydrogels and their application in removal of heavy metal ions. Polymer 44:1785–1793. https://doi.org/10.1016/s0032-3861(03)00033-8
Kemp KC, Seema H, Saleh M, Le NH, Mahesh K, Chandra V, Kim KS (2013) Environmental applications using graphene composites: water remediation and gas adsorption. Nanoscale 5:3149. https://doi.org/10.1039/c3nr33708a
Khan M, Lo IMC (2016) A holistic review of hydrogel applications in the adsorptive removal of aqueous pollutants: recent progress, challenges, and perspectives. Water Res 106:259–271. https://doi.org/10.1016/j.watres.2016.10.008
Kim H-S, Kim K-R, Kim H-J, Yoon J-H, Yang JE, Ok YS, Kim K-H (2015a) Effect of biochar on heavy metal immobilization and uptake by lettuce (*Lactuca sativa* L.) in agricultural soil. Environ Earth Sci 74:1249–1259. https://doi.org/10.1007/s12665-015-4116-1
Kim MK, Sundaram KS, Iyengar GI, Lee K-P (2015b) A novel chitosan functional gel included with multiwall carbon nanotube and substituted polyaniline as adsorbent for efficient removal of chromium ion. Chem Eng J 267:51–64. https://doi.org/10.1016/j.cej.2014.12.091
Krstic M, Stupar M, Đukic-Cosic D, Baralic K, Mracevic SD (2021) Health risk assessment of toxic metals and toxigenic fungi in commercial herbal tea samples from Belgrade, Serbia. J Food Compos Anal 104:104159. https://doi.org/10.1016/j.jfca.2021.104159
Lehmann J, Rillig MC, Thies J, Masiello CA, Hockaday WC, Crowley D (2011) Biochar effects on soil biota—a review. Soil Biol Biochem 43(9):1812–1836. https://doi.org/10.1016/j.soilbio.2011.04.022
Lim L, Rosli N, Ahmad I, Mat Lazim A, Mohd Amin M (2017) Synthesis and swelling behavior of pH-sensitive semi-IPN superabsorbent hydrogels based on poly(acrylic acid) reinforced with cellulose nanocrystals. Nanomaterials 7:399. https://doi.org/10.3390/nano7110399
Liu XL, Zeng ZX, Chen QW, Zou HL (2014) Effects of biochar and lime additives on non-point load of heavy metals in paddy soil. J Hydraul Eng 45:682–690
Liu Y, Kang Y, Huang D, Wang A (2012) Cu^{2+} removal from aqueous solution by modified chitosan hydrogels. J Chem Technol Biotechnol 87:1010–1016. https://doi.org/10.1002/jctb.3722
Liu Y, Su M, Fu Y, Zhao P, Xia M, Zhang Y, He B, He P (2018) Corrosive environments tolerant, ductile and self-healing hydrogel for highly efficient oil/water separation. Chem Eng J 354:1185–1196. https://doi.org/10.1016/j.cej.2018.08.071
Lucchini P, Quilliam RS, DeLuca TH, Vamerali T, Jones DL (2014) Does biochar application alter heavy metal dynamics in agricultural soil? Agric Ecosyst Environ 184:149–157. https://doi.org/10.1016/j.agee.2013.11.018
Lytle DA, Sorg T, Wang L, Chen A (2014) The accumulation of radioactive contaminants in drinking water distribution systems. Water Res 50:396–407. https://doi.org/10.1016/j.watres.2013.10.050
Mahmoud ME, Mohamed AK (2020) Novel derived pectin hydrogel from mandarin peel based metal-organic frameworks composite for enhanced Cr(VI) and Pb(II) ions removal. Int J Biol Macromol 164:920–931
Mahmud M, Daik R, Adam Z (2018) Influence of poly(ethylene glycol) on the characteristics of γ radiation-crosslinked poly(vinyl pyrrolidone)-low molecular weight chitosan network hydrogels. Sains Malays 47:1189–1197
Makhado E, Pandey S, Modibane KD, Kang M, Hato MJ (2020) Sequestration of methylene blue dye using sodium alginate poly (acrylic acid)@ZnO hydrogelnanocomposite: kinetic, Isotherm, and thermodynamic investigations. Int J Biol Macromol 162:60–73
Mansoor S, Kour N, Manhas S, Zahid S, Wani OA, Sharma V, Ahmad P (2021) Biochar as a tool for effective management of

drought and heavy metal toxicity. Chemosphere 271:129458. https://doi.org/10.1016/j.chemosphere.2020.12

Mirabedini M, Kassaee MZ, Poorsadeghi S (2016) Novel magnetic chitosan hydrogel film, cross-linked with glyoxal as an efficient adsorbent for removal of toxic Cr(VI) from water. Arab J Sci Eng 42:115–124. https://doi.org/10.1007/s13369-016-2062-1

Mishra S, Rath CC, Das AP (2019) Marine microfiber pollution: a review on present status and future challenges. Marine Pollut Bullet 140:188–197. https://doi.org/10.1016/j.marpolbul.2019.01.0

Mohamed AK, Mahmoud ME (2020a) Encapsulation of starch hydrogel and doping nanomagnetite onto metal-organic frameworks for efficient removal of fluvastatin antibiotic from water. Carbohydr Polym 245:116438

Mohamed AK, Mahmoud ME (2020b) Nanoscale Pisumsativum pods biochar encapsulated starch hydrogel: a novel nano sorbent for efficient chromium (VI) ions and naproxen drug removal. Bioresour Technol 308:123263. https://doi.org/10.1016/j.biortech.2020.12326

Moharrami P, Motamedi E (2020) Application of cellulose nanocrystals prepared from agricultural wastes for synthesis of starch-based hydrogel nanocomposites: efficient and selective nano adsorbent for removal of cationic dyes from water. Bioresour Technol 313:123661

Nagireddi S, Katiyar V, Uppaluri R (2017) Pd(II) adsorption characteristics of glutaraldehyde cross-linked chitosan copolymer resin. Int J Biol Macromol 94:72–84. https://doi.org/10.1016/j.ijbiomac.2016.09.08

Nassaj-Bokharaei S, Motesharezedeh B, Etesami H, Motamedi E (2021) Effect of hydrogel composite reinforced with natural char nanoparticles on improvement of soil biological properties and the growth of water deficit-stressed tomato plant. Ecotoxicol Environ Saf 223:112576. https://doi.org/10.1016/j.ecoenv.2021.112576

Ngwabebhoh FA, Erdem A, Yıldız U (2016) Synergistic removal of Cu (II) and nitrazine yellow dye using an eco-friendly chitosan-montmorillonite hydrogel: optimization by response surface methodology. J AppPolym Sci 133:43664. https://doi.org/10.1002/app.43664

Nieder R, Benbi DK, Reichl FX (2018) Reactive water-soluble forms of nitrogen and phosphorus and their impacts on environment and human health. In: Nieder R, Benbi DK, Reichl FX (eds.) Soil Compon Human Health, pp 223–255. https://doi.org/10.1007/978-94-024-1222-25

Ohemeng-Boahen G, Sewu DD, Woo SH (2019) Preparation and characterization of alginate-kelp biochar composite hydrogel bead for dye removal. Environ Sci Pollut Res 26:33030–33042. https://doi.org/10.1007/s11356-019-06421-2

Pal A, Gin KY-H, Lin AY-C, Reinhard M (2010) Impacts of emerging organic contaminants on freshwater resources: review of recent occurrences, sources, fate and effects. Sci Total Environ 408:6062–6069. https://doi.org/10.1016/j.scitotenv.2010.09.0

Paulino AT, Belfiore LA, Kubota LT, Muniz EC, Tambourgi EB (2011) Efficiency of hydrogels based on natural polysaccharides in the removal of Cd^{2+} ions from aqueous solutions. Chem Eng J 168:68–76. https://doi.org/10.1016/j.cej.2010.12.037

Qi C, Liu J, Jin Y, Xu L, Wang G, Wang Z, Wang L (2018) Photo-crosslinkable, injectable sericinhydrogel as 3D biomimetic extracellular matrix for minimally invasive repairing cartilage. Biomaterials 163:89–104. https://doi.org/10.1016/j.biomaterials.2018.0

Rees F, Sterckeman T, Morel JL (2016) Root development of non-accumulating and hyperaccumulating plants in metal-contaminated soils amended with biochar. Chemosphere 142:48–55. https://doi.org/10.1016/j.chemosphere.2015.03.068

Reza MS, Ahmed A, Caesarendra W, Abu Bakar MS, Shams S, Saidur R, Azad AK (2019) *Acacia Holosericea*: an invasive species for bio-char, bio-oil, and biogas production. Bioengineering 6:33. https://doi.org/10.3390/bioengineering6020033

Rich AL, Crosby EC (2013) Analysis of reserve pit sludge from unconventional natural gas hydraulic fracturing and drilling operations for the presence of technologically enhanced naturally occurring radioactive material (TENORM). New Solut 23(1):117–135. https://doi.org/10.2190/ns.23.1.h

Rodríguez-Hernández Á, Camacho M, Henríquez-Hernández LA, Boada LD, Ruiz-Suárez N, Valerón PF, Luzardo OP (2016) Assessment of human health hazards associated with the dietary exposure to organic and inorganic contaminants through the consumption of fishery products in Spain. Sci Total Environ 557–558:808–818. https://doi.org/10.1016/j.scitotenv.2016.03.0

Rutigliano FA, Romano M, Marzaioli R, Baglivo I, Baronti S, Miglietta F, Castaldi S (2014) Effect of biochar addition on soil microbial community in a wheat crop. Eur J Soil Biol 60:9–15. https://doi.org/10.1016/j.ejsobi.2013.10.007

Sanyang ML, Ghani WAWAK, Idris A, Ahmad MB (2014) Hydrogel biochar composite for arsenic removal from wastewater. Desalin Water Treat 57:3674–3688. https://doi.org/10.1080/19443994.2014.989412

Shafiq Z, Ajmal M, Kiran S, Zulfiqar S, Yasmeen G, Iqbal M, Farooqi ZH, Ahmad Z, Sahiner N, Mahmood K, Ahmad HB, Al-Harrasi A (2019) Facile synthesis of hydrogel-nickel nanoparticle composites and their applications in adsorption and catalysis. Pure Appl Chem 91:1567–1582. https://doi.org/10.1515/pac-2018-1201

Shan R, He Y, Zi T, Wang G, Liu X, Han Z, Zhu Y (2018) Immobilization of calcined layered double hydroxide into alginate hydrogel beads for PNP and PAP removal: kinetics, isotherms, thermodynamics, and mechanism. Water Air Soil Poll 229:329. https://doi.org/10.1007/s11270-018-3976-x

Shan X, Yang L, Zhao Y, Yang H, Xiao Z, An Q, Zhai S (2022) Biochar/Mg-Al spinel carboxymethyl cellulose-La hydrogels with cationic polymeric layers for selective phosphate capture. J Colloid Interface Sci 606:736–747. https://doi.org/10.1016/j.jcis.2021.08.078

Sim DHH, Tan IAW, Lim LLP, Hameed BH (2021) Encapsulated biochar-based sustained release fertilizer for precision agriculture: a review. J Cleaner Prod 303:127018. https://doi.org/10.1016/j.jclepro.2021.127018

Simeng L, Chen G (2018) Using hydrogel-biochar composites for enhanced cadmium removal from aqueous media. Material Sci Eng 2:294–298. https://doi.org/10.15406/mseij.2018.02.00073

Singh A, Sharma PK, Garg VK, Garg G (2010) Hydrogels: a review. Int J Pharm Sci Rev Res 4:97–105

Sinha V, Chakma S (2019) Advances in the preparation of hydrogels for wastewater treatment: a concise review. J Environ Chem Eng 7:103295. https://doi.org/10.1016/j.jece.2019.103295

Sirajuddin NA, MdJamil MS, Mat Lazim MAS (2014) Effect of cross-link density and the healing efficiency of self-healing poly (2-hydroxyethyl methacrylate) hydrogel. e-Polymers 14:289–294. https://doi.org/10.1515/epoly-2014-0036

Song XD, Xue XY, Chen DZ, He PJ, Dai XH (2014) Application of biochar from sewage sludge to plant cultivation: influence of pyrolysis temperature and biochar-to-soil ratio on yield and heavy metal accumulation. Chemosphere 109:213–220. https://doi.org/10.1016/j.chemosphere.2014.01

Steinbeiss S, Gleixner G, Antonietti M (2009) Effect of biochar amendment on soil carbon balance and soil microbial activity. Soil Biol Biochem 41:1301–1310. https://doi.org/10.1016/j.soilbio.2009.03.016

Tan X, Liu Y, Gu Y, Xu Y, Zeng G, Hu X, Li J (2016) Biochar-based nano-composites for the decontamination of wastewater: a review. Bioresour Technol 212:318–333. https://doi.org/10.1016/j.biortech.2016.04.09

Tang J, Zhu W, Kookana R, Katayama A (2013) Characteristics of biochar and its application in remediation of contaminated soil. J Biosci Bioeng 116:653–659. https://doi.org/10.1016/j.jbiosc.2013.05.035

Tang S, Yang J, Lin L, Peng K, Chen Y, Jin S, Yao W (2020) Construction of physically crosslinked chitosan/sodium alginate/calcium ion double-network hydrogel and its application to heavy metal ions removal. Chem Eng J 393:124728. https://doi.org/10.1016/j.cej.2020.124728

Thakur S, Govender PP, Mamo MA, Tamulevicius S, Mishra YK, Thakur VK (2017) Progress in lignin hydrogels and nanocomposites for water purification: future perspectives. Vacuum 146:342–355. https://doi.org/10.1016/j.vacuum.2017.08.011

Thakur S, Pandey S, Arotiba OA (2016) Development of a sodium alginate-based organic/inorganic superabsorbent composite hydrogel for adsorption of methylene blue. Carbohydr Polym 153:34–46. https://doi.org/10.1016/j.carbpol.2016.06.104

Tong X, Li J, Yuan J, Xu R (2011) Adsorption of Cu(II) by biochars generated from three crop straws. Chem Eng J 172:828–834. https://doi.org/10.1016/j.cej.2011.06.069

Utech S, Boccaccini AR (2015) A review of hydrogel-based composites for biomedical applications: enhancement of hydrogel properties by addition of rigid inorganic fillers. J Mater Sci 51:271–310. https://doi.org/10.1007/s10853-015-9382-5

Wang W, Jin X, Zhu Y, Zhu C, Yang J, Wang H, Lin T (2016) Effect of vapor-phase glutaraldehyde crosslinking on electrospun starch fibers. Carbohydr Polym 140:356–361. https://doi.org/10.1016/j.carbpol.2015.12.061

Wu Y, Brickler C, Li S, Chen G (2021) Synthesis of microwave-mediated biochar-hydrogel composites for enhanced water absorbency and nitrogen release. Polym Testing 93:106996. https://doi.org/10.1016/j.polymertesting.2020

Wu Z, Chen X, Yuan B, Fu M-L (2019) A facile foaming-polymerization strategy to prepare 3D MnO_2 modified biochar-based porous hydrogels for efficient removal of Cd(II) and Pb(II). Chemosphere 239:124745. https://doi.org/10.1016/j.chemosphere.2019.12

Wu Z, Zhang P, Zhang H, Li X, He Y, Qin P, Yang C (2022) Tough porous nanocomposite hydrogel for water treatment. J Hazard Mater 421:126754. https://doi.org/10.1016/j.jhazmat.2021.126754

Yao W, Yu S, Wang J, Zou Y, Lu S, Ai Y, Wang X (2017) Enhanced removal of methyl orange on calcined glycerol-modified nanocrystallined Mg/Al layered double hydroxides. Chem Eng J 307:476–486. https://doi.org/10.1016/j.cej.2016.08.117

Yin W, Yang X, Feng-Ling S, Yi-Qiong Y, Xiao-Dong Z (2016) Ultrasonic-assisted synthesis of a superabsorbent composite hydrogel for the responsive and removal properties of Pb (II). Acta Phys-Chim Sin 32:2563–2573. https://doi.org/10.3866/PKU.WHXB201607122

Zainal SH, Mohd NHH, Suhaili N, Anuar FH, Lazim AM, Othaman R (2020) Preparation of cellulose-based hydrogels: a review. J Mater Res Technol 10:935–952. https://doi.org/10.1016/j.jmrt.2020.12.012

Zhang H, Xiao R, Li R, Ali A, Chen A, Zhang Z (2020a) Enhanced aqueous Cr(VI) removal using chitosan-modified magnetic biochars derived from bamboo residues. Chemosphere 266:128963. https://doi.org/10.1016/j.chemosphere.2020.12

Zhang L, Lu H, Yu J, Fan Y, Yang Y, Ma J, Wang Z (2018) Synthesis of lignocellulose-based composite hydrogel as a novel biosorbent for Cu^{2+} removal. Cellulose 25:7315–7328. https://doi.org/10.1007/s10570-018-2077-8

Zhang L, Tang S, Guan Y (2020b) Excellent adsorption-desorption of ammonium by a poly(acrylic acid)-grafted chitosan and biochart composite for sustainable agricultural development. ACS Sustain Chem Eng 44:16451–16462. https://doi.org/10.1021/acssuschemeng.0c05070

Zhao Y, Ma L, Zeng R, Tu M, Zhao J (2016) Preparation, characterization and protein sorption of photo-crosslinked cell membrane-mimicking chitosan-based hydrogels. Carbohydr Polym 151:237–244. https://doi.org/10.1016/j.carbpol.2016.05.067

Zheng Y, Huang D, Wang A (2011) Chitosan-g-poly(acrylic acid) hydrogel with crosslinked polymeric networks for Ni^{2+} recovery. Anal Chim Acta 687:193–200. https://doi.org/10.1016/j.aca.2010.12.026

Production of Biochar-Based Nanocomposites from Chemical and Biological Methods

Lata Verma and Jiwan Singh

Abstract

Advanced nanomaterials, including nanobiochar, have provided long-term sustainable solutions for a number of contemporary challenges. Due to their special qualities and wide range of applications in industries including energy, agriculture, materials, and environment, particularly in the remediation of organic and inorganic pollutants. Biochar nanostructured materials have been produced as a result of integrated biochar technology and nano-biotechnology, revolutionizing the study of relevant subjects. Giving detailed and current information about the production procedure is the main aim of this chapter. This chapter also highlights the chemical and biological methods used for nanobiochar production and also provides insight into the application of nanobiochar in several environmental domains, such as soil amendment, removal of contaminants, catalysts, and fuel additives.

Keywords

Application • Biological techniques • Nanobiochar composites • Synthesis methods

1 Introduction

Biochar is an extremely porous, carbonaceous material produced under a low-oxygen environment from a variety of biomass feedstock, often either vegetal or animal waste. It possesses strong anti-decomposition and aromatization properties (Daful and Chandraratne 2020). The thermal degradation process is used to convert the abundantly present biomass into carbonaceous material (Singh et al. 2015). The pyrolysis temperature and feedstock source significantly influence biochar production (Huang et al. 2021). The biochar can be synthesized by the process of gasification, pyrolysis, and dry carbonization of waste biomass, including organic municipal waste, forest leftovers, animal dung, manure, agricultural waste, etc. (Xiang et al. 2020; Verma and Singh 2022). Biochar can have diverse qualities depending on the production technique, feedstock, temperature of pyrolysis, and some other pre- and post-processing factors, which can lead to a broad number of uses and great interest among researchers. For example, biomass gasification and pyrolysis between temperatures 800–1000 °C and 300–700 °C, respectively, result in biochar, whereas torrefaction at 200–300 °C temperature results in biochar with varied properties (Basu 2018). Residence time, method of synthesis, and feedstock attributes also contribute to some of the biochar characteristics such as yield of biochar, specific surface area, elemental composition, and porosity. On the other hand, pre- and post-modification strategies can be used to enhance the properties of biochar. Biochar is viewed as an appealing source for the synthesis of carbon-based materials due to its low cost of production, abundant availability, customizable characteristics, and sustainability (Soffian et al. 2022; Verma and Singh 2023). It has been established that CO_2 generated during biomass combustion, which includes plant wastes (fallen leaves, fruit shells, seeds, etc.), is ultimately consumed by plants, making biochar a carbon–neutral source of energy. The carbon in biochar is very resistant to further deterioration, and it stays for a long period of time in the soil when applied on agricultural land (Singh et al. 2015). Consequently, biochar is considered as a carbon-negative source of energy that not only produces energy but also sequesters carbon, subsequently. Biochar has become a potential technique in relation to manage and

L. Verma (✉) · J. Singh
Laboratory of Environmental Nanotechnology and Bioremediation, Department of Environmental Science, Babasaheb Bhimrao Ambedkar University, Lucknow, 226025, Uttar Pradesh, India
e-mail: lataverma2501@gmail.com

J. Singh
e-mail: jiwansingh95@gmail.com

D. Mishra et al. (eds.), *Biochar-Based Nanocomposites for Contaminant Management*, Advances in Science, Technology & Innovation, https://doi.org/10.1007/978-3-031-28873-9_4

mitigate several environmental problems such as climate change and land degradation (Timmons et al. 2017; Singh et al. 2019; Arif et al. 2020).

Developing nanobiochar composites by reducing biochar particle size is an effective approach that can enhance the surface area, porosity, mobility, and other properties of bulk biochar particles. This method is also effective in converting subpar biochar into high-quality nanobiochar (Chausali et al. 2021). The nanobiochar composites can be produced through physical, chemical, and biological techniques. This chapter emphasizes more about the chemical (acid and alkali treatment, infusion with metal oxides/hydroxide) and biological (inoculation of microorganisms) methods of nanobiochar composites production. The definition of nanobiochar is given by several researchers as biochar with <100 nm particle size. Additionally, nanobiochar is a term used to refer to biochar with typical diameters ranging from 4 to 5 to about 500 nm (Naghdi et al. 2019b). A larger surface volume ratio boosts adsorption potential, surface energy, and also biological efficacy. Further biochar particle size reduction to a nano-sized range up to 100 nm or lesser enhances their characteristics (Lonappan et al. 2016; Taheran et al. 2016). Nanobiochar is therefore trending right now because it combines the advantages of biochar technology with nanotechnology.

Nanobiochar has evolved to address the limitations of biochar by enhancing its chemical, physical, and structural properties. The particle size in the nano-scale range is controlled potentially by exfoliation or by pyrolysis temperature, as documented in the literature (Chen et al. 2017). Recently, Noreen and Abd-Elsalam (2021) noted that nanobiochar composites have an abundance of functional groups with better thermal and mechanical stability and an increased specific surface area because of their nano-size that makes them special and useful (Chausali et al. 2021). Additionally, reactive organic species, which have greater adsorption capacity than that of macrobiochar, may be produced by larger negative zeta potentials, lower hydrodynamic radii, and more oxygenated functional groups and carbon defects (Greco et al. 2019; Ramanayaka et al. 2020a). Due to its porous structure, chemical, and physical characteristics, nanobiochar has recently emerged as a promising option for soil amendment and an excellent adsorbent. The nanobiochar materials demonstrated enhanced adsorption capacity for a variety of contaminants like heavy metals, polycyclic aromatic hydrocarbons, dyes, polychlorinated biphenyls, herbicides, etc., due to their microporosity, specific surface area, surface functional groups, and surface hydrophobicity (Gámiz et al. 2017; Rajput et al. 2022). Furthermore, due to increased pH and alkalinity, nanobiochar may also be used to counteract soil acidity. Nanobiochar composites frequently showed amazing improvement in a number of soil properties including surface functional groups, porous structure, and active binding sites, which boosted catalytically desolation ability and separation also became simple and quick (Goswami et al. 2022). Zhang et al. (2021) stated that nanobiochar nanocomposites derived from cellulosic nanocrystals consist of active sites more than that of unmodified or pristine biochar materials and efficiently act as a potent photocatalyst for the degradation of phenol. In a study, nitrate along with the coexisting ions of chloride and nitro-benzene were efficiently removed from wastewater using nano zero-valent iron biochar composites (Ahmed et al. 2018; Wei et al. 2019). In this chapter, we have given insight on the production of nanobiochar composites by various chemical and biological methods, followed by the utilization scenario of these composites in diverse environmental applications.

2 Production of Nanobiochar Composites

2.1 Chemical Methods

The intrinsic characteristics of nanobiochar can be simply modified, and they serve as a framework for the building of several functional materials employed in a variety of applications. Consequently, various methods including animation, surface modification, oxidation, etc., were used for the functionalizing of biochar materials in order to improve their performance in various environmental applications (Anupama and Khare 2021). The process of chemical modification includes a one and two-step process modification. In one-step activation process, activation and carbonization are both accomplished simultaneously in the presence of a chemical activator. Whereas in the two-step activation method, raw biomass is pretreated with the chemical agent and then carbonized or the precursor is firstly carbonized and then the carbonized material is treated with a chemical agent for chemical activation (Godwin et al. 2019). Usually, for chemical modification biochar with a high surface area is used as a substrate for material deposition. Such a technique leads to the development of absolutely new biochar surface functional groups.

2.1.1 Oxidizing

This method of chemical modification can either be used before biomass carbonization (in-situ modification) or modification of biochar after pyrolysis (post-treatment method). Carbon precursor treatment or treatment of biochar after pyrolysis during the process of chemical modification shows improvement in the biochar performance as an adsorbent or amendment (Rajapaksha et al. 2016). Oxidation is mainly carried out by acid and base addition. Various chemicals like acids (H_2O_2, HCl, H_2SO_4, H_3PO_4), bases (NaOH, KOH), ozone (O_3), etc., have been used for the production of biochar/nanobiochar composites. In a study, it was reported that for nanobiochar functionalization pure acid or acid mixtures (H_2SO_4/HNO_3/HCl, HNO_3/HCL, H_2SO_4/HNO_3) are used that enhance the formation of the carboxylic group and also adsorbed laccase activity efficiently (Naghdi et al. 2017). As stated by Uchimiya et al. (2012), carboxylic group formation on biochar enhanced by acid treatment is suitable for removing Cu, Pd, Cd, oxytetracycline, and sulfamethazine, etc. (Vithanage et al. 2015). While treatment with base results development of more aromatic functional groups or graphitic carbon and also enhances π–π interaction, formation of complexes, and surface precipitation that helps in better removal of cadmium, arsenic, tannic acid, etc. (Panahi et al. 2020). The biochar is heated at high temperatures along with a very strong base like KOH under an inert atmosphere for its activation for several hours for its application as a super-capacitor. To develop a porous nanostructure having an ideal pore size across a broad surface area, the mixture is then put through a series of washing with chemical and drying stages (Gupta et al. 2015).

2.1.2 Chemical Impregnation/Coating

The surface area of biochar can be increased by the introduction of metal oxides or nanoparticles via chemical coating or chemical impregnation process to biochar at different phases of pyrolysis. The production of biochar-based nanocomposites combines the benefits of nanotechnology as well as biochar application. Both carbon sequestration and water treatment can be achieved by the valorization of low cost and abundantly present organic biomass as biochar with efficient physical and chemical properties with higher functionalities as suggested by Tan et al. (2016). On the basis of chemical impregnation and coating, nanobiochar composited fabrication is divided into three categories: (i) magnetic nanobiochar composites, (ii) functional nanobiochar, and (iii) oxide or hydroxide nanobiochar.

Magnetic Nanobiochar Composites

Magnetic biochar-based nanocomposites can be produced by two methods one is the pre-treating precursor with iron salt or chemically co-precipitating iron oxide on biochar. The magnetic nanobiochar materials solve the problem of their separation from the aqueous medium, and they easily get separated from the aqueous solution via the application of a magnetic field. It also enhances the adsorption capacity of biochar materials. These methods include the coating of magnetic iron oxide which are of nano-size, such as cobalt ferrite ($CoFe_2O_4$) nanoparticles, ferric oxide (c-FeO_3), and ferrous oxide (Fe_3O_4) on biochar surface, therefore, assuring the introduction of magnetic properties and active site availability for removal of pollutants (Chausali et al. 2021). The magnetic nanobiochar composites have demonstrated effective removal of pollutants because of their enhanced properties like active binding site availability, increased surface area, pore-size distribution, ease of separation, and catalytic degradation capacity. Several studies reported that magnetic nanobiochar composites had superior adsorption capacity and can effectively adsorb various contaminants from water (Yan et al. 2015).

Functional Nanoparticle-Coated Nanobiochar

Functional nanoparticles comprising chitosan, graphene, ZnS nanocrystals, layered double hydroxide, carbon nanotubes (CNT), Fe^0 nanoparticles, etc., can be coated on biochar to generate more efficient nanobiochar composites that are also very beneficial in removing a variety of contaminants (Wan et al. 2018).

Oxide or Hydroxide Nanobiochar

Mainly three methods are employed for the generation of oxide/hydroxide nanobiochar composites: (i) pre-treating biomass by metal oxides; (ii) target element amelioration by the process of bioaccumulation; and (iii) insertion of particles of nano-metal oxide afterward pyrolysis (Panahi et al. 2020).

2.2 Biological Techniques

Biological modification technique includes anaerobic digestion or bacterial conversion of biochar that improves the surface area, anion exchange capacity, hydrophobicity, and cation exchange capacity and adds a more negative charge on the surface. Microorganisms (like fungi/bacteria, etc.) due to their microscopic size can easily colonize the porous structure of biochar and form a biofilm on both the interior and exterior surfaces that can preclude them from washing out and improve the capacity of biochar to adsorb or break down certain organic pollutants very effectively (Fig. 1). For instance, a study was conducted by Frankel et al. (2016), in which the removal of naphthenic acid was enhanced from 30 to 87% because of the biofilm introduction to the biochar. Another current study came to the conclusion that biologically modified biochar might be a potential method for removing pharmacological drugs from on-site sewage systems (Dalahmeh et al. 2018).

The relevance of biogenic nanobiochar made with microorganisms has increased recently. They are an excellent contender for biosorption and biodegradation of emerging pollutants due to their distinctive features including strong catalytic reactivity and large surface area. They are non-toxic due to their biogenic origin and become a sustainable

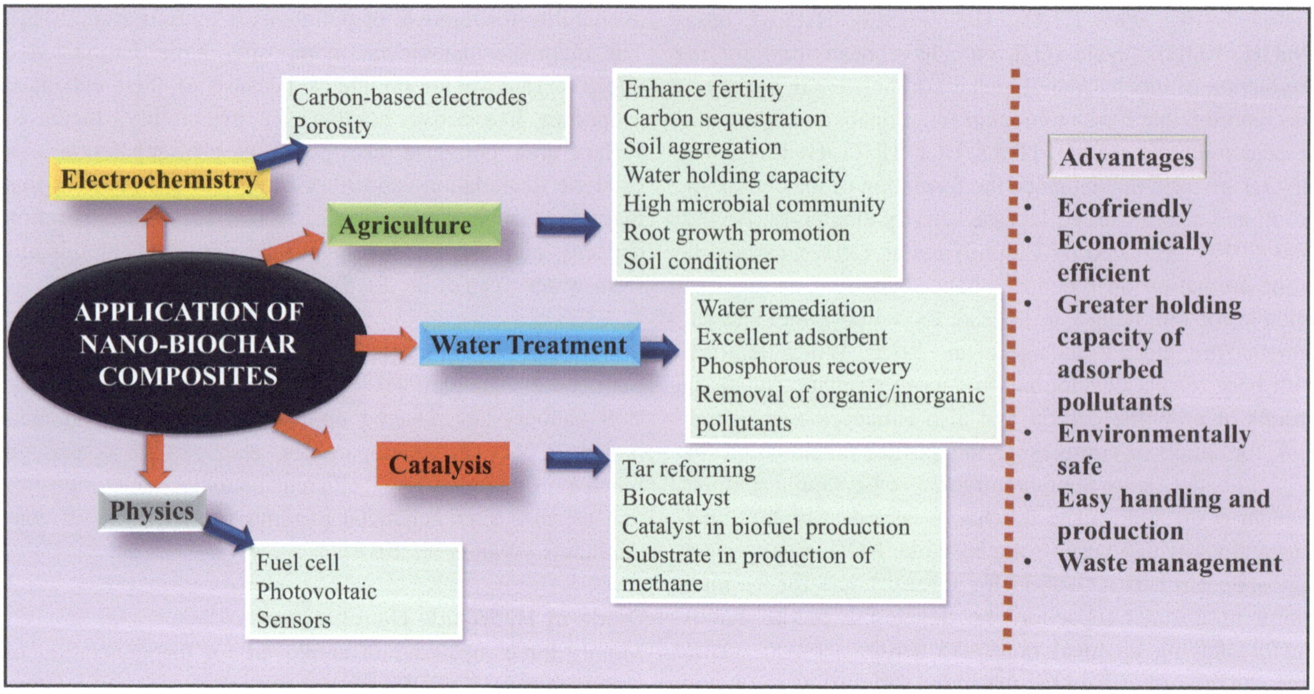

Fig. 1 Synthesis of biochar/nanobiochar via biological method

substitute for chemically produced nanomaterials (Mathur and Sanyal 2022; Sinha and Dahiya 2022). By settling on the large surface area of biochar and producing biofilms, microorganisms might alter inert biochar biochemically. This method of post-modification enables the creation of biochar composites (nanobiochar) with desired characteristic features. More specifically, it was possible to see the synchronous sorption of organic environmental pollutants and their breakdown by biochar /nanobiochar scaffold and the inoculated microorganisms. There are some microorganisms that are excellent biosorbents of various organic and inorganic pollutants (Panahi et al. 2020). Micro/nanobiochar has also been observed to advance sustainable agriculture operations by focusing on issues of global food security and reducing the emission of greenhouse gas (Ramadan and Abd-Elsalam 2020). The biomass residue generated after anaerobic digestion because of its appropriate redox potential and pH value can be effectively utilized for the fabrication of modified biochar. As discussed by Yao et al. (2018), biochar synthesized using anaerobically treated biomass exhibits excellent specific surface area, more negatively charged surface charge, and cation exchange capacity in comparison with pristine biochar and can be applied for the remediation of environmental pollutants.

Eltaweil et al. (2022) synthesized a novel silver nanobiochar composite using leaves of *Chenopodium ambrosioides* as a feedstock for biochar production, and its leaves extract as a reducing agent for the silver ions. The silver nanobiochar composite was then examined for its photocatalytic potential for the degradation of methylene blue dye from contaminated water. The method of green synthesis of nanobiochar composites has become a viable alternative to traditionally used physical and chemical methods by minimizing the secondary effects on the environment. Less energy consumption and the use of sustainable waste materials make this process more feasible (El-Maghrabi et al. 2021; Hosny et al. 2022).

3 Application of Nanobiochar Composites

Nanobiochar composites due to their chemical and physical properties have now been explored for a number of diverse environmental applications such as waste management, wastewater treatment, energy production, soil amendment, carbon sequestration, and phytoremediation and have been also used as electrode materials as illustrated in Fig. 2.

3.1 Nanobiochar as Adsorbent for Water Treatment

Nanobiochar has been extensively used for the removal of various organic and inorganic pollutants including heavy

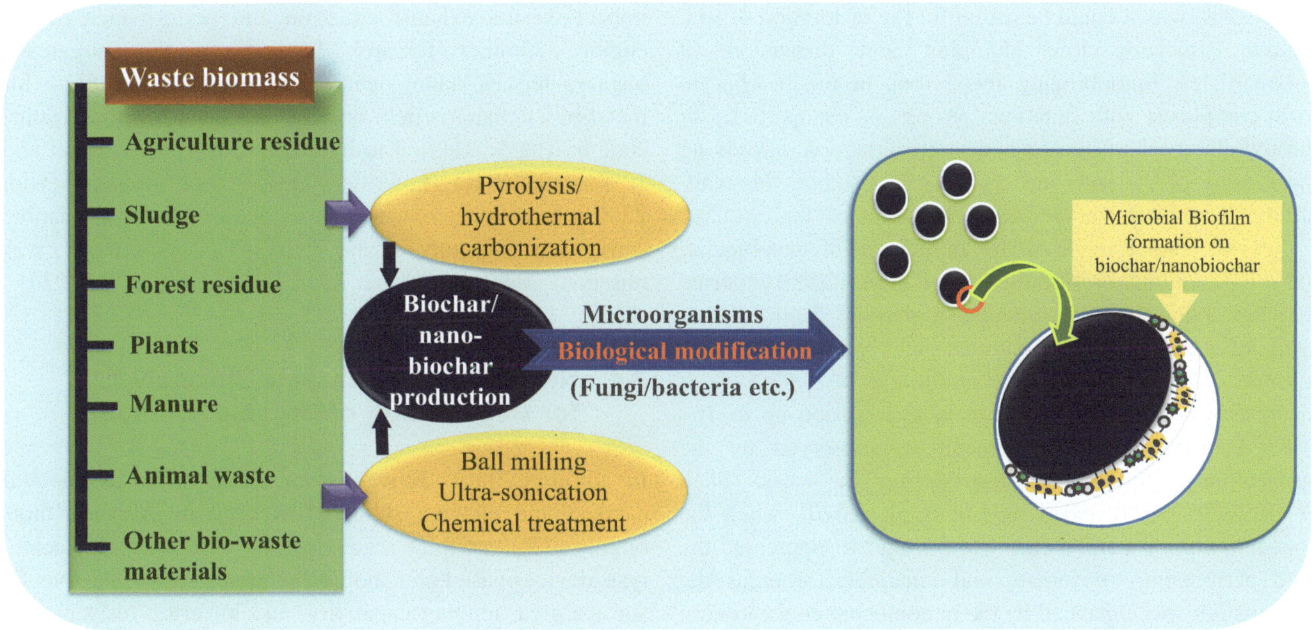

Fig. 2 Application of nanobiochar composites and their advantages

metals from wastewater as well as from contaminated groundwater by the process of adsorption. Due to the ease of use, efficacy, and accessibility of a variety of plant-based biosorbents, like biochar, fly ash, clay minerals, resins, etc., process of adsorption is frequently used to remove or reduce the pollutants from contaminated water bodies (Verma and Singh 2019; Verma et al. 2019). Nanobiochar composites provide excellent removal capacity for different variety of environmental contaminants including herbicides, pesticides, heavy metals, polychlorinated biphenyls (PCBs), and other organic/inorganic pollutants due to their high surface area, surface charge, and functional groups.

In a study conducted by Ramanayaka et al. (2020b), graphitic nanobiochar was prepared from waste material generated during the different processes in the thermal power plant and applied for the removal of Cr(VI) and Cd(II). The adsorption capacities of graphitic nanobiochar for Cd(II) and Cr(VI) were 520 mg/g and 922 mg/g, correspondingly. They also found that the Hills equation was best suited to the values of Cd(II) and Cr(IV) adsorption via nanobiochar material, demonstrating the cooperation of numerous binding sites of the synthesized nanobiochar. Cd(II), Pb(II), and Cu(II) were successfully adsorbed by engineered nanobiochar developed from cow bone meal with adsorption capacities of 166, 559, and 218 mg/g, respectively, as reported by Xiao et al. (2020). Ni(II) and dissolved arsenic (As) were removed from the aqueous solution with 1949 and 5987 mg/g adsorption capacities using nanobiochar (Lyu et al. 2018; Nath et al. 2019) while the removal of As with magnetic biochar showed lesser capacities as compared with the nanobiochar (Verma et al. 2021; Verma and Singh 2023). It was also reported that 95% removal of the pharmaceutical drug carbamazepine by nanobiochar was synthesized from solid pine wood (Naghdi et al. 2019b). Jia et al. (2020) reported the removal of glyphosate, an herbicide, through nanobiochar derived from phragmites with 269 mg/g adsorption capacity. Adsorption of sulfonamide antibiotics on nanobiochar derived from sugarcane bagasse, raw bamboo, and hickory chips is reported by Huang et al. (2020). Besides these, a number of research and reviews are reported on biochar-based nanocomposites or nanobiochar composites for contaminants removal from the environmental matrix (Tan et al. 2016).

3.2 Nanobiochar in Agriculture/Soil Amendment

It has been extensively documented that the use of biochar for soil amendment is highly efficient because it can enhance soil properties by restoring nutrients, preserving the biota of soil, and regulating bacterial activity (Singh et al. 2015, 2019). The intriguing properties (chemical and physical) of nanocomposites like enhanced porosity, surface area, etc., could greatly increase the nanocomposite's capacity to remove soil contaminants. The properties of nanobiochar composites, which could be useful for the restoration of soil, include allocating low clumping and dispersion of nanoparticles, immobilizing metal ions in the biochar to form complexes with functional groups of nanoparticles or through the mechanism of electrostatic attraction, improving the fertility of the soil, and checking by-products bioavailability (Rodriguez-Narvaez et al. 2019).

Several studies reported the application of nanobiochar for soil amendment/restoration. Chen et al. (2020) reported that the application of nanobiochar derived from coconut shells effectively reduced sediment yield by ~8–41%, with a runoff of about 39–74%, and nitrate loss of ~13–59%. It also contributes to the reduction in soil erosion up to 10–11%. Plant growth enhancement was observed by the application of nanobiochar derived from the wheat (Shen et al. 2020). As revealed by Liu et al. (2020), when the concentration of nanobiochar was 50 mg/l it suppressed the seed germination of tomato and cucumber, whereas the germination was enhanced by the nanobiochar concentration when it was increased to 100 mg/l. Tan et al. (2016) reported that the water-holding capacity of the soil was enhanced by nanobiochar incorporation. The area of leaf, plant height, and stem diameter of maize cultivated in soil amended with nanobiochar were noticeably higher than those cultivated in soil without nanobiochar (Yang et al. 2020). Feng et al. (2018), in their study, observed that when biochar is applied at a greater rate than nanobiochar, identical nitrate fixation rates were obtained. Thus, for nitrate fixation in soil and reducing nitrogen loss, use of nanobiochar could be an economically efficient option.

3.3 Nanobiochar Application in Fuel Additives

Nanobiochar has been employed as a sustainable approach in a variety of sectors, including energy, catalysis, biomedicine, fuel, and material science, in addition to uses in the environment and agriculture. Bio-nano enhancers are chosen over nano-metal because of their non-toxic characteristics since they play a significant role in lowering hazardous emissions of automobiles when utilized with fuels. To reduce the harmful emission of gases during the bioethanol blending with unadulterated diesel researchers focused on the development of metallic-nano additives but the metallic nanomaterials have carcinogenic in nature and are not environmentally friendly. Consequently, nanobiochar could be a great substitute for metal-based nanomaterials (Anupama and Khare 2021). In their investigation, Mirbagheri et al. (2020) discovered that adding nanobiochar to the blend of biodiesel, diesel, and ethanol had a favorable impact on the exhaust emission and performance of the engine. Another research has shown that sugarcane bagasse-derived nanobiochar is an excellent substitute for metal-based nanoparticle additives as gave terrific results. Nanobiochar's addition to hybrid fuel with 90% diesel and 10% of fossil oil shows less emission of pollutants with 2300 rpm engine speed and reduction in incombustible carbon and NOx up to 25% and 20.5%, respectively, was observed after nanobiochar addition (Ardebili et al. 2020).

3.4 Nanobiochar as Enzyme Biocatalyst for Degradation of Organic Pollutants

In addition, the effectiveness of nanobiochar as an acid catalyst and enzyme support has been investigated thoroughly. The variables regulating the efficiency of biocatalysts devices based on nanobiochar are by escalating specific surface area, loading capability, and lowering resistance to the transfer of mass. Because of nanobiochar materials' porous structure, more functional groups on the surface, and larger surface-to-volume ratio, these are demonstrated to be very effective as immobilization-sustaining materials in the enzyme system (Naghdi et al. 2018). A hydrogel technique to incorporate laccase between chitosan and nanobiochar was recently described by Naghdi et al. (2019a). They claimed that by attaching laccase to nanobiochar and chitosan, its stability under high temperature and storage conditions was improved. Several different organic contaminants have been degraded by using nanobiochar-held laccase. About 100% of diclofenac was removed from the wastewater by nanobiochar-supported laccase within 2 h.

3.5 Nanobiochar Application as Sensing Materials

In a number of electrochemical systems, nanobiochar is used as a sensing material. The toxin microcystin-LR was detected in water using a paper-based immunosensor that was built using conductive nanobiochar paper as a conducting layer because of its cheap cost (Yao et al. 2021). Mao et al. (2022) fabricated an ultrasensitive sensor (electrode) for monitoring Lead (Pb^{2+}) and cadmium (Cd^{2+}) using iron-imprinted nanobiochar with a high number of oxygenated function groups and good conductivity. The electrode can detect very low limits of Pb^{2+} and Cd^{2+} in the samples. The limit of detection for Cd^{2+} was 0.883 aM and 5.86 fM, respectively. This fabricated electrode with nanobiochar posed anti-interference to other organic substances and competitive ions, and it can be recycled up to seven times without sensing signal deterioration. There are an enormous number of possibilities for the application of nanobiochar that need to be investigated.

4 Futuristic Approach

- Despite their several benefits, the toxicity of nanobiochar materials on humans and the environment need to be addressed because only a few research or literature are available on their impact on different species of plants, soil biota, and a mammalian cell.
- Further research would concentrate on investigating methods for improving nanobiochar yield, electrode manufacture, production of super-capacitor, minimizing clumping, and application in the field of health care.
- Although there are a number of studies which have been performed on the development of modified biochar, much of these studies still focused on laboratory-scale application or at the theoretical stage. There are only a few reports on the performance of modified biochar and its large-scale application available. Researchers still need to provide greater assistance for the application procedure and commercial production of modified biochar. The marketing and use of engineered biochar are largely dependent on the expense of modification.
- More pertinent exploration will be needed further for the development and utilization of economically efficient and environmentally friendly organic chemicals for biochar modification because chemically (strong acids and alkali) modified biochar may impose the risk of secondary pollution in the environment. The possible harm that biochar might do to the ecosystems and the environment needs to get enough consideration.
- More research must be done on the regeneration and recovery of spent modified biochar under varied circumstances. Before choosing a technique and beginning large-scale manufacturing, it is necessary to evaluate its life cycle assessment and the environmental effects of the production of nanobiochar.
- It is recommended that the development of nanobiochar should be combined with gaseous and liquid biofuel production for the proper valorization of biomass. Nanobiochar production which requires more machinery and handling increases its production cost; therefore, its viability has to be further examined in terms of energy balance and technological economics.

5 Conclusions

Biochar is regarded as superior carbon material due to its affordability, ease of production, environmentally friendly, and adjustable physicochemical characteristics. Compared to other carbon-based nanomaterials, the production of biochar-based nanocomposites (e.g., nanobiochar) might be an economical and ecologically benign technique. As a result, its distinctive physicochemical characteristics, abundant surface functional groups, and easy surface modification have made nanobiochar a promising material with a wide range of their application. Chemical impregnation, coating of functional nanoparticles, and surface oxidation were employed for the nano-scaling of biochar. The process of impregnating biochar with nanoparticles extends the range of environmental applications of both biochar and nanotechnology. Chemical and biological methods are discussed briefly as some chemically modified nanobiochar have a harmful impact on humans and the environment because of the use of the synthetic chemicals. Thus, in this reference, biological techniques are more safe and sustainable technique as compared to chemical modification. The process of biological modification has been less studied and needs to be explored in the future.

References

Ahmad M, Ahmad M, Usman AR, Al-Faraj AS, Abduljabbar AS, Al-Wabel MI (2018) Biochar composites with nano zerovalent iron and eggshell powder for nitrate removal from aqueous solution with coexisting chloride ions. Environ Sci Pollut Res 25(26):25757–25771. https://doi.org/10.1007/s11356-017-0125-9

Ardebili SMS, Taghipoor A, Solmaz H, Mostafaei M (2020) The effect of nano-biochar on the performance and emissions of a diesel engine fueled with fusel oil-diesel fuel. Fuel 268:117356. https://doi.org/10.1016/j.fuel.2020.117356

Arif M, Jan T, Riaz M, Fahad S, Adnan M, Ali K, Rasul F (2020) Biochar; a remedy for climate change. In: Environment, climate, plant and vegetation growth. Springer, Cham, pp 151–171

Anupama, Khare P (2021) A comprehensive evaluation of inherent properties and applications of nano-biochar prepared from different methods and feedstocks. J Clean Prod 320:128759. https://doi.org/10.1016/j.jclepro.2021.128759

Basu P (2018) Biomass gasification, pyrolysis and torrefaction: practical design and theory. Academic Press

Chausali N, Saxena J, Prasad R (2021) Nanobiochar and biochar-based nanocomposites: advances and applications. J Agric Food Inf 5:100191. https://doi.org/10.1016/j.jafr.2021.100191

Chen L, Chen XL, Zhou CH, Yang HM, Ji SF, Tong DS, Zhong ZK, Yu WH, Chu MQ (2017) Environmental-friendly montmorillonite-biochar composites: facile production and tunable adsorption-release of ammonium and phosphate. J Clean Prod 648–659. https://doi.org/10.1016/j.jclepro.2017.04.050

Chen X, Zhou B, Wang Q, Tao W, Lin H (2020) Nano-biochar reduced soil erosion and nitrate loss in sloping fields on the Loess Plateau of China. CATENA 187:104346. https://doi.org/10.1016/j.catena.2019.104346

Daful AG, Chandraratne MR (2020) Biochar production from biomass waste-derived material. Encyclopedia of Renewable and Sustainable Materials 4:370-378. https://doi.org/10.1016/B978-0-12-803581-8.11249-4

Dalahmeh S, Ahrens L, Gros M, Wiberg K, Pell M (2018) Potential of biochar filters for onsite sewage treatment: adsorption and biological degradation of pharmaceuticals in laboratory filters with active, inactive and no biofilm. Sci Total Environ 612:192–201. https://doi.org/10.1016/j.scitotenv.2017.08.178

El-Maghrabi N, El-Borady OM, Hosny M, Fawzy M (2021) Catalytic and medical potential of a phyto-functionalized reduced graphene oxide-gold nanocomposite using willow-leaved knotgrass. ACS Omega 6(50):34954–34966. https://doi.org/10.1021/acsomega.1c05596

Eltaweil AS, Abdelfatah AM, Hosny M, Fawzy M (2022) Novel biogenic synthesis of a Ag@ biochar nanocomposite as an antimicrobial agent and photocatalyst for methylene blue degradation. ACS Omega 7(9):8046–8059. https://doi.org/10.1021/acsomega.1c07209

Feng Y, Sun H, Xue L, Wang Y, Yang L, Shi W, Xing B (2018) Sawdust biochar application to rice paddy field: reduced nitrogen loss in floodwater accompanied with increased NH_3 volatilization. Environ Sci Pollut Res 25(9):8388–8395. https://doi.org/10.1007/s11356-017-1059-y

Frankel ML, Bhuiyan TI, Veksha A, Demeter MA, Layzell DB, Helleur RJ, Hill JM, Turner RJ (2016) Removal and biodegradation of naphthenic acids by biochar and attached environmental biofilms in the presence of co-contaminating metals. Bioresour Technol 216:352–361. https://doi.org/10.1016/j.biortech.2016.05.084

Gámiz B, Cox L, Hermosín MC, Spokas K, Celis R (2017) Assessing the effect of organoclays and biochar on the fate of abscisic acid in soil. J Agric Food Chem 65(1):29–38. https://doi.org/10.1021/acs.jafc.6b03668

Godwin PM, Pan Y, Xiao H, Afzal MT (2019) Progress in preparation and application of modified biochar for improving heavy metal ion removal from wastewater. J Bioresour Bioprod 4(1):31–42. https://doi.org/10.21967/jbb.v4i1.180

Goswami L, Kushwaha A, Singh A, Saha P, Choi Y, Maharana M, Patil SV, Kim BS (2022) Nano-biochar as a sustainable catalyst for anaerobic digestion: a synergetic closed-loop approach. Catalysts 12(2):186. https://doi.org/10.3390/catal12020186

Greco C, Cosentino U, Pitea D, Moro G, Santangelo S, Patanè S, D'Arienzo M, Fiore M, Morazzoni F, Ruffo R (2019) Role of the carbon defects in the catalytic oxygen reduction by graphite nanoparticles: a spectromagnetic, electrochemical and computational integrated approach. Phys Chem Chem Phys 21(11):6021–6032

Gupta RK, Dubey M, Kharel P, Gu Z, Fan QH (2015) Biochar activated by oxygen plasma for supercapacitors. J Power Sourc 274:1300–1305. https://doi.org/10.1016/j.jpowsour.2014.10.169

Hosny M, Fawzy M, El-Fakharany EM, Omer AM, Abd El-Monaem EM, Khalifa RE, Eltaweil AS (2022) Biogenic synthesis, characterization, antimicrobial, antioxidant, antidiabetic, and catalytic applications of platinum nanoparticles synthesized from polygonum salicifolium leaves. J Environ Chem Eng 10(1):106806. https://doi.org/10.1016/j.jece.2021.106806

Huang J, Zimmerman AR, Chen H, Gao B (2020) Ball milled biochar effectively removes sulfamethoxazole and sulfapyridine antibiotics from water and wastewater. Environ Pollut 258:113809. https://doi.org/10.1016/j.envpol.2019.113809

Huang WH, Lee DJ, Huang C (2021) Modification on biochars for applications: a research update. Bioresour Technol 319:124100. https://doi.org/10.1016/j.biortech.2020.124100

Jia D, Liu M, Xia J, Li C (2020) Effective removal of aqueous glyphosate using $CuFe_2O_4$@ biochar derived from phragmites. J Chem Technol Biotechnol 95(1):196–204. https://doi.org/10.1002/jctb.6221

Liu W, Li Y, Feng Y, Qiao J, Zhao H, Xie J, Fang Y, Shen S, Liang S (2020) The effectiveness of nanobiochar for reducing phytotoxicity and improving soil remediation in cadmium-contaminated soil. Sci Rep 10(1):1–10. https://doi.org/10.1038/s41598-020-57954-3

Lonappan L, Rouissi T, Das RK, Brar SK, Ramirez AA, Verma M, Surampalli RY, Valero JR (2016) Adsorption of methylene blue on biochar microparticles derived from different waste materials. J Waste Manag 49:537–544. https://doi.org/10.1016/j.wasman.2016.01.015

Lyu H, Gao B, He F, Zimmerman AR, Ding C, Huang H, Tang J (2018) Effects of ball milling on the physicochemical and sorptive properties of biochar: experimental observations and governing mechanisms. Environ Pollut 233:54–63. https://doi.org/10.1016/j.envpol.2017.10.037

Mao D, Hu J, Duan P, Qin C, Piao Y (2022) Ultrasensitive and highly reusable electrochemical sensor with ion imprinted nanobiochar. Sens Actuators B: Chem 371:132490. https://doi.org/10.1016/j.snb.2022.132490

Mathur P, Sanyal D (2022) Removal of emerging contaminants from pharmaceutical waste through application of bio nanotechnology. In: Desalination water treat. Elsevier, pp 485–500. https://doi.org/10.1016/B978-0-323-85583-9.00028-4

Mirbagheri SA, Ardebili SMS, Kiani MKD (2020) Modeling of the engine performance and exhaust emissions characteristics of a single-cylinder diesel using nano-biochar added into ethanol-biodiesel-diesel blends. Fuel 278:118238. https://doi.org/10.1016/j.fuel.2020.118238

Naghdi M, Taheran M, Brar SK, Kermanshahi-Pour A, Verma M, Surampalli RY (2018) Pinewood nanobiochar: a unique carrier for the immobilization of crude laccase by covalent bonding. Int J Biol Macromol 115:563–571. https://doi.org/10.1016/j.ijbiomac.2018.04.105

Naghdi M, Taheran M, Brar SK, Kermanshahi-Pour A, Verma M, Surampalli RY (2017) Immobilized laccase on oxygen functionalized nanobiochars through mineral acids treatment for removal of carbamazepine. Sci Total Environ 584:93–401. https://doi.org/10.1016/j.scitotenv.2017.01.021

Naghdi M, Taheran M, Brar SK, Kermanshahi-Pour A, Verma M, Surampalli RY (2019a) Fabrication of nanobiocatalyst using encapsulated laccase onto chitosan-nanobiochar compositeInt. J Biol Macromol 124:530–536. https://doi.org/10.1016/j.ijbiomac.2018.11.234

Naghdi M, Taheran M, Pulicharla R, Rouissi T, Brar SK, Verma M, Surampalli RY (2019b) Pine-wood derived nanobiochar for removal of carbamazepine from aqueous media: adsorption behavior and influential parameters. Arab J Chem 12(8):5292–5301. https://doi.org/10.1016/j.arabjc.2016.12.025

Nath BK, Chaliha C, Kalita E (2019) Iron oxide Permeated Mesoporous rice-husk nanobiochar (IPMN) mediated removal of dissolved arsenic (As): chemometric modelling and adsorption dynamics. J Environ Manag 246:397–409. https://doi.org/10.1016/j.jenvman.2019.06.008

Noreen S, Abd-Elsalam KA (2021) Biochar-based nanocomposites: a sustainable tool in wastewater bioremediation. In: Aquananotechnology. Elsevier, pp 185–200. https://doi.org/10.1016/B978-0-12-821141-0.00023-9

Panahi HKS, Dehhaghi M, Ok YS, Nizami AS, Khoshnevisan B, Mussatto SI, Aghbashlo M, Tabatabaei M, Lam SS (2020) A comprehensive review of engineered biochar: production, characteristics, and environmental applications. J Clean Prod 270:122462. https://doi.org/10.1016/j.jclepro.2020.122462

Rajapaksha AU, Chen SS, Tsang DC, Zhang M, Vithanage M, Mandal S, Gao B, Bolan NS, Ok YS (2016) Engineered/designer biochar for contaminant removal/immobilization from soil and water: potential and implication of biochar modification. Chemosphere 148:276–291. https://doi.org/10.1016/j.chemosphere.2016.01.043

Rajput VD, Minkina T, Ahmed B, Singh VK, Mandzhieva S, Sushkova S, Bauer T, Verma KK, Shan S, Hullebusch ED, Wang B (2022) Nano-biochar: a novel solution for sustainable agriculture and environmental remediation. Environ Res 210:112891. https://doi.org/10.1016/j.envres.2022.112891

Ramadan MM, Abd-Elsalam KA (2020) Micro/nano biochar for sustainable plant health: present status and future prospects. In: Abd-Elsalam KA (ed.) Carbon Nanomater Agri-Food Environ Appl. pp. 323–357. https://doi.org/10.1016/B978-0-12-819786-8.00016-5

Ramanayaka S, Tsang DC, Hou D, Ok YS, Vithanage M (2020a) Green synthesis of graphitic nanobiochar for the removal of emerging contaminants in aqueous media. Sci Total Environ 706:135725. https://doi.org/10.1016/j.scitotenv.2019.135725

Ramanayaka S, Vithanage M, Alessi DS, Liu WJ, Jayasundera AC, Ok YS (2020b) Nanobiochar: production, properties, and multifunctional applications. Environ Sci Nano 7(11):3279–3302. https://doi.org/10.1039/D0EN00486C

Rodriguez-Narvaez OM, Peralta-Hernandez JM, Goonetilleke A, Bandala ER (2019) Biochar-supported nanomaterials for environmental applications. J Ind Eng Chem 78:21–33. https://doi.org/10.1016/j.jiec.2019.06.008

Shen Y, Tang H, Wu W, Shang H, Zhang D, Zhan X, Xing B (2020) Role of nano-biochar in attenuating the allelopathic effect from Imperata cylindrica on rice seedlings. Environ Sci Nano 7(1):116–126. https://doi.org/10.1039/C9EN00828D

Singh R, Kumar R, Babu JN, Srivastava P, Singh P, Raghubanshi AS (2015) Multifaceted application of crop residue biochar as a tool for sustainable agriculture: an ecological perspective. Ecol Eng 77:324–347. https://doi.org/10.1016/j.ecoleng.2015.01.011

Singh R, Srivastava P, Singh P, Sharma AK, Singh H, Raghubanshi AS (2019) Impact of rice-husk ash on the soil biophysical and agronomic parameters of wheat crop under a dry tropical ecosystem. Ecol Ind 105:505–515. https://doi.org/10.1016/j.ecolind.2018.04.043

Sinha N, Dahiya P (2022) Removal of emerging contaminants from pharmaceutical wastewater through application of bionanotechnology. In: Shah M, Rodriguez-Couto S, Biswas J (eds.) Development in Wastewater Treatment Research and Processes. pp. 247–264. https://doi.org/10.1016/B978-0-323-85583-9.00004-1

Soffian MS, Halim FZA, Aziz F, Rahman MA, Amin MAM, Chee DNA (2022) Carbon-based material derived from biomass waste for wastewater treatment. Environ Adv 9:100259. https://doi.org/10.1016/j.envadv.2022.100259

Taheran M, Naghdi M, Brar SK, Knystautas EJ, Verma M, Ramirez AA, Surampalli RY, Valéro JR (2016) Adsorption study of environmentally relevant concentrations of chlortetracycline on pinewood biochar. Sci Total Environ 571:772–777. https://doi.org/10.1016/j.scitotenv.2016.07.050

Tan XF, Liu YG, Gu YL, Xu Y, Zeng GM, Hu XJ, Liu SB, Wang X, Liu SM, Li J (2016) Biochar-based nano-composites for the decontamination of wastewater: a review. Bioresour Technol 212:318–333. https://doi.org/10.1016/j.biortech.2016.04.093

Timmons D, Lema-Driscoll A, Uddin G (2017) The economics of biochar carbon sequestration in Massachusetts. University of Massachusetts, UMass Clean Energy Extension

Uchimiya M, Bannon DI, Wartelle LH (2012) Retention of heavy metals by carboxyl functional groups of biochars in small arms range soil. J Agric Food Chem 60(7):1798–1809. https://doi.org/10.1021/jf2047898

Verma L, Azad A, Singh J (2021) Performance of a novel iron infused biochar developed from Raphanus sativus and Artocarpus heterophyllus refuse for trivalent and pentavalent arsenic adsorption from an aqueous solution: mechanism, isotherm and kinetics study. Int J Phytoremed 2021:1–14. https://doi.org/10.1080/15226514.2021.1985078

Verma L, Siddique MA, Singh J, Bharagava RN (2019) As (III) and As (V) removal by using iron impregnated biosorbents derived from waste biomass of Citrus limmeta (peel and pulp) from the aqueous solution and ground water. J Environ Manag 109452. https://doi.org/10.1016/j.jenvman.2019.109452

Verma L, Singh J (2019) Synthesis of novel biochar from waste plant litter biomass for the removal of Arsenic (III and V) from aqueous solution: a mechanism characterization, kinetics and thermodynamics. J Environ Manage 248:109235. https://doi.org/10.1016/j.jenvman.2019.07.006

Verma L, Singh J (2022) Arsenic adsorption from aqueous solution and groundwater using monometallic (Fe) and bimetallic (Fe/Mn) Tectona biochar synthesized from plant refuse: mechanism, isotherm, and kinetic study. Environ Eng Res 28(3):220110. https://doi.org/10.4491/eer.2022.110

Verma L, Singh J (2023) Removal of As (III) and As (V) from aqueous solution using engineered biochar: batch and fixed-bed column study. Int J Environ Sci Technol 20:1961–1980. https://doi.org/10.1007/s13762-022-03920-7

Vithanage M, Rajapaksha AU, Zhang M, Thiele-Bruhn S, Lee SS, Ok YS (2015) Acid-activated biochar increased sulfamethazine retention in soils. Environ Sci Pollut Res 22(3):2175–2186. https://doi.org/10.1007/s11356-014-3434-2

Wan S, Wu J, Zhou S, Wang R, Gao B, He F (2018) Enhanced lead and cadmium removal using biochar-supported hydrated manganese oxide (HMO) nanoparticles: Behavior and mechanism. Sci Total Environ 616:1298–1306. https://doi.org/10.1016/j.scitotenv.2017.10.188

Wei G, Zhang J, Luo J, Xue H, Huang D, Cheng Z, Jiang X (2019) Nanoscale zero-valent iron supported on biochar for the highly efficient removal of nitrobenzene. Front Environ Sci Eng 13(4):1–11. https://doi.org/10.1007/s11783-019-1142-3

Xiang W, Wan Y, Zhang X, Tan Z, Xia T, Zheng Y, Gao B (2020) Adsorption of tetracycline hydrochloride onto ball-milled biochar: governing factors and mechanisms. Chemosphere 255:127057. https://doi.org/10.1016/j.chemosphere.2020.127057

Xiao J, Hu R, Chen G (2020) Micro-nano-engineered nitrogenous bone biochar developed with a ball-milling technique for high-efficiency removal of aquatic Cd (II), Cu (II) and Pb (II). J Hazard Mater 387:121980. https://doi.org/10.1016/j.jhazmat.2019.121980

Yan L, Kong L, Qu Z, Li L, Shen G (2015) Magnetic biochar decorated with ZnS nanocrytals for Pb(II) removal. ACS Sustain Chem Eng 3(1):125–132. https://doi.org/10.1021/sc500619r

Yang Y, Zhou B, Hu Z, Lin H (2020) The effects of nano-biochar on maize growth in northern Shaanxi Province on the Loess Plateau. Appl Ecol Environ Res 18:2863–2877. https://doi.org/10.15666/aeer/1802_28632877

Yao L, He L, Yang Y, Zhang Y, Liu Z, Liang L, Piao Y (2021) Nanobiochar paper based electrochemical immunosensor for fast and ultrasensitive detection of microcystin-LR. Sci Total Environ 750:141692. https://doi.org/10.1016/j.scitotenv.2020.141692

Yao Y, Zhang Y, Gao B, Chen R, Wu F (2018) Removal of sulfamethoxazole (SMX) and sulfapyridine (SPY) from aqueous solutions by biochars derived from anaerobically digested bagasse. Environ Sci Pollut Res 25(26):25659–25667. https://doi.org/10.1007/s11356-017-8849-0

Zhang Y, Zhao G, Xuan Y, Gan L, Pan M (2021) Enhanced photocatalytic performance for phenol degradation using ZnO modified with nano-biochar derived from cellulose nanocrystals. Cellulose 28(2):991–1009. https://doi.org/10.1007/s10570-020-03581-0

Comparative Investigation of Biochar-Based Nanocomposites Over Pristine Biochar: An Overview

Disha Mishra, Shilpi Jain, Puja Khare, and Rishikesh Singh

Abstract

The role of biochar has been intensified currently into various fields due to its advanced physicochemical features. Their role has currently extended many advanced research aspects like energy storage, fuel cells, transportation activities, and membrane development. However, the synthesis of biochar for their extended applications requires specific reaction conditions, which are sometimes compromised during synthesis. Additionally, due to application in smaller grinded powder form, their separations and recyclability are always a matter of concern. To overcome these obstacles for the smooth application, the modifications of biochar into the nanocomposites would be a viable and application-oriented strategy. The preparations of biochar nanocomposites through the impregnation of various chemical entities (nanometal/metal oxides, hydroxides, magnetic biochar, or other functional groups) will enhance their sorption potential. This chapter will present an overview about the comparison of the pristine biochar with the biochar-based nanocomposites along with summarizing the ongoing research on this field. Further, the chapter will discuss the potential of both pristine biochar and their nanocomposites in various applications and finally finding of the current loopholes that must be addressed in future research experiments.

D. Mishra (✉) · P. Khare
Central Institute of Medicinal and Aromatic Plants, Lucknow, 226015, India
e-mail: mishra.disha1@gmail.com

P. Khare
e-mail: pujakhare@cimap.res.in

S. Jain
Maharaja Sayajirao University of Baroda, Gujarat, 390002, India
e-mail: shilpijain55@gmail.com

R. Singh
Department of Botany, Panjab University, Chandigarh, 160014, India
e-mail: rishikesh.iesd@gmail.com

Keywords

Nanocomposites • Nanometal • Magnetic • Sorption • Recyclability

1 Introduction

The growing population and rapid industrialization have created immense pressure on environmental sustainability in terms of resource degradation, contamination, global food security, and severe climate change (Singh et al. 2019). There is an alarming situation now for the remediation and management of these globally recognized issues. For this, biochar has bloomed with its extraordinary and multidisciplinary properties regarding environmental management. Biochar is a recalcitrant carbonaceous matrix obtained via pyrolysis of biomass, and it was primarily introduced as a soil conditioner and carbon sequestration material. In the pyrolysis of the organic materials, the simpler compounds and gaseous products are produced along with the char. The origin of biochar was documented from high fertility soil called *terra preta* ("dark earth") situated in the Amazon Basin (Lehmann et al. 2003). Due to their extraordinary features like higher surface area, functional groups, zeta potential, porosity, recalcitrant carbon, and nutrient-rich nature, biochar can be used for agricultural farming, reduction of greenhouse gas emissions, remediation of pollutants, soil quality improvement, and energy production (Singh et al. 2015). Moreover, the production of biochar is economic and sustainable as it can convert waste biomass into biochar at a very low cost. Recently, many studies have documented the application of biochar for agriculture, environmental, engineering, energy, and carbon sequestration at a very low amount. Therefore, the application of biochar in a small amount for targeted applications can improve the economics and sustainable use of biochar production (Wang et al. 2020). The properties of the biochar

D. Mishra et al. (eds.), *Biochar-Based Nanocomposites for Contaminant Management*, Advances in Science, Technology & Innovation, https://doi.org/10.1007/978-3-031-28873-9_5

such as size, water holding capacity, nutrient amount, cation exchange capacity, zeta potential, aromaticity, and surface chemistry entirely depend on the feedstock and pyrolysis operating parameters. Generally rising pyrolysis temperature is responsible for higher C content, surface area, and porosity, while lower O and H content and bulk density (Liu et al. 2015).

The production of bulk biochar (size 0.04–20 mm) is mainly applied for the agriculture and environmental application, but decreasing the particle size (10–600 μm) further has scaled up the properties of biochar. However, the applicability of biochar for various applications such as adsorption, recyclability, and output efficiency is sometimes restricted due to a higher load of contaminants or interaction with other molecules in the matrices. Also, the powder form of the biochar makes it difficult to separate from the medium (Zhao et al. 2021). Therefore, to avoid these limitations of biochar, current research on biochar has been diverted to the production of the engineered surface of biochar depending upon the application suitability. The reduction of the biochar into nanosize or construction of their nanocomposites (up to 100 nm or below it) has turned out an excellent option due to increased surface area and energy, adsorption potential, and consequently biological effectiveness (Chausali et al. 2021; Khare 2021). Hence, the engineering technologies for the production of nanocomposites or nanobiochar amalgamate through the benefits of nanotechnology present an advanced formulation of biochar for various applications (Tan et al. 2016). The modifications of biochar structures through various techniques have generated novel and efficient biochar materials. Production of biochar with other nanomaterials has created revolutionary multiple advantages such as waste management, low cost, carbon sequestration, pollutant management, and energy production (Chausali et al. 2021). Thus, the benefits of biochar-based nanocomposites over pristine biochar are quite high and engineering allowed them to gain optimum characteristics for specific applications.

Therefore, current chapter will present the fabrication, properties, and performance of biochar-based nanocomposites, and further their comparison with the pristine biochar will be discussed. In addition to this, the chapter will also throw a light on the current ongoing research scenario and existing knowledge gaps for further improvement.

2 Production of Biochar and Their Nanocomposites

The elucidation of production methods involved for biochar is a crucial step as the major features of the produced biochar lie on the production routes. However, biochar production can be done through many methods such as gasification, slow and fast pyrolysis, torrefaction, and hydrothermal carbonization (Kambo and Dutta 2015).

In the slow pyrolysis method, the biochar production occurred over a broad temperature range (300–800 °C), with a slow heating rate (5–7 °C/min) for more than one hour. Generated biochar has more yield (30%) in slow pyrolysis than that in fast pyrolysis (12%) or gasification (10%) (Inyang and Dickenson 2015). During the pyrolysis, the organic compounds of the feedstocks thermally break down and generated three major fractions, i.e., in solids: biochar, in volatile liquids: bio-oil, and non-condensable gases (CO, CO_2, CH_4, and H_2) (Ashwath et al. 2018). The physical reactions occurring during pyrolysis are dependent on the type of feedstocks, heating rates, temperatures, and reactor conditions. The judicious conversion of major elements like carbon, hydrogen, and oxygen into gases and other volatile compounds during pyrolysis directly governs the aromaticity, degradability, and polarity of the biochar while pyrolysis temperature correlates yield, surface functionalities, area, pore volume, and carbon content. Due to the decrease in the reaction temperature, the molecular orientation varied as the interplanar distances decreases (Weber and Quicker 2018).

On the other side, the construction of biochar-based nanocomposites offers improved characteristics due to the impregnation of foreign chemical entities in the pristine biochar structures, thus expanding their applicability. Biochar-based nanocomposites can be divided into the following types, i.e., nanometal/oxides/hydroxides based, magnetic biochar, and other types of functionalized biochar (Leng et al. 2021). Their fabrication can be done through the impregnation of metal oxides/hydroxides of (Ca, Zn, Mg, Mn, Ce, Ti, Al, Ni, Fe) or the nanoparticles (chitosan, cellulose, carbon nanotubes (CNT), layered double hydroxides (LDHs), nanoscale zero-valent iron (nZVI), graphene, or other biogenic nanoparticles) (Sizmur et al. 2017). However, they can be decorated on the surface of the biochar during pre- or post-treatment of biomass (Fig. 1). The resultant composites showed significant improvement in the properties like surface area, porosity, pore size distribution, catalytic ability, chemically reactive, and recyclability than pristine biochar. Many studies have shown the excellent adsorption ability and catalytic degradation ability of modified biochar for various organic contaminants (Rajapaksha et al. 2016; Huang et al. 2019; Chausali et al. 2021). Thus, this practice has now become crucial for extension in the specificity and applicability of biochar. However, the associated toxicity with the use of nanomaterials cannot be ignored. The coming section will describe the method for the construction of biochar-based nanocomposites briefly, as the preparation of biochar-based nanocomposites has been elaborated in various earlier chapters of the book.

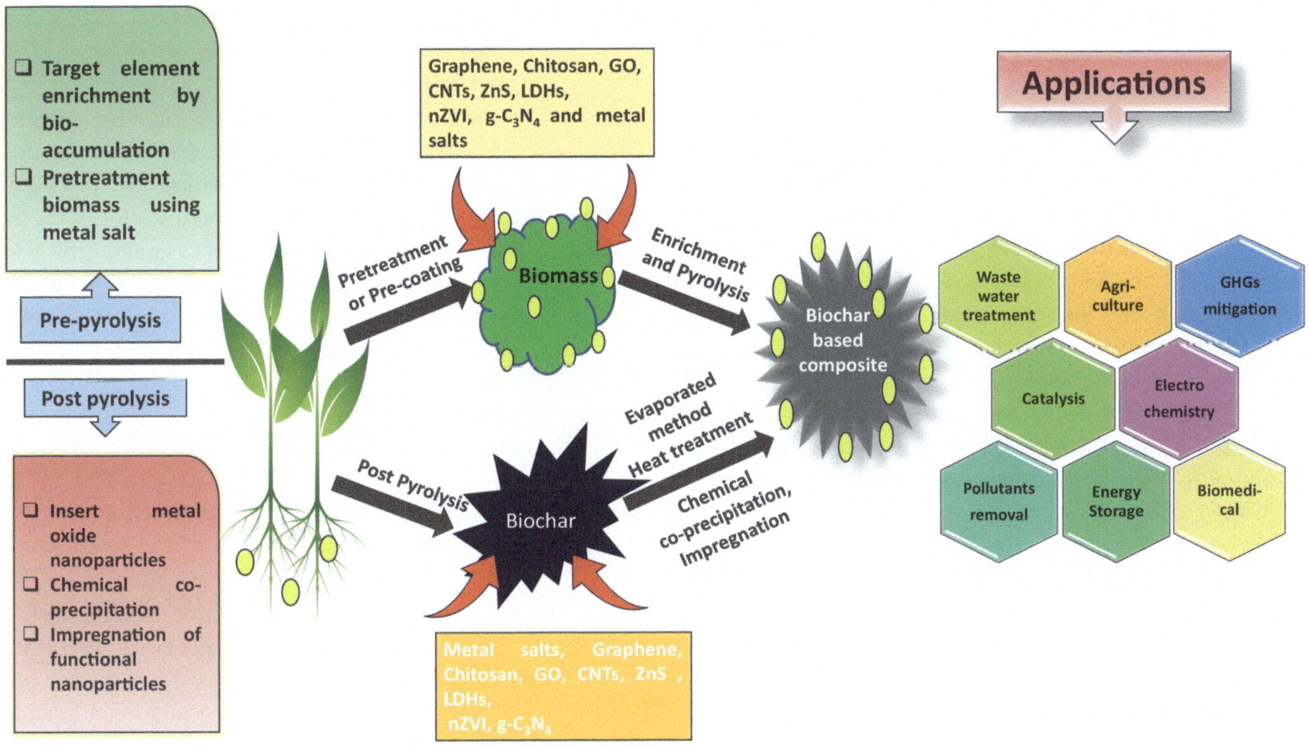

Fig. 1 Various production routes and application of biochar-based composites (yellow circles are showing inserted functionalities on the biochar surface) CNT: carbon nanotubes, LDH: layered double hydroxides, GO: Graphene oxides, nZVI: nano-zero-valent iron

2.1 Nanometal Oxide/Hydroxide-Biochar Composites

The impregnation of various metal oxides/hydroxides on the biochar can significantly improve the biochar functionality. The preparation of these types of composites can be done via three routes: (i) pre-treatment of the biomass by metal salt, (ii) pre-pyrolysis bioaccumulation of target element, and (iii) post-pyrolysis metal oxide nanoparticles loading (Huang et al. 2019). In the former one, the biomass is treated with some metal salts before pyrolysis which results into metal ions enrichment onto or interior of the biochar surface. These metal salts get converted into metal oxides/hydroxides after pyrolysis, and thus, generating biochar-based nanocomposites. The transformed surface of biochar was prepared through the introduction of some of the common metal salts where the addition of $AlCl_3$ (into aluminum oxide and aluminum hydroxide) (Tran et al. 2021), dual $CaCl_2$ and $ZnCl_2$ (into $CaCO_3$/ZnO) (Chen et al. 2022), $FeCl_3$ transformation into FeOOH, Fe_3O_4 and Fe_2O_3 (Yan et al. 2021), $KMnO_4$ (into MnO) (Deng et al. 2022), $MgCl_2$ (into MgO) (Shi et al. 2021), $MnCl_2$ (into MnO) (Liao et al. 2022), etc. Due to these modifications, the surface area and active functional sites of biochar get increased, which further helps in the removal of various contaminants. Recently, in a study, the adsorption efficiency of iron-modified biochar (FeYBC) was compared with pristine biochar (YBC) for Cr(VI) and phenol. The result showed that FeYBC was 310 and 16 times more efficient than YBC for Cr(VI) and phenol removal, respectively. They were more than 90% efficient for the adsorptive removal of those contaminants. The generation of several O-containing groups such as Fe_2O_3, FeO, FeOOH, and Fe_3O_4 on the FeYBC surface removed Cr(VI) through ion exchange and surface complexation while π–π interaction and electron donor–acceptor complex were mainly responsible for phenol removal, suggesting the more efficient nature of modified biochar composites (Dong et al. 2022).

While for bioaccumulation, the insertion of specific (target) chemical elements was performed inside the biomass during its growth. Then, the collected metal element could be converted into hydroxide/oxide during pyrolysis. This method can also turn out to be an innovative method for the remediation of contaminants through accumulation in the tissue of the hyperaccumulator plants (Panahi et al. 2020). These plants can be further used for the synthesis of biochar. However, for the post-pyrolysis methods, heat treatments, evaporative methods, direct hydrolysis, and conventional wet impregnation were used.

2.2 Magnetic Biochar

Inspired by traditional magnetic force for the isolation of the compounds, researchers have introduced magnetically activated biochar surfaces for the remediation of contaminants and ease of separation by applying magnetic fields. Magnetic biochar can be prepared through the pre-treatment of biomass and post-loading of the magnetic entity. Generally, chemical modification is practiced during magnetic modification to avoid any loss of porous surface due to magnetic medium. Various chemical groups of nitrogen (NH_2 and $CO\text{-}NH_2$), oxygen (COOH, OH), and sulfur (SH, and $S(=O)_2$) can be introduced for further chelation with contaminants (Zhu et al. 2018a). A novel magnetic biochar with zirconium hydroxide showed highest adsorption quantity of 107.6 mg/g and 40.8 mg/g at pH 6.5 and 8.5, for As(III) and As(V), respectively, owing to complexation reaction and electrostatic attraction onto ZBC (Peng et al. 2022). In another study, magnetic biochar composites with amino groups demonstrated efficient phosphorus (P) removal. The kinetics of adsorption has suggested a pseudo-second-order kinetic model with the adsorption of 53.56 and 62.06 mg P/g for modified biochar. However, the recovery from the swine water was 88.2% and 96.5% by MTBC-2 N and MTBC-4N, respectively (Zhang et al. 2022). The review has also discussed the importance of magnetic-modified biochar for the adsorption of heavy metals and organic pollutants, including the numerous mechanisms involved and the role of functional group grafting. They have also highlighted a deep insight into the future application of magnetic-modified biochar (Qu et al. 2022).

2.3 Other Types of Functionalized Biochar

Production of functional biochar can also be done through the coating of functional nanoparticles either through in situ or post-treatment. Till now, many engineered nanoparticles such as chitosan, cellulose nanofibers, nanocrystals, graphene, clay, carbon nanotubes (CNTs), nZVI, LDHs, and biogenic nanoparticles can be inserted onto the biochar surface. They are predominantly highly advantageous due to the synergistic properties of foreign functional entities and existing modified biochar (Tan et al. 2016). The new groups can be coated on the biochar before pyrolysis where they can be enriched onto biomass directly through the mixing and drying method. Other routes for the functional nanoparticles are embedded in the matrix post-pyrolysis. This method is successful and produces a superior hybrid than pristine biochar due to the introduction of more accurate active sites for selective removal of the contaminants. The N-doped carbon nanotubes composites supported on porous biochar (Ni@N-K-C) have ensured the mass diffusion of Cr(VI) ions through surface adsorption reactions (Zhu et al. 2021). The review published by Fang et al. (2021) has also discussed the importance of LDH/biochar composites as adsorbents and catalysts for the pollutants removal with superior qualities like chemical stability, more surface functional groups, excellent anion exchangeability, and remarkable electronic properties. Recently, biochar surface modification with the low-cost adsorbent such as clay, bentonite, montmorillonite, and kaolinite has been practiced either in situ or post-pyrolysis (Chu et al. 2020). These kinds of sustainable and environment-friendly approaches can be turned into fruitful options for future research for pollutant removal through biochar-based nanocomposites.

3 Pristine Biochar Versus Biochar-Based Nanocomposites

The characteristics of biochar are the major deciding factors for their further applications in various fields. However, the physicochemical properties generally varied as per the feedstock and pyrolysis conditions. Some important features of the pristine biochar like pH, porosity, surface area, organo-chemical characteristics, pH, and elemental compositions are discussed here. pH value is an important factor of biochar for their agricultural application. The biomass is either slightly acidic or basic (pH: 5.0–7.5), and during pyrolysis mainly acidic functional groups (−COOH, −CHO) get detached, turning the whole biochar basic in nature. Also, the amount of ash in the biochar is a deciding factor for pH of biochar. However, the high ash content in biochar limits their industrial application. The parent feedstocks are responsible for the final ash content of biochar. The major components of the ash are SiO_2, CaO, and K_2O with a little amount of P_2O_5 (especially in animal residues), Al_2O_3, and MgO (Gunamantha and Widana 2018).

The surface area and porosity of biochar are interrelated, and their distribution of the surface decides the heterogeneity of biochar. During the pyrolysis, the dehydration results into the loss of volatiles leaving the pore spaces. Slow pyrolysis has generated more pores in woody biochar owing to the slow degradation of the lignin content (Weber and Quicker 2018). Certain rise in the temperature (>400 °C) results into the thermal condensation of organic matter and the generation of micropores (He et al. 2021). The drastic temperature rise (400–500 °C) turns the amorphous carbons into crystalline carbons via condensation, followed by volatile loss, and thus, more voids generation turn to the formation of pores (Keiluweit et al. 2010). Also, the distribution of pores, shapes, and surface to volume ratio directly affect the application of biochar. However, the appropriate combination of both biomass and pyrolysis temperature influences the surface characteristics of the biochar. The alkaline nature of the biomass is helpful in the self-activation for pore development (Leng et al. 2021). The impact of the biomass on the surface area was shown as the lotus stems-derived biochar (1610 m^2/g) has more surface area than lotus leaves-based biochar (1039 m^2/g), due to the presence of K ion which acted as an activator for pore formation (Zhang et al. 2017). Thus, tuning in the surface properties should be according to the target sorbate molecules. It was observed that in co-pyrolysis method, the interaction of more than one type of biomass leads to distinct pore structures. A study has explored that covering one biomass with other biomass resulted in pore blockage, thereby lowering porosity and surface area (Wang et al. 2018). The biomass having high lignin content is always stable thereby preserving the porous structures (Sun et al. 2014). Another crucial factor for the biochar performance is the morphology which is also closely related to pore connectivity and shapes and feedstocks. The biochar derived from lignocellulosic biomass is more porous than that of animal based (Kumar et al. 2020). However, the correlation between biomass, particle size, and porosity cannot be defined. Various components of biomass like lignocellulose, fats, phytosterols, and phenolics, and inorganics, minerals play a prominent role in the nutrient and elemental composition of the final biochar. The heating rate also affects the surface area and total pore volume in a zig-zag manner. An optimum heating rate is always required for the maximum surface area and porosity of pristine biochar. Similarly, the residence time is also a component that needs to be taken care in the pyrolysis process and required an optimum residence time for the maximum value of the surface area and total pore volume. Therefore, the tuning in the biochar characteristic through various surface modifications method further extends their application in wastewater treatment, soil conditioner, catalytic degradation, energy storage, and carbon captures.

There are plenty of documents available regarding the relationship of the surface area and sorption capacity of biochar after surface modification (Nartey and Zhao 2014; Ambaye et al. 2021; Lawal et al. 2021; Qiu et al. 2022). Addition of nanometals to the biochar is helpful in increasing the sorptive sites, reaction mechanism, surface area, structures, and thermal stability. Past researchers have successfully concluded the advanced performance of the modified biochar through the addition of metal oxides of Mg, Mn, Al, Cr, Cu, Fe, Zn, etc. The addition of Fe_3O_4 to the biochar depicted changed surface properties and the presence of FeO(OH) in the biochar (Yu et al. 2022). The rise in surface area and pore volume of modified biochar (444 m^2/g and 0.317 cm^3/g) than that of pristine biochar (68.3 m^2/g and 0.0516 cm^3/g), respectively, were found after Cu doping (Wan et al. 2019).

The loading of metal oxides (nZVI), nano-iron oxyhydroxide (nHIO) with the surface tuning through oxidation/amination of biochar influenced the development of oxygen-containing functional groups and helped in the development of better pore structures. The modification leads to 1.3 times rise in the biochar than pristine biochar and also favored the adsorption of 2,4-D on modified biochar by forming π-π interaction, chemisorptions, and hydrogen bonding (Zhu et al. 2018b). The production of magnetic biochar along with the surface modification (thioamine and aminofunctionalized) of the biochar has also shown a high sorption capacity (61.25 mg/g Pb(II) and 53.75 mg/g Cd(II)) along with economic, green, easy recyclable, and high-performance method (Zahedifar et al. 2021). Synthesis of biogenic silver-loaded biochar nanocomposites has shown its photocatalytic degradation potential for the methylene blue (up to 80% removal efficiency) along with the 100% antimicrobial actions against *Escherichia coli*. Biochar has also provided high surface area for the photoluminescent action of Ag (Eltaweil et al. 2022). However, in limited research, the addition of heteroatoms can also lower the surface area and porosity after blocking the macropores (Feng et al. 2020). The effect of Ce-modified biochar positively affected trice growth along with the improvement in soil physicochemical properties (Servin et al. 2017). The comparison of the fluoride removal potential of lanthanum oxide nanoparticle-coated biochar (AcLaO) and graphene oxide-coated biochar (SDGO) showed a better removal capacity of the farmer one (Chakraborty and Das 2022). Development of biochar (BR) and hydroxyapatite-loaded biochar nanocomposites (BR/HAP) accelerated their U(VI) adsorption capacities with (q_e) of 110.56 mg/g and 428.25 mg/g for BR and BR/HAP, respectively (Ahmed et al. 2021). Wood-waste modification of biochar through ZnCr LDH/carbon nanotube (CNT) showed an exceptional improvement in the 100% degradation of rifampicin, and the addition of biochar has helped for rising the surface area up to 7.0 m^2/g than that of CNT (5 m^2/g) and ZnCr LDH (1.2 m^2/g), respectively (Rad et al. 2022). The fabrication of highly efficient and reusable bentonite clay@biochar@-magnetic nanocomposites showed the 98% removal of Hg^{2+} along with the reusability up to 5 cycles. Additionally, the bentonite@biochar@magnetic nanocomposites showed extended removal capacity for COD (37.5%), BOD_5 (28.9%), Hg^{2+}(65%), As^{3+}(60.5%), and As^{5+} (50%), suggesting the incredible performance (Alali et al. 2022).

Thus, sufficient reports have proven that biochar acquires tremendous superior qualities, i.e., porous, functional, mechanically strong, recalcitrant, and with various chemical characteristics. However, their production largely depends on the right selection and optimization of feedstocks as well as operating parameters. Additionally, the pristine biochar can be upgraded to more active, efficient, and engineered surfaces through the impregnation of various chemical entities depending on the application strategies. Thus, the formation of biochar-based nanocomposites after modification of pristine biochar could be a fruitful option for various environmental applications.

4 Application Suitability of Pristine Biochar Versus Biochar-Based Nanocomposites

4.1 For Sustainable Agriculture

The role of biochar as an amendment for sustainable agriculture is well established. Biochar is primarily used for enhancing soil fertility, improving soil nutritional status, alteration of enzymatic dynamics, and remediation of polluted soil (Singh et al. 2015). Nutrient-rich biochar amendment has played an important role in improving soil physiochemical features such as organic matter, porosity, water retention capacity, soil cation exchange capacity, and bulk density. The implementation of biochar can modify the microbial community, improving soil enzymatic functions, and nutrient retention. Even, biochar can be applied to problematic soils (soil facing the issues like poor aggregation, high salinity, extreme pH levels, low microbial activity, and low production) for promoting plant growth and increasing productivity. Biochar's porous structures provide a suitable habitat for microbial proliferation and propagation and help in fixing nitrogen. They are always helpful in increasing the bioavailability and translocation of nutrients directly to the plants (Hossain et al. 2020). Most of the biochar is alkaline hence quite helpful in managing the pH of the soil depending on the characteristics of the soil. However, biochars produced from animal residues are mostly acidic, and the biochar with higher ash content has basic nature (Hassen-Trabelsi et al. 2014). Aging of the biochar is another important factor to be considered as it promotes cation exchange capacity (CEC) of the biochars as 15 months old biochar showed more CEC (173 mmolc/kg) than that of fresh biochar (26.2 mmolc/kg) (Mukherjee et al. 2014). Aged biochars are better soil CEC enhancers than fresh biochars. Li et al. (2019) have compared the properties of aged biochar after freeze–thaw, chemical (i.e., oxidation), and in natural field conditions through meta-analysis. Currently, the application of biochar-based nanocomposites on the soil is being studied, and the improved properties of the engineered biochar are more helpful in promoting soil health than that of pristine biochar (Gupta et al. 2022).

Synthesis of new core–shell P-laden biochar/ZnO/g–C_3N_4 composite (Pbi–ZnO–g–C_3N_4) was used for photocatalytic degrading atrazine and also as sustained release fertilizer simultaneously. The formulation showed removal efficiency of 85.3% after 260 min. However, the composite system revealed extended P-releasing performance with a total P release of 216.40 g/L in 260 min (An et al. 2022). The effect of two modified biochar, i.e., iron, zinc, and their combination loaded on biochar was compared with pristine biochar, and results showed that the composite system led to improved soil physicochemical parameters like pH, CEC, exchangeable sodium percentage, nutrients, the activity of H^+-ATPase and H^+-PPase. The biochar composites system was able to alleviate soil toxicity including plant growth by decreasing Na uptake and increasing nutrient uptake (Rahimzadeh and Ghassemi-Golezani 2022).

Cellulose nanofiber/montmorillonite-modified composites showed improvement in the soil water holding capacity by 7.8% and water retention capacity by 3.0% after 25 days of the application in the soil. They have also acted as a slow P release fertilizers which has reduced 43.9% of P leaching within 80 days of the incubation (Kassem et al. 2022). Thus, bio-based cellulosic biochar composites are good candidate for improving soil physiological features. The engineered composites of biochar were also tested for the remediation of various inorganic and organic contaminants in soil. A biogenic novel Hg-volatilizing *Lecythophora* sp. fungus (DC-F1) loaded on the biochar which has volatilized >86% of Hg(II) from liquid medium with an initial feed value of 7.0 mg/L within 16 h. The level of Hg(II) in soils and lettuce shoots were reduced by 13.3–26.1% and 49.5–67.7%, respectively, after composite application of over 56 days of incubation. The composite system also alleviated the soil fungal communities (Chang et al. 2019). Addition of Mg and Mn oxide-loaded biochar altered the phenology, morphology, and productivity of safflower plants (*Carthamus tinctorius* L.) under salt stress. They were able to raise pH and CEC of soil, plant growth, chlorophyll content, and stability, K, Mg, Mn levels, duration of reproductive stage, plant biomass, seed and oil yields of safflower under salt stress (Golezani et al. 2021). Thus, biochar-based nanocomposites are able to manage the soil quality through the introduction of precision and smart methods, thereby decreasing environmental and economic burdens and promoting sustainable agriculture.

4.2 For Pollutants Removal

Biochar is an extraordinary material to act as a sorbent for various ranges of pollutants present in different environmental matrices due to its better adsorption capacity and porous characteristics. Moreover, the application of modified or nanocomposites of the biochar provides an additional benefit by increasing surface area, porosity, and surface functional groups for the attachment of the pollutants (Fig. 2). However, the use of the grinded biochar creates problem in their separation step owing to its small particle size and density. To overcome these limitations, various surface tuning has been adopted for the manipulation in

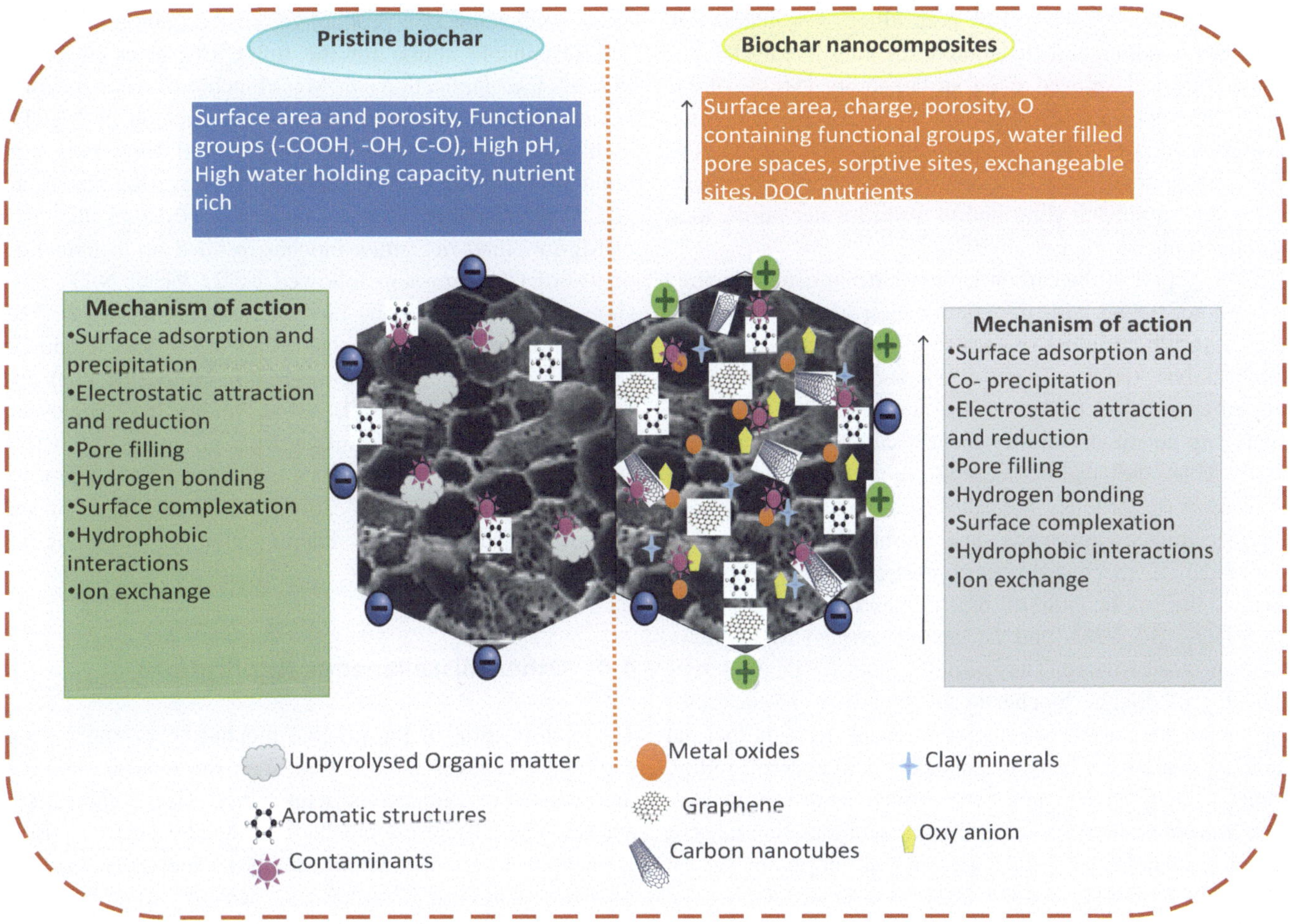

Fig. 2 Comparative features and action of pristine biochar vs biochar-based nanocomposites for contaminant remediation

surface features and easy separation (Pahani et al. 2020). The utilization of magnetic biochar in the medium led to easy separations. A used magnetic biochar with adsorbed Cu^{2+} (Cu–Fe@BRC) was tried for the permonosulfate activator for norfloxacin degradation, and they resulted into 91.47% of norfloxacin degradation with low Fe and Cu leaching. The composites were able to run for a long duration, stable, had no harmful by-products and were cost-effective. These magnetic composites can easily be separated from the medium using an external magnet (Pan et al. 2021). Due to the oxide formation on the biochar surface after pyrolysis, various O-containing groups provided extra binding of pollutants. The activation of biochar through the formation of MgO-biochar nanocomposites caused significant changes, i.e., variation in the surface area, pore volume, surface charge, and chemical properties of biochars which led to rise in the NH_4^+ and PO_4^{3-} adsorption capacity (Tran et al. 2022). Thus, they can also be helpful in minimizing the eutrophication risks of waste streams through the adsorption of P by the application of various metal oxides-loaded biochar-based nanocomposites (Tran et al. 2021). A zinc oxide/betaine-modified biochar showed removal of phosphates (10 mg/L: 3 g/L phosphate vs nanocomposite dosage) in in vitro conditions (Nakarmi et al. 2020).

Ca-magnetic biochar was able to turn Cd and As into their stable form (bi-dentate chelate and ternary surface complexes) thus decreasing the bioavailability and improving the pH and CEC of the respective soil (Wu et al. 2020). Engineered biochar has been extensively explored for binding of the organic contaminants through various mechanisms like pore filling, *p-p* interaction, hydrogen bonding, electrostatic attraction, cation exchange, hydrophobic interactions, complexes adsorption, and partitioning (Dai et al. 2019). The application of hydroxyapatite/biochar-based nanocomposites (HAP/BC-NCs) for the elimination of Cu (II) from aqueous media showed that the cation exchange and the formation of inner-sphere surface complexes were the potential mechanisms involved for the adsorption of Cd (Jung et al. 2019). A novel magnetically modified metal–organic framework ZIF-8/biochar (MZBC) was utilized for the sorption of ketamine (KET), and they demonstrated higher removal than that of pristine biochar which was almost 24.8 times more than that of without organic framework addition to the biochar. The whole adsorption mechanisms were governed by pore filling, π-π interaction, hydrogen bonding, and chelation. Even after five cycles, the nanocomposites showed good sustainability (>90% of the initial adsorption capacity) (Liu et al. 2021a, b). Therefore, biochar can be used for the construction of various biochar-supported filters, barriers, reactive films, and membranes for significant contribution toward water purification systems (Palansooriya et al. 2020).

Being a part of the carbon family, biochar can be applied as a photocatalyst with the combination of other nanomaterials, resulting into the formation of biochar-supported photocatalyst (BSP). These BSPs showed commendable properties such as better surface area, porosity, improved catalyzing potential, stability, and reusability (Mian and Liu 2018). Thus, they can be successfully used for the removal of various hazardous molecules such as dyes, phenols, pharmaceutical compounds, and antibiotics through photodegradation mechanism in a cost-effective way. In a review, the application of biochar-based nanocomposites (BC–TiO_2, BC–ZnO, and BC-spinel ferrites) as a photocatalysts was explored. They acquire some special properties than that of pristine biochar to be acted as a photocatalyst such as photocatalytic activity, mechanical strength, thermal stability, chemically inert, magnetically active, permeability, reduced energy band gaps, renewability, and easy separation. Moreover, they also attain lower recombination rates of the electron–hole pairs which are essential for photocatalytic applications (Ahmaruzzaman 2021; Fito et al. 2022). The biochar-supported composites of Pbi–ZnO–g–C_3N_4 showed photocatalytic performance due to the formation of Z-shaped heterojunction and where biochar serves as an electron-transfer bridge (An et al. 2022).

4.3 Carbon Sequestration

Soil also emits carbon dioxide (CO_2) into the atmosphere through soil respiration and also releases greenhouse gasses (GHGs) (CO_2, CH_4, N_2O). Their emission can be controlled through biochar amendments owing to changes in soil physicochemical properties. In the presence of the biochar, there will be a decrease in CO_2 flux than that of without it (Palviainen et al. 2018). Likewise, the emission of CH_4 from soil can be controlled due to the activation of methanotrophs and methanogens after biochar amendments. The application of chemical fertilizers is also helpful in climate change mitigation, reducing chemical fertilizer application rate, enhancing crop productivity, and improving water retention (Tauqeer et al. 2022). The co-application of chemical fertilizer and biochar has proven as a novel strategy for carbon sequestration, minimized dependency of chemical fertilizers and increase crop yield. The algae derived from biochar have high potential of CO_2 capturing as well as rich in nutrients. Their application is a productive, cost economic, efficient, and sustainable option for the future (Mona et al. 2021). Further the application of MgO/biochar-based nanocomposites showed the potential for the formation of $MgCO_3$ with soil CO_2, and reduced CO_2 emission. Thus, they captured soil derived CO_2 in an effective way (Ibrahim et al. 2022). In an another research, the GHGs reduction efficiency of Mg-modified rice straw biochar resulted up to reduction of 9% of CO_2 emission followed by 22.3% of N_2O emissions than that of pristine biochar (Khan et al. 2022). The application of modified biochar with $CaCO_3$, kaolin, and Ca $(H_2PO_4)_2$ showed carbon sequestration up to 29% than that of pristine biochar (Li et al. 2014). Studies have suggested that the introduction of nitrogen-rich groups at the active sites of biochar significantly curtailed their CO_2 capturing potential. Thus, the fortification and modification of the pristine biochar can be a valuable option for enhancing the GHGs emission control.

4.4 Other Miscellaneous Applications

The modification of the pristine biochar is responsible for significant changes like surface area, mechanical strength, thermal stability, tunable structures, pore size, conductance, and the reactivity of the biochar and thereby extending their application in energy storage, catalysis, fuel cells, material sciences, supporting structures, sensors, and biomedical fields. A composite of biochar with NiO showed electrochemical properties with a specific capacitance of 10.66 F/g in an asymmetric configuration, suggesting a possible alternative for energy storage devices (Endler et al. 2020). In another study, a composite system of bio-$CuWO_4$ nanocomposites showed their photocatalytic potential for ciprofloxacin and electrocatalyst for the oxidation of methanol simultaneously. They have degraded up to 97% of ciprofloxacin aqueous suspension after 90 min of visible light exposure along with the higher oxidation rate of methanol (Thiruppathi et al. 2020). Since the role of bio-nano fuel additives is currently being adapted owing to their non-toxicity. Therefore, a nickel ferrite ($NiFe_2O_4$) nanorod/poly(3,4-ethylenedioxythiophene) (PEDOT) composite was used as an anode in microbial fuel cells (MFCs). MFC with PEDOT/$NiFe_2O_4$/BC showed a peak power of 1200 mW/m^2 because of the unique catalytic nanostructures, compact biofilm formation, continuous electron conduction paths, and advanced structural and chemical stabilities of PEDOT/$NiFe_2O_4$ (Senthilkumar et al. 2019). The role of biochar as an enzyme support and acid catalyst is well known. Biochar provides good immobilization potential owing to its superior properties. A novel ZnO-MoO_3-C-based nanocomposite was used for sensing of acetaminophen as an electrochemical biosensor. In a real-time testing, the sensors have proved recovery in the range from 94.7 to 101.0% of acetaminophen, suggesting their application in bioanalysis (Liu et al. 2021a, b). The application of Ag-embedded biochar composite showed better specific capacitance of composite (655 F/g in 1M Na_2SO_4) at a current density of 0.35 A/g. The electrodes have shown profound cyclic stability of 79.3% after 5000 charge/discharge cycles, high energy, and 98% adsorption of Congo Red dye after 5 min (Kaushal et al. 2019). Due to their higher porous nature, functional group, and surface area, the application of chitosan-coated biochar was tested for cellulase immobilization. The supported composites retained 67% activity of free cellulase at pH = 4 and 60 °C and approximately 90.8% glucose productivity even after 10 cycles (Mo et al. 2020). Further, the promotion of active sites on the modified biochar surface added the catalytic devastation ability as well as easy separation. Magnetic biochar nanocomposites demonstrated Fenton like catalytic potential as compared to pristine structures due to dispersed crystals of loaded Fe_3O_4 and increased O-containing functional groups (Tu et al. 2020). Thus, the potential of biochar-based nanocomposite is well proven into diverse applications; however, there is still a dire need of extensive research to be carried out with comprehensive assessment.

5 Conclusions and Future Perspectives

The literature presented here has justified the proven potential of biochar-based nanocomposites over the pristine biochar. The superior qualities of biochar-based nanocomposites have successfully demonstrated "the biochar effect" in various fields. Also, the sustainable management of waste biomass can be easily operated through biochar production followed by their pre/post-pyrolysis modification strategies. However, being a pyrogenic carbon material, their aging for a long time generates the threat of aromaticity followed by leaching of bound contaminants and eventually ecotoxicity. The used biochar for contaminants adsorption could be a prominent reason behind the secondary pollution, and also, their presence can cause toxicity at different trophic levels. Overall, the production of biochar-based nanocomposites is an advantageous option for creating new avenues for future research.

Despite of excessive progressive research on the biochar-based nanocomposites, there are still few knowledge gaps identified, and future work should be focused on the following points:

- An elaborative testing should be performed regarding their toxicity, resistance, PAHs profiles, trace metals presence, and mechanism of actions of the biochar-based nanocomposites.
- More diversion should be given toward scaled-up field applications along with the knowledge about the effect of additives, binders, and nanometals on the toxicity of mammalian cells, plant cells, and different environmental matrices.
- The application-related issues like in situ activation, processing, separation, agglomeration, and reusability for a long duration should be addressed.

References

Ahmaruzzaman M (2021) Biochar based nanocomposites for photocatalytic degradation of emerging organic pollutants from water and wastewater. Mat Res Bull 140:111262

Ahmed W, Núñez-D A, Mehmood S, Ali S, Qaswar M, Shakoor A, Chen DY (2021) Highly efficient uranium (VI) capture from aqueous solution by means of a hydroxyapatite-biochar nanocomposite: adsorption behavior and mechanism. Environ Res 201:111518

Alali AF, Almojil SF, Almohana AI, Almoalimi KT (2022) Highly reusable bentonite clay@ biochar@ magnetic nanocomposite for Hg (II) removal from synthetic and real wastewater. pp 1–31. https://doi.org/10.21203/rs.3.rs-2097581/v1

Ambaye TG, Vaccari M, van Hullebusch ED, Amrane A, Rtimi S (2021) Mechanisms and adsorption capacities of biochar for the removal of organic and inorganic pollutants from industrial wastewater. Int J Environ Sci Tech 10:3273–3294

An X, Wang H, Dong C, Jiang P, Wu Z, Yu B (2022) Core-shell P-laden biochar/ZnO/g-C_3N_4 composite for enhanced photocatalytic degradation of atrazine and improved P slow-release performance. J Colloid Interface Sci 608:2539–2548

Ashwath N, Nam H, Capareda SC (2018) Optimising pyrolysis conditions for thermal conversion of beauty leaf tree (*Calophyllum inophyllum L.*) press cake. In: Application of thermo-fluid processes in energy systems. Springer, Singapore, pp 267–280

Chakraborty V, Das P (2022) Investigation on efficiency of synthesized lanthanum oxide–coated biochar and graphene oxide–coated biochar on removal of fluoride: batch and fixed bed continuous reactor performance modelling. Biomass Convers Biorefin 12:1–4

Chang J, Duan Y, Dong J, Shen S, Si G, He F, Yang Q, Chen J (2019) Bioremediation of Hg-contaminated soil by combining a novel Hg-volatilizing *Lecythophora* sp. fungus, DC-F1, with biochar: Performance and the response of soil fungal community. Sci Tot Environ 671:676–684

Chausali N, Saxena J, Prasad R (2021) Nanobiochar and biochar based nanocomposites: advances and applications. J Agri Food Res 5:100191

Chu Q, Xu S, Xue L, Liu Y, Feng Y, Yu S, Yang L, Xing B (2020) Bentonite hydrochar composites mitigate ammonia volatilization from paddy soil and improve nitrogen use efficiency. Sci Tot Environ 718:137301

Dai Y, Zhang N, Xing C, Cui Q, Sun Q (2019) The adsorption, regeneration and engineering applications of biochar for removal organic pollutants: a review. Chemosphere 223:12–27

Deng H, Zhang J, Huang R, Wang W, Meng M, Hu L, Gan W (2022) Adsorption of Malachite Green and Pb^{2+} by $KMnO_4$-modified biochar: insights and mechanisms. Sustainability 14(4):2040

Dong FX, Yan L, Zhou XH, Huang ST, Liang JY, Zhang WX, Guo ZW, Guo PR, Qian W, Kong LJ, Chu W (2022) Simultaneous adsorption of Cr (VI) and phenol by biochar-based iron oxide

composites in water: Performance, kinetics and mechanism. J Hazard Mat 416:125930
Eltaweil AS, Abdelfatah AM, Hosny M, Fawzy M (2022) Novel biogenic synthesis of a Ag@ Biochar nanocomposite as an antimicrobial agent and photocatalyst for methylene blue degradation. ACS Omega 7(9):8046–8059
Endler LW, Wolfart F, Mangrich AS, Vidotti M, Marchesi LF (2020) Facile method to prepare biochar–NiO nanocomposites as a promisor material for electrochemical energy storage devices. Chem Papers 74(5):1471–1476
Fang Q, Ye S, Yang H, Yang K, Zhou J, Gao Y, Lin Q, Tan X, Yang Z (2021) Application of layered double hydroxide-biochar composites in wastewater treatment: recent trends, modification strategies, and outlook. J Hazard Mat 420:126569
Feng D, Zhang Y, Zhao Y, Sun S, Wu J, Tan H (2020) Mechanism of in-situ dynamic catalysis and selective deactivation of H_2O-activated biochar for biomass tar reforming. Fuel 279:118450
Fito J, Kefeni KK, Nkambule TT (2022) The potential of biochar-photocatalytic nanocomposites for removal of organic micropollutants from wastewater. Sci Total Environ 16:154648
Ghassemi-Golezani K, Farhangi-Abriz S (2021) Biochar-based metal oxide nanocomposites of magnesium and manganese improved root development and productivity of safflower (*Carthamus tinctorius* L.) under salt stress. Rhizosphere 19:100416
Gunamantha IM, Widana GA (2018) Characterization the potential of biochar from cow and pig manure for geoecology application. IOP Conf Ser Earth Environ Sci (IOP Publishing) 131(1):0120552018
Gupta I, Singh R, Batish, DR, Singh HP, Raghubanshi AS, Kohli RK (2022) Engineered biochar as soil fertilizer. In: Ramola S, Mohan D, Masek O, Méndez A, Tsubota T (eds) Engineered biochar. Springer, Singapore, pp 197–221. https://doi.org/10.1007/978-981-19-2488-0_11
Hassen-Trabelsi AB, Kraiem T, Naoui S, Belayouni H (2014) Pyrolysis of waste animal fats in a fixed-bed reactor: Production and characterization of bio-oil and bio-char. Waste Manag 34(1):210–218
He M, Xu Z, Sun Y, Chan PS, Lui I, Tsang DC (2021) Critical impacts of pyrolysis conditions and activation methods on application-oriented production of wood waste-derived biochar. Biores Technol 341:125811
Hossain MZ, Bahar MM, Sarkar B, Donne SW, Ok YS, Palansooriya KN, Kirkham MB, Chowdhury S, Bolan N (2020) Biochar and its importance on nutrient dynamics in soil and plant. Biochar 2:379–420
Huang Q, Song S, Chen Z, Hu B, Chen J, Wang X (2019) Biochar-based materials and their applications in removal of organic contaminants from wastewater: state-of-the-art review. Biochar 1(1):45–73
Ibrahim MM, Guo L, Wu F, Liu D, Zhang H, Zou S, Xing S, Mao Y (2022) Field-applied biochar-based MgO and sepiolite composites possess CO_2 capture potential and alter organic C mineralization and C-cycling bacterial structure in fertilized soils. Sci Total Environ 813:152495
Inyang M, Dickenson E (2015) The potential role of biochar in the removal of organic and microbial contaminants from potable and reuse water: a review. Chemosphere 134:232–240
Jung KW, Lee SY, Choi JW, Lee YJ (2019) A facile one-pot hydrothermal synthesis of hydroxyapatite/biochar nanocomposites: adsorption behavior and mechanisms for the removal of copper (II) from aqueous media. Chem Eng J 369:529–541
Kambo HS, Dutta A (2015) A comparative review of biochar and hydrochar in terms of production, physico-chemical properties and applications. Renew Sustain Energy Rev 45:359–378
Kassem I, Ablouh EH, El Bouchtaoui FZ, Hannache H, Ghalfi H, Sehaqui H, El Achaby M (2022) Cellulose nanofibers/engineered biochar hybrid materials as biodegradable coating for slow-release phosphate fertilizers. ACS Sustain Chem Eng 10(46):15250–15262
Kaushal I, Saharan P, Kumar V, Sharma AK, Umar A (2019) Superb sono-adsorption and energy storage potential of multifunctional Ag-Biochar composite. J Alloys Compds 785:240–249
Keiluweit M, Nico PS, Johnson MG, Kleber M (2010) Dynamic molecular structure of plant biomass-derived black carbon (biochar). Environ Sci Technol 44(4):1247–53
Khan MN, Li D, Shah A, Huang J, Zhang L, Núñez-Delgado A, Han T, Du J, Ali S, Sial TA, Lan Z (2022) The impact of pristine and modified rice straw biochar on the emission of greenhouse gases from a red acidic soil. Environ Res 208:112676
Khare P (2021) A comprehensive evaluation of inherent properties and applications of nano-biochar prepared from different methods and feedstocks. J Cleaner Prod 320:128759
Kumar H, Cai W, Lai J, Chen P, Ganesan SP, Bordoloi S, Liu X, Wen Y, Garg A, Mei G (2020) Influence of in-house produced biochars on cracks and retained water during drying-wetting cycles: comparison between conventional plant, animal, and nano-biochars. J Soils Sediments 4:1983–1996
Lawal AA, Hassan MA, Zakaria MR, Yusoff MZ, Norrrahim MN, Mokhtar MN, Shirai Y (2021) Effect of oil palm biomass cellulosic content on nanopore structure and adsorption capacity of biochar. Biores Technol 332:125070
Lehmann J, Pereira da Silva Jr J, Steiner C, Nehls T, Zech W, Glaser B (2003) Nutrient availability and leaching in an archaeological Anthrosol and a Ferralsol of the Central Amazon basin: fertilizer, manure and charcoal amendments. Plant Soil 249:343–357
Leng L, Xiong Q, Yang L, Li H, Zhou Y, Zhang W, Jiang S, Li H, Huang H (2021) An overview on engineering the surface area and porosity of biochar. Sci Total Environ 763:144204
Li F, Cao X, Zhao L, Wang J, Ding Z (2014) Effects of mineral additives on biochar formation: carbon retention, stability, and properties. Environ Sci Technol 48(19):11211–11217
Li H, Lu X, Xu Y, Liu H (2019) How close is artificial biochar aging to natural biochar aging in fields? A meta-analysis. Geoderma 352:96–103
Liao X, Chen C, Liang Z, Zhao Z, Cui F (2022) Selective adsorption of antibiotics on manganese oxide-loaded biochar and mechanism based on QSPR method. Bioresour Technol 367:128262.https://doi.org/10.2139/ssrn.4223123
Liu C, Zhang N, Huang X, Wang Q, Wang X, Wang S (2021a) Fabrication of a novel nanocomposite electrode with ZnO-MoO_3 and biochar derived from mushroom biomaterials for the detection of acetaminophen in the presence of DA. Microchem J 161:105719
Liu WJ, Jiang H, Yu HQ (2015) Development of biochar-based functional materials: toward a sustainable platform carbon material. Chem Rev 115(22):12251–12285
Liu Y, Cao S, Xi C, Su H, Chen Z (2021b) A new nanocomposite assembled with metal organic framework and magnetic biochar derived from pomelo peels: a highly efficient adsorbent for ketamine in wastewater. J Environ Chem Eng 9(5):106207
Mian MM, Liu G (2018) Recent progress in biochar-supported photocatalysts: synthesis, role of biochar, and applications. RSC Adv 8(26):14237–14248
Mo H, Qiu J, Yang C, Zang L, Sakai E, Chen J (2020) Porous biochar/chitosan composites for high performance cellulase immobilization by glutaraldehyde. Enzyme Microb Technol 138:109561
Mona S, Malyan SK, Saini N, Deepak B, Pugazhendhi A, Kumar SS (2021) Towards sustainable agriculture with carbon sequestration, and greenhouse gas mitigation using algal biochar. Chemosphere 275:129856
Mukherjee A, Zimmerman AR, Hamdan R, Cooper WT (2014) Physicochemical changes in pyrogenic organic matter (biochar) after 15 months of field aging. Solid Earth 5(2):693–704

Nakarmi A, Bourdo SE, Ruhl L, Kanel S, Nadagouda M, Alla PK, Pavel I, Viswanathan T (2020) Benign zinc oxide betaine-modified biochar nanocomposites for phosphate removal from aqueous solutions. J Environ Manag 272:111048

Nartey OD, Zhao B (2014) Biochar preparation, characterization, and adsorptive capacity and its effect on bioavailability of contaminants: an overview. Adv Mat Sci Eng 2014:1–12

Palansooriya KN, Yang Y, Tsang YF, Sarkar B, Hou D, Cao X, Meers E, Rinklebe J, Kim KH, Ok YS (2020) Occurrence of contaminants in drinking water sources and the potential of biochar for water quality improvement: a review. Crit Rev Environ Sci Technol 50(6):549–611

Palviainen M, Berninger F, Bruckman VJ, Köster K, de Assumpção CR, Aaltonen H, Makita N, Mishra A, Kulmala L, Adamczyk B, Zhou X (2018) Effects of biochar on carbon and nitrogen fluxes in boreal forest soil. Plant Soil 425(1):71–85

Pan J, Gao B, Duan P, Guo K, Xu X, Yue Q (2021) Recycling exhausted magnetic biochar with adsorbed Cu^{2+} as a cost-effective permonosulfate activator for norfloxacin degradation: Cu contribution and mechanism. J Hazard Mat 413:125413

Panahi HK, Dehhaghi M, Ok YS, Nizami AS, Khoshnevisan B, Mussatto SI, Aghbashlo M, Tabatabaei M, Lam SS (2020) A comprehensive review of engineered biochar: production, characteristics, and environmental applications. J Cleaner Prod 270:122462

Peng Y, Azeem M, Li R, Xing L, Li Y, Zhang Y, Guo Z, Wang Q, Ngo HH, Qu G, Zhang Z (2022) Zirconium hydroxide nanoparticle encapsulated magnetic biochar composite derived from rice residue: application for As (III) and As (V) polluted water purification. J Hazard Mat 423:127081

Qiu B, Shao Q, Shi J, Yang C, Chu H (2022) Application of biochar for the adsorption of organic pollutants from wastewater: modification strategies, mechanisms and challenges. Sep Purification Technol 11:121925

Qu J, Shi J, Wang Y, Tong H, Zhu Y, Xu L, Wang Y, Zhang B, Tao Y, Dai X, Zhang H (2022) Applications of functionalized magnetic biochar in environmental remediation: a review. J Haz Mat 434:128841

Rad TS, Khataee A, Rad SS, Arefi-Oskoui S, Gengec E, Kobya M, Yoon Y (2022) Zinc-chromium layered double hydroxides anchored on carbon nanotube and biochar for ultrasound-assisted photocatalysis of rifampicin. Ultrasonics Sonochem 82:105875

Rahimzadeh S, Ghassemi-Golezani K (2022) Biochar-based nutritional nanocomposites altered nutrient uptake and vacuolar H^{+}-pump activities of dill under salinity. J Soil Sci Plant Nut 22(3):3568–3581

Rajapaksha AU, Chen SS, Tsang DC, Zhang M, Vithanage M, Mandal S, Gao B, Bolan NS, Ok YS (2016) Engineered/designer biochar for contaminant removal/immobilization from soil and water: potential and implication of biochar modification. Chemosphere 148:276–291

Senthilkumar N, Pannipara M, Al-Sehemi AG (2019) PEDOT/$NiFe_2O_4$ nanocomposites on biochar as a free-standing anode for high-performance and durable microbial fuel cells. New J Chem 43(20):7743–7750

Servin AD, De la Torre-Roche R, Castillo-Michel H, Pagano L, Hawthorne J, Musante C, Pignatello J, Uchimiya M, White JC (2017) Exposure of agricultural crops to nanoparticle CeO_2 in biochar-amended soil. Plant Physiol Biochem 110:147–157

Shi Q, Zhang H, Shahab A, Zeng H, Zeng H, Nabi I, Siddique J, Ullah H (2021) Efficient performance of magnesium oxide loaded biochar for the significant removal of Pb^{2+} and Cd^{2+} from aqueous solution. Ecotoxic Environ Saf 221:112426

Singh R, Kumar R, Babu JN, Srivastava P, Singh P, Raghubanshi AS (2015) Multifaceted application of crop residue biochar as a tool for sustainable agriculture: an ecological perspective. Ecol Eng 77:324–347

Singh R, Singh H, Raghubanshi AS (2019) Challenges and opportunities for agricultural sustainability in changing climate scenarios: a perspective on Indian agriculture. Trop Ecol 60:167–185

Sizmur T, Fresno T, Akgül G, Frost H, Moreno-Jiménez E (2017) Biochar modification to enhance sorption of inorganics from water. Biores Technol 246:34–47

Sun J, Lian F, Liu Z, Zhu L, Song Z (2014) Biochars derived from various crop straws: characterization and Cd (II) removal potential. Ecotoxic Environ Saf 106:226–231

Tan X, Shaaban M, Yang J, Cai Y, Wang B, Peng QA (2021) Efficient removal of hexavalent chromium from an aquatic system using nanoscale zero-valent iron supported by ramie biochar. Nanomaterials 11(10):2698

Tan XF, Liu YG, Gu YL, Xu Y, Zeng GM, Hu XJ, Liu SB, Wang X, Liu SM, Li J (2016) Biochar-based nano-composites for the decontamination of wastewater: a review. Biores Technol 212:318–333

Tauqeer HM, Turan V, Farhad M, Iqbal M (2022) Sustainable agriculture and plant production by virtue of biochar in the era of climate change. Managing plant production under changing environment. Springer, Singapore, pp 21–42

Thiruppathi M, Leeladevi K, Ramalingan C, Chen KC, Nagarajan ER (2020) Construction of novel biochar supported copper tungstate nanocomposites: a fruitful divergent catalyst for photocatalysis and electrocatalysis. Mat Sci Semicond Pro 106:104766

Tran DT, Pham TD, Dang VC, Pham TD, Nguyen MV, Dang NM, Ha MN, Nguyen VN, Nghiem LD (2022) A facile technique to prepare MgO-biochar nanocomposites for cationic and anionic nutrient removal. J Water Process Eng 47:102702

Tran TC, Nguyen TP, Nguyen TT, Tran TN, Nguyen TA, Tran QB, Nguyen XC (2021) Enhancement of phosphate adsorption by chemically modified biochars derived from *Mimosa pigra* invasive plant. Case Stud Chem Environ Eng 4:100117

Tu Y, Peng Z, Huang J, Wu X, Kong L, Liang Z, Yang L, Lin Z (2020) Preparation and characterization of magnetic biochar nanocomposites via a modified solvothermal method and their use as efficient heterogeneous fenton-like catalysts. Indus Eng Chem Res 59 (5):1809–1821

Wan Z, Sun Y, Tsang DC, Iris KM, Fan J, Clark JH, Zhou Y, Cao X, Gao B, Ok YS (2019) A sustainable biochar catalyst synergized with copper heteroatoms and CO_2 for singlet oxygenation and electron transfer routes. Green Chem 21(17):4800–4814

Wang L, Ok YS, Tsang DC, Alessi DS, Rinklebe J, Wang H, Mašek O, Hou R, O'Connor D, Hou D (2020) New trends in biochar pyrolysis and modification strategies: feedstock, pyrolysis conditions, sustainability concerns and implications for soil amendment. Soil Use Manag 36(3):358–386

Wang Z, Shu X, Zhu H, Xie L, Cheng S, Zhang Y (2018) Characteristics of biochars prepared by co-pyrolysis of sewage sludge and cotton stalk intended for use as soil amendments. Environ Technol 41:1347–1357

Weber K, Quicker P (2018) Properties of biochar. Fuel 217:240–261

Wu J, Li Z, Huang D, Liu X, Tang C, Parikh SJ, Xu J (2020) A novel calcium-based magnetic biochar is effective in stabilization of arsenic and cadmium co-contamination in aerobic soils. J Hazard Mat 387:122010

Yan L, Dong FX, Lin X, Zhou XH, Kong LJ, Chu W, Diao ZH (2021) Insights into the removal of Cr (VI) by a biochar–iron composite from aqueous solution: reactivity, kinetics and mechanism. Environ Technol Innovat 24:102057

Yu Y, Zhong Z, Guo H, Yu Y, Zheng T, Li H, Chang Z (2022) Biochar–goethite composites inhibited/enhanced degradation of triphenyl phosphate by activating persulfate: insights on the mechanism. Sci Total Environ 858:159940

Zahedifar M, Seyedi N, Shafiei S, Basij M (2021) Surface-modified magnetic biochar: highly efficient adsorbents for removal of Pb (II) and Cd (II). Mat Chem Phys 271:124860

Zhang Y, Liu S, Zheng X, Wang X, Xu Y, Tang H, Kang F, Yang QH, Luo J (2017) Biomass organs control the porosity of their pyrolyzed carbon. Adv Funct Mat 27(3):1604687

Zhang Y, Shi G, Wu W, Ali A, Wang H, Wang Q, Xu Z, Qi W, Li R, Zhang Z (2022) Magnetic biochar composite decorated with amino-containing biopolymer for phosphorus recovery from swine wastewater. Colloids Surf A: Physicochem Eng Asp 634:127980

Zhao M, Ma D, Ye Y (2021) Adsorption, separation and recovery properties of blocky zeolite-biochar composites for remediation of cadmium contaminated soil. Chin J Chem Eng 54:272–279. https://doi.org/10.1016/j.cjche.2021.10.020

Zhu D, Shao J, Li Z, Yang H, Zhang S, Chen H (2021) Nano nickel embedded in N-doped CNTs-supported porous biochar for adsorption-reduction of hexavalent chromium. J Hazard Mat 416:125693

Zhu L, Zhao N, Tong L, Lv Y, Li G (2018b) Characterization and evaluation of surface modified materials based on porous biochar and its adsorption properties for 2, 4-dichlorophenoxyacetic acid. Chemosphere 210:734–744

Zhu Y, Li H, Zhang G, Meng F, Li L, Wu S (2018a) Removal of hexavalent chromium from aqueous solution by different surface-modified biochars: Acid washing, nanoscale zero-valent iron and ferric iron loading. Bioresour Technol 261:142–150. https://doi.org/10.1016/j.biortech.2018.04.004

Application of Biochar-Based Nanocomposites for Remediation of Emerging Contaminants from the Environment

Biochar-Based Nanocomposites for Separation of Inorganic Contaminants from the Environment

Sushmita Mahour and Shalini Srivastava

Abstract

The contamination of the water ecosystem by various inorganic pollutants has aroused a serious global concern. The excess concentration of these pollutants in environment exhibits a harmful effect on the living being. Thus, remediation of such pollutants using an economical and environmentally friendly material such as biochar has emerged as a substantial solution. The utilization of biochar for heavy metal sequestration reduces the overall cost of the process since it can easily be prepared using an agricultural by-product. Despite various merits, biochar shows low adsorption efficiency. Thus, manipulating the biochar using nanoscale reinforcement is an attractive pathway to enhance its stability, surface area, active adsorption sites, and surface functionalities. These biochar nanohybrids/ nanocomposites have improved dispersity and a higher potential for adsorption of inorganic pollutants. Hence, the application of biochar nanocomposites (BNCs) for the separation of contaminants has emerged as an advanced, sustainable, and revolutionary solution from the water/wastewater. The chapter will focus on the separation/adsorption of the non-metals, metal ions, trace metals, metalloids, and radionuclei from the wastewater, drinking water, soil, and sediments. Further, chapter will also cover the technical and engineered aspects of biochar nanocomposites for the removal of anionic pollutants (nitrate, ammonia, and phosphate) generated through agricultural runoff, municipal wastewater, or industrial discharges. The detailed mechanism of sorption, interaction, immobilization, degradation, and mineralization of emerging inorganic pollutants in presence of biochar nanocomposites will be highlighted in this chapter.

S. Mahour · S. Srivastava (✉)
Department of Chemistry, Faculty of Science, Dayalbagh Educational Institute, Agra, 282005, India
e-mail: dei.shalinisrivastava@gmail.com

Keywords

Adsorption • Mechanism • Remediation • Sustainable • Toxic pollutants

1 Introduction

Technological progressions and rapid economic advancement have significantly enhanced the quality of life. However, the subsequent negative effect of development has aggravated the environmental concerns (Bilal et al. 2019; Amusat et al. 2021). Due to innumerable anthropogenic actions, the quality of the soil and water system is deteriorating. In developing countries, almost 80% of industrial wastewater and municipal water is released into the water system or on the wasteland without adequate treatment. This wastewater consists of a huge quantity of biological (bacteria, parasites, and viruses), organic (dyes, drugs, pesticides, biphenyl, etc.), and inorganic (heavy metal ions, chlorides, fluorides, nitrates, bromates, and phosphates) contaminants that have posed severe threats to living beings (Singh et al. 2020; Khan et al. 2021). Thus, there is an indispensable need for the remediation of these possible toxic contaminants. The ample treatment of wastewater stream will provide the recycled clean water. Moreover, the nutrient and metal ions contaminants present in these streams can also be recovered and used in economical advantages (van der Hoek et al. 2016; Yu et al. 2019). Different technical approaches are utilized for treating the wastewater and to achieve considerable recovery of resources. But, the application of these technologies on a wider scale is constrained due to extensive energy requirements, expensive infrastructure and the release of secondary pollutants. Recently, researchers are encouraging the approach of zero-liquid discharge (ZLD) which is regarded as an emerging technology to recover resources and reduce waste. This fundamental technology of ZLD is an adsorption-based separation method that is

D. Mishra et al. (eds.), *Biochar-Based Nanocomposites for Contaminant Management*, Advances in Science, Technology & Innovation, https://doi.org/10.1007/978-3-031-28873-9_6

sighted as an attractive futuristic approach (Yang et al. 2020a). For this purpose, the use of natural biomass, agricultural, and industrial by-product-based adsorbents is gaining the attention of the scientific community over the conventional adsorbent.

Nowadays, biochar has appeared as an exemplary adsorbent owing to its superior physicochemical properties like environmental friendliness, considerable porosity, large surface area, high mineral content, and presence of numerous oxygen-containing functional groups (Zhang et al. 2018; Shaheen et al. 2019; Liang et al. 2021). It can easily be fabricated by controlled heating of agricultural/industrial by-products that are low-cost and abundantly available. However, ordinary biochar has been associated with certain disadvantages like low density, smaller particle size, and low selectivity toward pollutants (Zhang et al. 2020). Thus, manipulating the biochar using nanoscale reinforcement is an attractive pathway to augment its properties. The biochar can be modified using various agents like metal oxide/hydroxide (Van Vinh et al. 2015; Yu et al. 2016; Zheng et al. 2020), magnetic particles (Wang et al. 2015), layered double hydroxides (Zhang et al. 2021), carbon nanotubes (Inyang et al. 2015), graphene oxide (Zhang et al. 2018), etc., to form biochar nanocomposites (BNCs) which possess the combined advantages of biochar as well as nano-materials. These BNC materials exhibit enhanced stability, surface area, pore properties, active adsorption sites, and surface functionalities (Li et al. 2018a; Hasan et al. 2020). They also have improved dispersity, electron transfer capacity, and admirable abilities to adsorb a wide range of contaminants from the environment (Wang et al. 2019a; Zhao et al. 2021). The present chapter aims to summarize: (1) different type of engineered/designed BNCs, (2) the application of BNCs for remediation of different contaminants from water/wastewater and soil, (3) to discuss the underlying mechanisms, and (4) put forward the prospects of future researches in BNCs.

2 Biochar-Based Nanocomposites (BNCs)

On the basis of various nano-materials utilized to functionalize biochar, such as magnetic iron oxide, nano-metal oxide, nano-metal hydroxide, nanoscale zero-valent iron, chitosan, layered double hydroxides, carbon nanotubes, graphene, graphene oxide, ZnS nanocrystals, and graphitic C_3N_4, BNCs can be categorized as magnetic nano-biochar composites, nano-metal oxide/hydroxide biochar composites, and functional nanoparticles-coated biochar. All these composites can be fabricated using different methods as shown in Fig. 1.

2.1 Magnetic Nano-Biochar Composites

Magnetic nano-biochar can be synthesized by the pre-treatment of the biomaterials with iron ion and chemical co-precipitation of iron oxides (Fe_3O_4, γ-Fe_2O_3 and $CoFe_2O_4$) onto biochar (Baig et al. 2014; Reddy and Lee 2014; Wang et al. 2015).

2.1.1 Pre-treatment of Biomaterial Using Iron Ions

This strategy involves the impregnation or co-precipitation of Fe^{3+}/Fe^{2+} onto the biomass surface before the pyrolysis. Subsequently, it is treated by microwave heating or pyrolyzing in different situations. Yin et al. (2018) modified the rice straw by the one-step process to prepare magnetic

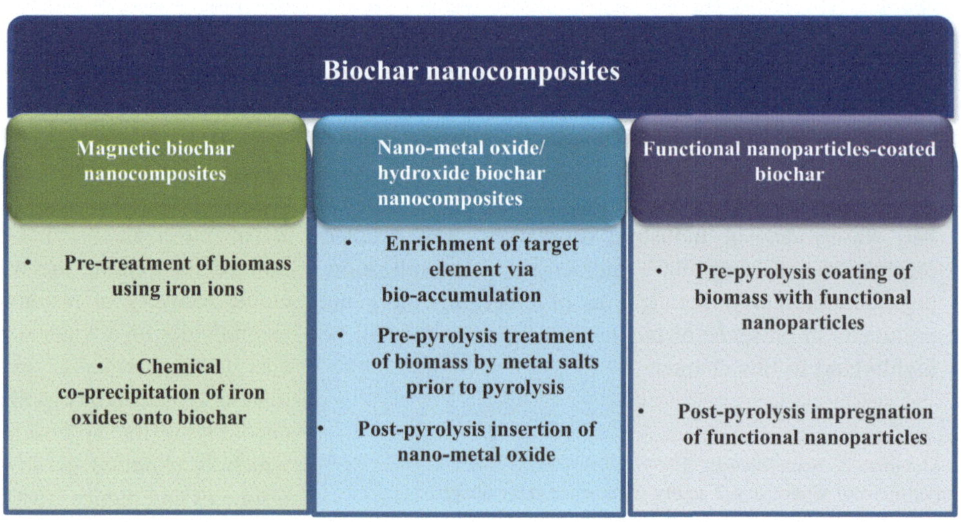

Fig. 1 Different types of biochar nanocomposites

biochar composite. The rice straw was treated with $ZnCl_2$ and $FeCl_3.6H_2O$ followed by the pyrolysis at different temperatures. The fabricated magnetized biochar acts as an effective adsorbent for Cu(II) ions. Nyamunda et al. (2019) prepared magnetic biochar via chemical co-precipitation method using $FeCl_2$ and $FeCl_3$ mixture on water hyacinth and then pyrolyzed. Wang et al. (2015) produced magnetized biochar by treating the raw biomass with hematite minerals. The modified biochar possesses stronger magnetic properties and the iron oxide particles behave as active adsorption sites for arsenic ions.

2.1.2 Chemical Co-precipitation of Iron Oxides onto Biochar

Chemical co-precipitation is the most widely used method for the preparation of magnetic biochar. This method takes lesser time to complete the process, reaction conditions are simple, and a high purity product is obtained. Formerly, the pristine biochar is prepared and undergoes co-precipitation of magnetic particles onto the biochar matrix. Lomenech et al. (2021) prepared humin-derived biochar by heating it at 250 °C for one hour and then its magnetic composite was formed by dissolving it in $FeCl_3$ and $FeSO_4$ solutions. The magnetic biochar is suspended in NaOH solution for the precipitation of iron oxide nanoparticles on its surface. Khan et al. (2020) synthesized magnetic biochar from corn straw. The corn straw was pyrolyzed at 600 and 800 °C for 2 h under a nitrogen atmosphere. Further, it was treated with ferric nitrate and subjected to sonication. The solution was dried by evaporating the solution which undergoes calcination at different temperatures. The synthesized magnetic biochar possesses a greater surface area (313.9 m^2/g) and increased adsorption efficiency for As(III). Baig et al. (2014) studied the pre-pyrolysis co-precipitation and post-pyrolysis co-precipitation of iron oxides onto biochar. The results indicate that biochar synthesized via pre-pyrolysis followed by co-precipitation has better physicochemical and magnetization properties as compared to the other method.

2.2 Nano-Metal Oxide/Hydroxide Biochar-Nanocomposites

Nano-metal oxide/hydroxide BNCs are prepared by three main methods, viz. (i) enrichment of target element via bio-accumulation, (ii) pre-treatment of biomaterial by metal salt, and (iii) post-pyrolysis addition of nano-metal oxide into biomaterial. The first two processes involve the pre-pyrolysis impregnation of metal onto the biomaterial, while in the third process, nano-metal oxide/hydroxide is directly introduced into biochar produced after pyrolysis (Tan et al. 2016; Chausali et al. 2021).

2.2.1 Enrichment of Target Element via Bio-accumulation

The target element can be enriched into the plant-based biomass using the bio-accumulation phenomenon. These target elements enriched biomass is converted into nano-metal oxide/hydroxide biochar composites after thermal treatment. Yao et al. (2013) enrich Mg in a tomato plant by irrigating it with Hoagland solution two times a week. Afterward, the biomass is fabricated into biochar by slow pyrolysis. The produced biochar surface possesses nanoscale MgO and $Mg(OH)_2$ particles that show greater adsorption efficiency for phosphate. Therefore, bio-accumulation of elements into the biomass can work as a novel approach for producing engineered biochar composite showing the greater potential for future application.

2.2.2 Pre-pyrolysis Treatment of Biomass by Metal Salts

This method involves the pre-treatment of biomass by chemical reagents prior to the pyrolysis process. The biomass is immersed in metal salts solution which leads to the attachment of metal ions onto the biomass surface or insertion into the inner surface. After this, biomass is pyrolyzed to form nano-metal oxide/hydroxide biochar composite. ZnO (Tho et al. 2021), $ZnCl_2$ (Yin et al. 2018), $AlCl_3$ (Lawrinenko et al. 2017), $CaCl_2$ (Liu et al. 2016), $KMnO_4$ (Sun et al. 2019), and MnO_2 (Liang et al. 2017) are the most commonly used metal salts resulting into the formation of ZnO, Al_2O_3, AlOOH, CaO, and MnO_x nanoparticles onto the surface of biochar. The metal nanocomposites have a greater number of reactive sites and surface area to attract the inorganic contaminants.

2.2.3 Post-pyrolysis Insertion of Nano-Metal Oxide

This method involves the formation of pristine biochar by the pyrolysis of biomass. Post-pyrolysis metal oxide nanoparticles are introduced onto the biochar surface. Impregnation method (Maneechakr and Mongkollertlop 2020), direct hydrolyzes (Hu et al. 2015), heat treatment (Yu et al. 2016), and evaporative method (Cope et al. 2014) are some frequently used methods for this purpose. Yu et al. (2016) prepared MnO_x-loaded biochar for which pristine biochar was treated with $KMnO_4$ solution then undergoes heat treatment. The obtained biochar composite shows higher adsorption for Pb(II). Hu et al. (2015) directly hydrolyzed iron salt to synthesize iron (Fe)-impregnated biochar onto hickory biochar and applied for the removal of As ions. Cope et al. (2014) used the evaporative method to amend the iron oxide onto the biochar. Maneechakr and Mongkollertlop (2020) prepared the biochar using palm kernel cake residue and insert the MnO_2 nanoparticles through the impregnation method. The prepared biochar composites show enhanced adsorption properties for Cd, Cr, Hg, and Pb ions.

2.3 Functional Nanoparticles-Coated Biochar

The preparation of these composites includes the impregnation of functional nanoparticles (carbon nanotubes (CNTs), graphene oxides, layered double hydroxides, etc.) onto the biochar before or after pyrolysis.

2.3.1 Pre-pyrolysis Coating of Biomass with Functional Nanoparticles

Before pyrolysis, the biomass is treated with various types of functional nanoparticles. Generally, dip-coating strategy is followed to convert the biomass into biochar functional matrices. Functional nanoparticles are suspended in deionized water and homogenized using ultrasonication. The raw biomass is dipped into the functional nanoparticle suspension, dried and pyrolyzed to produce the biochar composite. Yang et al. (2020b) dip-coated chestnut shell in the carbon nanotubes suspension and pyrolyzed at 600 °C to synthesized carbon nanotube-biochar composite. Zhang et al. (2018) engineered a new biochar composite with graphene oxide nanosheets and corn straw. A mixture of corn straw and graphene oxides (GO) nanosheets was stirred for 1 h, dried and slowly pyrolyzed at 600 °C for 1 h in inert condition. GO nanosheets were coated onto biochar surface through π–π interactions. The prepared composite severed in the co-adsorption of Pb(II) and atrazine. Inyang et al. (2015) synthesized multi-walled carbon nanotube (CNT)-coated biochars via dip-coating the biomass into different concentrations of carboxyl-functionalized CNT solutions before the pyrolysis.

2.3.2 Post-pyrolysis Impregnation of Functional Nanoparticles

The process occurs by pyrolyzing raw biomass to form biochar on which impregnation of functional nanoparticles is carried out to get biochar composites. These composites have the combined advantages of biochar and functional nanoparticles. Chitosan (Chin et al. 2021), layered double hydroxide (Wang et al. 2020a), zero-valent iron (ZVI) (Wang et al. 2019a; Hasan et al. 2020), and ZnS nanocrystals (Yan et al. 2014) are commonly applied in the fabrication process. These functional nano-materials get attached onto the porous matrix of biochar that served as the active sites to adsorb contaminants from aqueous solutions. Sometimes, the deposition of these functional particles partially blocks the pores present onto the surface of biochar. The resulted composites have superior functions and properties inherited from both functional nanoparticles and biochar. Fortunately, the superior characteristics of functional nanoparticles overcome this demerit. Hasan et al. (2020) prepared a zero-valent iron modified biochar nanocomposites that shows exceptional ability to remove Cu, Cd, and Zn ions. Jia et al. (2019) synthesized magnetic biochar/Mg-Fe-layered double hydroxide composites using the hydrothermal method. Firstly, magnetic biochar was prepared by pyrolyzing $FeCl_2$ treated oil tea camellia shells. Further, proceeded with hydrothermal process in presence of $Fe(NO_3)_3$, $Mg(NO_3)_2$ and Na_2CO_3. Zhang et al. (2021) functionalized the biochar derived from forestry wastes and magnetized using $CoFe_2O_4$. This magnetic biochar is further decorated with Mg/Al-LDH that acts as an effective adsorbent.

3 Applications of Biochar-Based Nanocomposites for Removing the Contaminants from the Environment

Biochar-based nanocomposites show great potential for remediation of contaminants from the environment. Nowadays, many inorganic contaminants such as heavy metal ions, metalloids, radioactive ions, phosphorus, nitrates, and ammonium ions are polluting the soil and water system. These pollutants enter the environment through the release of industrial wastewater, municipal water, stormwater, and agricultural runoff. BNCs added value to agricultural/industrial by-products through their modification into composite materials and help in waste management. BNCs exhibit extraordinary adsorption capacity due to which they can eliminate the various pollutants from the soil and water systems. It acts as a great adsorbent and is widely used in wastewater treatment.

3.1 Removal of Heavy Metals

The adsorption of heavy metals onto the BNCs helps to remediate the water and soil. The large surface area and presence of functionalized active sites onto these nanocomposites amplify the separation of metal ions onto BNCs surface. Recently, researchers are applying different biochar-nanocomposites to capture the metal ions (Table 1). Yan and Li (2022) investigated the removal of Pb(II) using magnetite/graphene nanocomposites. To upgrade the performance of biochar composite, steam activation was carried out which results in the activation of oxygen functionalities and nanosizing of graphene sheets. The synthesized BNC exhibits adsorption capacity of 153.2 mg/g. Quan et al. (2022) prepared loofah sponge biochar-attapulgite nanocomposites via the mechano-chemical process. The prepared composite has greater Si–O–C and oxygen-containing moieties resulting in

Table 1 Removal of heavy metals by adsorption with biochar-based nanocomposites

Biomass	Modifying agent	Target pollutant (adsorption performance)	Sample	Reference
Rice husk	Mg–Al/LDH	Cd (II): 125.34 mg/g Cu (II): 104.34 mg/g	Wastewater	Li et al. (2022)
Pine wood	Graphene/ Fe nanoparticle	Pb (II): 153.2 mg/g	Synthetic water	Yan and Li (2022)
Loofah sponge	Attapulgite	Cd (II): 102.26 mg/g	Synthetic water	Quan et al. (2022)
Banana peel	Fe_3O_4/ZIF-67	Cd (II): 50.78 mg/g	Synthetic water	Foroutan et al. (2022)
Cassava root	ZnO	Pb (II): 44.27 mg/g Cd (II): 42.05 mg/g As (III): 39.52 mg/g Cr (VI): 28.37 mg/g	Synthetic water	Tho et al. (2021)
Palm fiber powder	H_2O_2–HAc and ZVI	Cd (II): 46.4 mg/g	Irrigation water	Liu et al. (2020)
Kenaf bar	ZVI and $FeSO_4{\cdot}7H_2O$	Cd (II): 10 mg/kg	Soil	Diao et al. (2020)
Pine sawdust	Si and Mg salts	Cu (II): 214.7 mg/g Zn (II): 277.3 mg/g	Synthetic water	Deng et al. (2020)
Rice husk	KOH	Pb (II): 3952.71 mg/kg Cd (II): 136.77 mg/kg	Soil	Wang et al. (2020b)
Poplar wood chips	Thiol modified and ball-milled biochar	Hg (II): 320.1 mg/g	Synthetic water	Lyu et al. (2020)
Wheat straw	Goethite-modification $Fe(NO_3)_3$	Cd (II): 1.01 mg/kg	Soil	Zhu et al. (2020b)
Rice straw	$CaCO_3$ and FeO	Cd (II): 6.89 μg/g	Soil	Wu et al. (2020a)
Rice husk	3D MnO_2	Cd (II): 84.76 mg/g Pb (II): 70.90 mg/g	Synthetic water	Wu et al. (2020b)
Corncob	MgO	Pb (II): 3410 mg/kg	Soil	Shen et al. (2019)
Oil-tea and camellia shells	$FeCl_2$, $Fe(NO_3)_3$, $Mg(NO_3)_2$ and MgFe LDH	Pb (II)	Synthetic water	Jia et al. (2019)
Corn straw	$Fe(NO_3)_3$, $Mn(NO_3)_2$ and $MnFe_2O_4$	Pb (II) Cd (II)	Synthetic water	Zhang et al. (2019)
Undaria pinnatifida roots	Hydroxyapatite nanoparticles	Cu(II)	Synthetic water	Jung et al. (2019)
Pineapple peel	$FeCl_3$, ZVI and γ-Fe_2O_3	Cr (VI)	Synthetic water	Shakya and Agarwal (2019)
Rice straw	$ZnCl_2$ and $FeCl_{3.}6H_2O$	Cu (II)	Synthetic water	Yin et al. (2018)
Corn stover	ZnO/ZnS	Pb (II): 135.8 mg/g Cu (II): 91.2 mg/g Cr (VI): 24.5 mg/g	Synthetic water	Li et al. (2018a, b)
Rice straw	KBH_4, $FeSO_4$ and ZVI	Cr (VI)	Synthetic water	Qian et al. (2018)
Hickory wood chips	Ball-milled biochar	Ni (II): 230–650 mmol/kg	Soil	Lyu et al. (2018)
Agricultural wastes	ZVI	Cd(II), Co(II), Zn(II) and Pb(II)	Synthetic water	Kolodynska et al. (2017)
Eucalyptus wood and poultry litter	$FeCl_3$, $FeCl_2$ and Fe_3O_4	Cd (II), Cu (II), Zn (II), Pb (II)	Soil	Lu et al. (2017)
Rice husk	MnO_x	Pb (II): 86.5 mg/g	Synthetic water	Yu et al. (2016)
Hickory chips and sugarcane bagasse	CNTs	Pb (II)	Synthetic water	Inyang et al. (2015)

effective capturing of Cd(II). Similarly, Vishnu et al. (2020) fabricated a novel nanocomposites for the eradication of Cu (II) and Pb(II) ions from wastewater. The biochar was derived from groundnut waste and modified with super paramagnetic nanoparticles of Co/Fe/Mn. The novel adsorbent has an adsorption capacity of 436.8 mg/g and 318.9 mg/g for Cu(II) and Pb(II), respectively. The adsorbent can be regenerated by acid washing and used for 10 cycles. Li et al. (2018b) used ZnO/ZnS nano-biochar to remove Cr(VI), Cu(II), and Pb(II) from aqueous solutions. The ZnO/ZnS decorated biochar has a greater surface area (397.4 m^2/g) as compared to raw biochar (102.9 m^2/g). The ZnO/ZnS biochar matrix has surface hydroxyl groups that form complexes with metal ions and also show ion exchange process.

3.2 Removal of Radioactive Ions and Metalloids

Metalloids are elements that have properties intermediate to metal and non-metals. They are also bio-accumulated into living beings and show severe noxious effects. Similarly, exposure to radioactive elements is highly hazardous for the environment and humans. These elements have a very long half-life and reside in the environment for many years. They enter the environment during the extraction of radioactive materials, testing of nuclear missiles, accidents at nuclear plants, and mishaps during transportation and disposal of nuclear matter. Removal of these pollutants is highly essential, for that many researchers are using various biochar-based nanocomposites (Hu et al. 2018; Lyu et al. 2021; Zhu et al. 2022). Table 2 represents the various biochar

Table 2 Removal of metalloids and radioactive elements using biochar nanocomposites

Biomass	Modifying agent	Target pollutant	Sample	Reference
Soya bean pod	Prussian blue nanoparticles	Cs^+: 207.20 mg/g	Synthetic water	Feng et al. (2022)
Bamboo	ZVI	As (III): 129.24 mg/g As (V): 127.15 mg/g	Synthetic water	Xu et al. (2022)
Scallion sheathings	Hydrated ferric oxide	As (III): 104.55 mg/g	Wastewater	Zhu et al. (2022)
Sugarcane bagasse	$FeCl_3$	As (V): 12.8 mg/g	Wastewater	Bai et al. (2022)
Rice straw	$CaCl_2.2H_2O$ $(NH_4)_2HPO_4$	U(VI): 110.56 mg/g	Synthetic water	Ahmed et al. (2021)
Bamboo	Mg–Al–LDH	U (VI)	Soil	Lyu et al. (2021)
Ginkgo leaves	$MnFe_2O_4$	As: 90 mg/g	Synthetic water	Wen et al. (2021)
Wheat straw	Goethite-modification with KOH and Fe $(NO_3)_3$	As: 67.48 mg/kg	Soil	Zhu et al. (2020b)
Rice straw	$CaCO_3$ and FeO	As: 3. 01 μg/g	Soil	Wu et al. (2020a)
Lufacylindrica fiber	MnO_2	U(VI): 904 mg/g	Synthetic water	Ioannou et al. (2019)
Palm fiber powder	H_2O_2-HAc and ZVI	As: 181.5 mg/g	Irrigation water	Liu et al. (2020)
Corn straw	$Fe(NO_3)_3/FeSO_4$and $KMnO_4$	As: 87.97 mg/kg	Soil	Lin et al. (2019a)
Phragmites australis	$CeCl_{3.}7H_2O$, $FeCl_{3.}6H_2O$ and $FeCl_{2.}4H_2O$	Sb (V)	Synthetic water	Wang et al. (2019b)
Siderite and Rice husk	Fe_3O_4	U (VI): 52.63 mg/g	Synthetic water	Li et al. (2019a)
Ficus microcarpa aerial root	MnO_2	U (VI): 24.6 mg/g	Synthetic water	Li et al. (2019b)
Phragmites australis	$LaCl_3$, $FeCl_{3.}6H_2O$ and $FeCl_{2.}4H_2O$	Sb (V)	Synthetic water	Wang et al. (2018)
Hydrophyte biomass	Fe_3O_4	U (VI)	Synthetic water	Hu et al. (2018)
Pinewood, natural hematite	Hematite	As (V): 429 mg/kg	Soil	Wang et al. (2015)

nanocomposites utilized for abatement of metalloids and radioactive elements. Feng et al. (2022) synthesized a new composite material for efficient abatement of Cs^+. The soybean pod surface was fabricated with nanoparticles of Prussian blue. The nanoparticles were uniformly dispersed onto the biochar matrix enhancing the adsorption capability. Wen et al. (2021) investigated the decontamination of metalloids from wastewater/drinking water. Porous biochar was derived from ginkgo leaves and modified by $MnFe_2O_4$ nanoparticle using the co-precipitation method. The biochar composite can adsorb 90 mg/g of As(V) species. Lin et al. (2019a) used Fe–Mn oxide biochar composite to eradicate As from water.

3.3 Removal of Nutrients

The inorganic ions like nitrate, phosphate, and ammonium are known as nutrients as they are required by plants and animals for proper growth. But the presence of excess nutrients in soil and water leads to eutrophication. Biochar-based nanocomposites can efficiently absorb these elements. This is because these nanocomposites carry a positive charge and are capable of attracting anionic ions via electrostatic force of attraction (Yao et al. 2011; Chen et al. 2017). Tran et al. (2022) examined the efficiency of MgO-biochar nanocomposites for removing NH_4^+ and PO_4^{3-} from water. Biochar is prepared from rice husk and corn cob which was further fabricated with MgO particles under an inert atmosphere at 400 oC and 500 oC. The modified biochar has a larger surface area, greater surface charge, and superior chemical properties which amplifies its adsorption performance. The adsorption of NH_4^+ and PO_4^{3-} onto biochar-based nanocomposites followed both chemisorption and physisorption. Zhu et al. (2018) prepared a porous nano-biochar composite via impregnating α-Fe_2O_3/Fe_3O_4 nanoparticles into bamboo biochar. The obtained material was utilized to remove phosphorus from sewage. Feng et al. (2017) removed phosphorus from soil using CeO functionalized nano-biochar synthesized through impregnation. He et al. (2017) studied that nano-Fe oxide fabricated biochar has good adsorption efficiency for nutrients. A Fe_3O_4-biochar nanocomposite with photosynthetic bacteria (PSB) was prepared through the chemical co-precipitation method and applied for bioremediation of wastewater. The loading of Fe_3O_4 nanoparticles onto biochar increases the surface area of nanomaterial and effectively removed 92.1% PO_4^{3-} and 87.5% NH_4^+ from wastewater. Other modified biochar nanocomposites used for the remediation are given in Table 3.

4 Mechanisms Involved

The scavenging of pollutants by the biochar-based nanocomposites is based on the adsorption with underlying several mechanisms: surface adsorption, electrostatic interactions, ion exchange, precipitation, complexation, and reduction (Qiu et al. 2020; Gholizadeh and Hu 2021). The adsorption of pollutants through biochar nanocomposites is facilitated by its specific surface area, cation exchange capacity, surface active functional groups, and nano-metal oxide. Since various factors affect the adsorption process, therefore, multiple mechanisms occur simultaneously during the adsorption of the ions (Fig. 2). The process of surface adsorption essentially includes the accumulation of pollutants onto the biochar nanocomposites. The pollutants are either diffusing into the cavities/ pore of BNCs or adsorb on the surface through Van der Waals forces. Since the surface adsorption is mainly governed by the intermolecular forces of attraction, the adsorption process is found to be reversible and adsorption affinity between the adsorbate and adsorbent is weak. The ion exchange phenomenon involves the exchange of ions between the charged surface ions present on the biochar surface and target pollutants in the solution. The size of pollutants and the surface characteristics of BNCs highly affect the ion exchange process. The complexation phenomenon occurs via the formation of inner- or outer-sphere complexes between the target pollutants and surface functional groups of the biochar nanocomposites. The BNCs contain various functional moieties such as carboxyl, hydroxyl, carbonyl, amine, and amide groups. The electron pair on either oxygen atom or nitrogen atom of these functional groups form coordinate bonds with the pollutants to form complexes (Qiu et al. 2021).

Binding the pollutants through electrostatic force of attraction is another adsorption mechanism. The presence of charged functional groups and metal/metal oxide nanoparticles onto the biochar surface enables the adsorption of charged pollutants through electrostatic interactions. The pH of the solutions dominates the extent of electrostatic interactions and point of zero charge of biochar-based nanocomposites. Another adsorption mechanism is precipitation, during this process, the pollutants precipitate onto the surface of biochar nanocomposites. Many researchers studied the adsorption of several pollutants on the biochar nanocomposites to elucidate the actual mechanism. Lin et al. (2019b) adsorbed arsenic ions onto Fe–Mn oxide biochar. Due to the impregnation of Fe–Mn oxide nanoparticles, the surface area of the adsorbent enhances. Arsenic ions diffuse to micro-pores of composite leading to an uneven surface and

Table 3 Removal of nutrients using biochar-based nanocomposites

Biomass	Modifying agent	Target pollutant	Sample	Reference
Coconut shell and rice husk	MnO_2	Li^+: 89 mg/g	Synthetic water	Kamran and Park (2022)
Rice husk and corn cob	MgO	NH_4^+: 21.32 mg/g PO_4^{3-}: 117.77 mg/g	Synthetic water	Tran et al. (2022)
Jackfruit	Iron oxide	Nitrate: 5.26 mg/g PO_4^{3-}: 7.94 mg/g	Wastewater and simulated water	Nayak et al. (2021)
Corn stover	α-Fe_2O_3 and γ-Fe_2O_3	PO_4^{3-}	Soil	Bakshi et al. (2021)
Dewatered sludge	$MnAl_2O_4$	PO_4^{3-}: 28.20 mg/g	Synthetic water	Peng et al. (2020)
Corn straw	MgO	PO_4^{3-}: 60.95 mg/g	Synthetic water	Zhu et al. (2020a)
Douglas fir	α-Fe_2O_3/Fe_3O_4	Nitrate: 15 mg/g	Synthetic water	Dewage et al. (2018)
Maize straw	Nanocerium oxide	PO_4^{3-}: 78 mg/g	Synthetic water	Feng et al. (2017)
Agricultural and forestry residues	FeO	NH_4^+: 87.5% PO_4^{3-}: 92.1%	Synthetic water	He et al. (2017)
Undaria pinnatifida roots	$MgFe_2O_4$	PO_4^{3-}	Synthetic water	Jung et al. (2017)
Water hyacinth	Fe	PO_4^{3-}	Synthetic water	Cai et al. (2017)
Ramie stem	CaO	PO_4^{3-}: > 100 mg/g	Synthetic water	Liu et al. (2016)
Vermiculite treated algal biomass	SiO_2	PO_4^{3-}: 159.4 mg/g	Synthetic water	Wang et al. (2016)
Sugarcane bagasse, cotton woods, sugar beet tailings, peanut shells, pine woods	MgO nanoflakes	Nitrate: 95 mg/g PO_4^{3-}: 835 mg/g	Synthetic water	Zhang et al. (2012)

reduction in surface area. The adsorption occurs through the physisorption process. Yan et al. (2021) suggested that the adsorption of Cr(VI) by the biochar–iron composite (FBC) involves multiple mechanistic pathways. Cr(VI) is absorbed on nanocomposite by a combination of complexation and surface precipitation, and then reduction process. Cr (VI) is reduced to Cr(III) through the reduction process. The iron oxide present on biochar favors the adsorption of Cr(VI). Wang et al. (2021) observed that the adsorption of nitrate on biochar nanocomposite produced from woody waste and modified with Mg/Al-LDHs (ABM) follows the route of both physical and chemical adsorption. ABM has a mesoporous surface with a mean pore size of 6.25 nm and the hydrated ion of NO_3^- has an ionic radius of 0.335 nm which enables the penetration of NO_3^- ions into the pores. The surface of ABM is positively charged owing to the fabrication of Mg/Al-LDHs, which in turn attracts the negatively charged NO_3^- ions via electrostatic force of attraction. In addition, chemical adsorption occurs by forming outer-sphere complex, metal bond bridges, and ion exchange. The occurrence of the above chemisorption phenomenon has been confirmed by XPS and FTIR techniques. Priyadarshni et al. (2020) applied iron and copper nanoparticles impregnated biochar for the removal of arsenite and arsenate contaminants. It was observed that arsenic was adsorbed by electrostatic and covalent type interactions between Fe/Cu–OH/OH_2^+ groups of the composites and As(III)/As(V) species. Qiu and Duan (2019) used biochar composite for the abatement of phosphorus from water. It was observed that phosphorus transfer to the mesopore of the adsorbent. Dong et al. (2021) found that Ca/A-LDH and hydroxyapatite decorated *Solidago canadensis*-derived biochar composite (BC@LDH@HAP) is highly effective in separating Eu(III) ions from wastewater. The spectra of XRD, XPS, and FTIR reveal that adsorption of Eu(III) onto adsorbent involved precipitation, ion exchange, and complexation.

5 Future Prospects

Removal of toxic pollutant is very crucial for guarding the environmental quality and sustaining human health. The emerging biochar-based nanocomposites have a great potential to remediate inorganic pollutants, but this field still

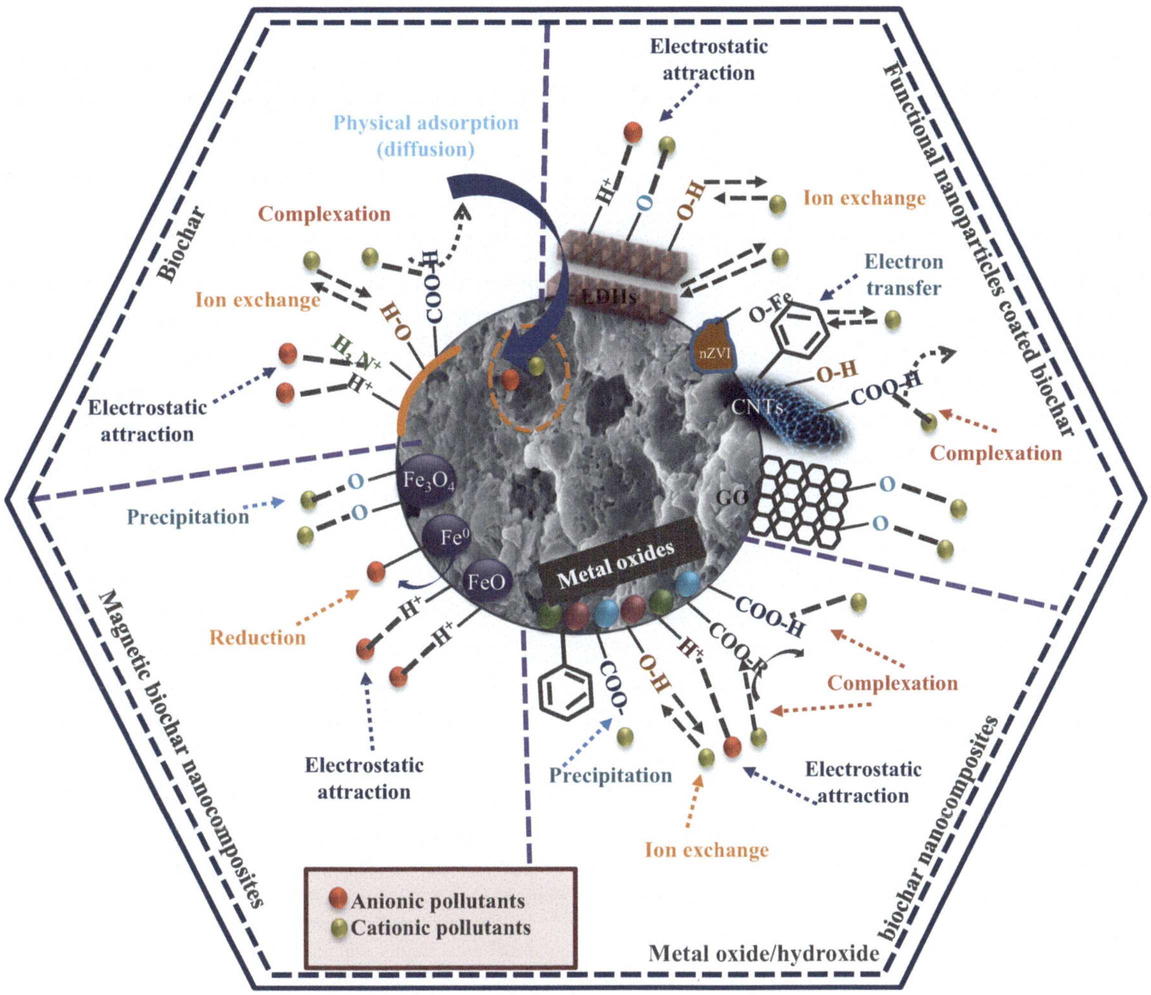

Fig. 2 Mechanism involves in adsorption of pollutants using biochar-based nanocomposites

needs extensive explorations and particular consideration should be paid to the following aspects in the future investigations:

- Inorganic pollutants have contiminated the environment so extensively that no single adsorbent eithier raw or modified can simultaneously treat these pollutanta from water/wastewater and soil. Hence, application of combination of different modified adsorbent and treatment methods can be utilized to remediate the environment from these pollutants.
- Utilization of agricultural residues, industrial by-products, and forest waste as feedstock for the production of biochar that will reduce the environmental waste and provide a valued added product that can be used for the removal of contaminants and promote the concept of "waste controlled by waste".
- Considering the use of raw materials which are rich in some specific elements to prepare self-doped biochar for target pollutants remediation. Furthermore, the different modification methods should integrate with other advance techniques like ultrasonic assistance, ball milling, and microwave assistance to form composite matrices.
- The influence of modifying agents and modification techniques on the physicochemical properties of biochar should be thoroughly studied in the future.
- Research on dealing with the noxious and unsafe substances produced during the fabrication of raw biochar should also gain the attention of scientific community.
- Applying a relatively greener nanotechnology to develop biochar-based nanocomposites that balance between the production costs and optimizing the overall performance with a higher number of regeneration, so that can be utilized on a commercial level is the need of future research.

6 Conclusions

The biochar derived from various natural/agricultural/ industrial byproducts and their derivatives is undeniably a new class of cost effective adsorbent that needs to be exploited. Biochar is specifically preferred as they have unique physicochemical properties and structural features. They are biodegradable, non-toxic, biocompatible, renewable, abundantly available, and undergo modification with ease. The modified biochar nanocomposites demonstrate superior characteristics such as greater surface area, better porosity, higher stability, and exemplary adsorption efficiency toward potentially toxic inorganic pollutants including heavy metals, metalloids, phosphate, nitrate, ammonium, and other radioactive elements. The main mechanism involved in the removal process includes surface adsorption, electrostatic interactions, ion exchange, precipitation, complexation, and reduction process. In spite of the vast practice of modification on biochar and its application for removal of toxic pollutants, it is still in its growing stage and the long-term influence of biochar before and after use requires more effective and reliable information.

References

Ahmed W, Núñez-Delgado A, Mehmood S, Ali S, Qaswar M, Shakoor A, Chen DY (2021) Highly efficient uranium (VI) capture from aqueous solution by means of a hydroxyapatite-biochar nanocomposite: adsorption behavior and mechanism. Environ Res 201:111518. https://doi.org/10.1016/j.envres.2021.111518

Amusat SO, Kebede TG, Dube S, Nindi MM (2021) Ball-milling synthesis of biochar and biochar–based nanocomposites and prospects for removal of emerging contaminants: a review. J Water Process Eng 41:101993. https://doi.org/10.1016/j.jwpe.2021.101993

Bai L, Huang A, Feng J, Su X, Mo W, Ma S, Lin H (2022) One-step microwave pyrolysis synthesis of bagasse biochar/ferrites nanocomposite and synergistic effect on As (V) adsorption in water. Mater Chem Phys 283:126035. https://doi.org/10.1016/j.matchemphys.2022.126035

Baig SA, Zhu J, Muhammad N, Sheng T, Xu X (2014) Effect of synthesis methods on magnetic Kans grass biochar for enhanced As (III, V) adsorption from aqueous solutions. Biomass Bioenergy 71:299–310. https://doi.org/10.1016/j.biombioe.2014.09.027

Bakshi S, Laird DA, Smith RG, Brown RC (2021) Capture and release of orthophosphate by Fe-modified biochars: mechanisms and environmental applications. ACS Sustain Chem Eng 9(658):668. https://doi.org/10.1021/acssuschemeng.0c06108

Bilal M, Adeel M, Rasheed T, Zhao Y, Iqbal HMN (2019) Emerging contaminants of high concern and their enzyme-assisted biodegradation–a review. Environ Int 124:336–353. https://doi.org/10.1016/j.envint.2019.01.011

Cai R, Wang X, Ji X, Peng B, Tan C, Huang X (2017) Phosphate reclaim from simulated and real eutrophic water by magnetic biochar derived from water hyacinth. J Environ Manage 187:212–219. https://doi.org/10.1016/j.jenvman.2016.11.047

Chausali N, Saxena J, Prasad R (2021) Nanobiochar and biochar based nanocomposites: advances and applications. J Agric Food Res 5:100191. https://doi.org/10.1016/j.jafr.2021.100191

Chen L, Chen XL, Zhou CH, Yang HM, Ji SF, Tong DS, Zhong ZK, Yu WH, Chu MQ (2017) Environmental-friendly montmorillonite-biochar composites: facile production and tunable adsorption-release of ammonium and phosphate. J Clean Prod 156:648–659. https://doi.org/10.1016/j.jclepro.2017.04.050

Chin JF, Heng ZW, Teoh HC, Chong WC, Pang YL (2021) Recent development of magnetic biochar crosslinked chitosan on heavy metal removal from wastewater–modification, application and mechanism. Chemosphere 291:133035. https://doi.org/10.1016/j.chemosphere.2021.133035

Cope CO, Webster DS, Sabatini DA (2014) Arsenate adsorption onto iron oxide amended rice husk char. Sci Total Environ 488:554–561. https://doi.org/10.1016/j.scitotenv.2013.12.120

Deng J, Li X, Wei X, Liu Y, Liang J, Song B, Shao Y, Huang W (2020) Hybrid silicate-hydrochar composite for highly efficient removal of heavy metal and antibiotics: coadsorption and mechanism. Chem Eng J 387:124097. https://doi.org/10.1016/j.cej.2020.124097

Dewage NB, Liyanage AS, Pittman CU, Mohan D, Mlsna T (2018) Fast nitrate and fluoride adsorption and magnetic separation from water on α-Fe_2O_3 and Fe_3O_4 dispersed on Douglas fir biochar. Bioresour Technol 263:258–265. 1016/j.biortech.2018.05.001

Diao ZH, Yan L, Dong FX, Qian W, Deng QH, Kong LJ, Yang JW, Lei ZX, Du JJ, Chu W (2020) Degradation of 2, 4-dichlorophenol by a novel iron based system and its synergism with Cd (II) immobilization in a contaminated soil. Chem Eng J 379:122313. https://doi.org/10.1016/j.cej.2019.122313

Dong L, Li S, Jin Y, Hu B, Sheng G (2021) Enhanced adsorption of Eu (III) from wastewater using Solidagocanadensis-derived biochar functionalized by Ca/Al-LDH and hydroxyapatite. Appl Surf Sci 567:150794. https://doi.org/10.1016/j.apsusc.2021.150794

Feng S, Cao X, Zheng W, Yue X, Li X, Li S, Wang X, Feng S (2022) In-situ formed Prussian blue nanoparticles supported by porous biochar as highly efficient removal of cesium ions. J Environ Chem Eng 24:107972. https://doi.org/10.1016/j.jece.2022.107972

Feng Y, Lu H, Liu Y, Xue L, Dionysiou DD, Yang L, Xing B (2017) Nano-cerium oxide functionalized biochar for phosphate retention: preparation, optimization and rice paddy application. Chemosphere 185:816–825. https://doi.org/10.1016/j.chemosphere.2017.07.107

Foroutan R, Peighambardoust SJ, Mohammadi R, Peighambardoust SH, Ramavandi B (2022) Cadmium ion removal from aqueous media using banana peel biochar/Fe_3O_4/ZIF-67. Environ Res 211:113020. https://doi.org/10.1016/j.envres.2022.113020

Hasan MS, Geza M, Vasquez R, Chilkoor G, Gadhamshetty V (2020) Enhanced heavy metal removal from synthetic stormwater using nanoscalezerovalent iron–modified biochar. Water Air Soil Pollut 231:220. https://doi.org/10.1007/s11270-020-04588-w

He S, Zhong L, Duan J, Feng Y, Yang B, Yang L (2017) Bioremediation of wastewater by iron oxide-biochar nanocomposites loaded with photosynthetic bacteria. Front Microbiol 8:823. https://doi.org/10.3389/fmicb.2017.00823

Hu Q, Zhu Y, Hu B, Lu S, Sheng G (2018) Mechanistic insights into sequestration of U(VI) toward magnetic biochar: batch, XPS and EXAFS techniques. J Environ Sci (china) 70:217–225. https://doi.org/10.1016/j.jes.2018.01.013

Hu X, Ding Z, Zimmerman AR, Wang S, Gao B (2015) Batch and column sorption of arsenic onto iron-impregnated biochar synthesized through hydrolysis. Water Res 68:206–216. https://doi.org/10.1016/j.watres.2014.10.009

Inyang M, Gao B, Zimmerman A, Zhou Y, Cao X (2015) Sorption and cosorption of lead and sulfapyridine on carbon nanotube-modified biochars. Environ Sci Pollut Res 22:1868–1876. https://doi.org/10.1007/s11356-014-2740-z

Ioannou K, Hadjiyiannis P, Liatsou I, Pashalidis I (2019) U (VI) adsorption by biochar fiber–MnO_2 composites. J Radioanal Nucl Chem 320:425–432. https://doi.org/10.1007/s10967-019-06479-9

Jia Y, Zhang Y, Fu J, Yuan L, Li Z, Liu C, Zhao D, Wang X (2019) A novel magnetic biochar/MgFe-layered double hydroxides composite removing Pb^{2+} from aqueous solution: Isotherms, kinetics and thermodynamics. Colloids Surf a: Physicochem Eng Asp 567:278–287. https://doi.org/10.1016/j.colsurfa.2019.01.064

Jung KW, Lee S, Lee YJ (2017) Synthesis of novel magnesium ferrite ($MgFe_2O_4$)/biochar magnetic composites and its adsorption behavior for phosphate in aqueous solutions. Bioresour Technol 245:751–759. https://doi.org/10.1016/j.biortech.2017.09.035

Jung KW, Lee SY, Choi JW, Lee YJ (2019) A facile one-pot hydrothermal synthesis of hydroxyapatite/biochar nanocomposites: adsorption behavior and mechanisms for the removal of copper(II) from aqueous media. Chem Eng J 369:529–541. https://doi.org/10.1016/j.cej.2019.03.102

Kamran U, Park SJ (2022) Hybrid biochar supported transition metal doped MnO_2 composites: efficient contenders for lithium adsorption and recovery from aqueous solutions. Desalination 522:115387. https://doi.org/10.1016/j.desal.2021.115387

Khan S, Bhardwaj U, Iqbal HM, Joshi N (2021) Synergistic role of bacterial consortium to biodegrade toxic dyes containing wastewater and its simultaneous reuse value as an added. Chemosphere 284:131273. https://doi.org/10.1016/j.chemosphere.2021.131273

Khan ZH, Gao M, Qiu W, Qaswar M, Islam M, Song Z (2020) The sorbed mechanisms of engineering magnetic biochar composites on arsenic in aqueous solution. Environ Sci Pollut Res 27:41361–41371. https://doi.org/10.1007/s11356-020-10082-x

Kołodyńska D, Bąk J, Kozioł M, Pylychuk LV (2017) Investigations of heavy metal ion sorption using nanocomposites of iron-modified biochar. Nanoscale Res Lett 12:1–3. https://doi.org/10.1186/s11671-017-2201-y

Lawrinenko M, Jing D, Banik C, Laird DA (2017) Aluminum and iron biomass pretreatment impacts on biochar anion exchange capacity. Carbon 118:422–430. https://doi.org/10.1016/j.carbon.2017.03.056

Li A, Zhang Y, Ge W, Zhang Y, Liu L, Qiu G (2022) Removal of heavy metals from wastewaters with biochar pyrolyzed from MgAl-layered double hydroxide-coated rice husk: mechanism and application. Bioresour Technol 347:126425. https://doi.org/10.1016/j.biortech.2021.126425

Li C, Zhang L, Gao Y, Li A (2018a) Facile synthesis of nano ZnO/ZnS modified biochar by directly pyrolyzing of zinc contaminated corn stover for Pb(II), Cu(II) and Cr(VI) removals. Waste Manag 79:625–637. https://doi.org/10.1016/j.wasman.2018.08.035

Li M, Liu H, Chen T, Dong C, Sun Y (2019a) Synthesis of magnetic biochar composites for enhanced uranium (VI) adsorption. Sci Total Environ 651(1020):1028. https://doi.org/10.1016/j.scitotenv.2018.09.259

Li N, Yin ML, Tsang DCW, Yang ST, Liu J, Li X, Song G, Wang J (2019b) Mechanisms of U(VI) removal by biochar derived from Ficusmicrocarpaaerial root: a comparison between raw and modified biochar. Sci Total Environ 697:134115. https://doi.org/10.1016/j.scitotenv.2019.134115

Li R, Wang JJ, Gaston LA, Zhou B, Li M, Xiao R, Wang Q, Zhang Z, Huang H, Liang W, Huang H (2018b) An overview of carbothermal synthesis of metal–biochar composites for the removal of oxyanion contaminants from aqueous solution. Carbon 129:674–687. https://doi.org/10.1016/j.carbon.2017.12.070

Liang J, Li X, Yu Z, Zeng G, Luo Y, Jiang L, Yang Z, Qian Y, Wu H (2017) Amorphous MnO_2 modified biochar derived from aerobically composted swine manure for adsorption of Pb (II) and Cd (II). ACS Sustain Chem Eng 5:5049–5058. https://doi.org/10.1021/acssuschemeng.7b00434

Liang L, Xi F, Tan W, Meng X, Hu B, Wang X (2021) Review of organic and inorganic pollutants removal by biochar and biochar-based composites. Biochar 3:255–281. https://doi.org/10.1007/s42773-021-00101-6

Lin L, Li Z, Liu X, Qiu W, Song Z (2019a) Effects of Fe-Mn modified biochar composite treatment on the properties of As-polluted paddy soil. Environ Pollut 244:600–607. https://doi.org/10.1016/j.envpol.2018.10.011

Lin L, Song Z, Huang Y, Khan ZH, Qiu W (2019b) Removal and oxidation of arsenic from aqueous solution by biochar impregnated with Fe-Mn oxides. Water Air Soil Pollut 230:105. https://doi.org/10.1007/s11270-019-4146-5

Liu K, Li F, Cui J, Yang S, Fang L (2020) Simultaneous removal of Cd (II) and As(III) by graphene-like biochar-supported zero-valent iron from irrigation waters under aerobic conditions: synergistic effects and mechanisms. J Hazard Mater 395:122623. https://doi.org/10.1016/j.jhazmat.2020.122623

Liu SB, Tan XF, Liu YG, Gu YL, Zeng GM, Hu XJ, Wang H, Zhou L, Jiang LH, Zhao BB (2016) Production of biochars from Ca impregnated ramie biomass (Boehmerianivea (L.) Gaud.) and their phosphate removal potential. RSC Adv 6:5871–5880. https://doi.org/10.1039/C5RA22142K

Lomenech C, Hurel C, Messina L, Schembri M, Tosi P, Orange F, Georgi F, Mija A, Kuzhir P (2021) A humins-derived magnetic biochar for water purification by adsorption and magnetic separation. Waste Biomass Valori 12:6497–6512. https://doi.org/10.1007/s12649-021-01481-3

Lu HP, Li ZA, Gascó G, Méndez A, Shen Y, Paz-Ferreiro J (2017) Use of magnetic biochars for the immobilization of heavy metals in a multi-contaminated soil. Sci Total Environ 622–623:892–899. https://doi.org/10.1016/j.scitotenv.2017.12.056

Lyu H, Gao B, He F, Zimmerman AR, Ding C, Huang H, Tang J (2018) Effects of ball milling on the physicochemical and sorptive properties of biochar: experimental observations and governing mechanisms. Environ Pollut 233:54–63. https://doi.org/10.1016/j.envpol.2017.10.037

Lyu H, Xia S, Tang J, Zhang Y, Gao B, Shen B (2020) Thiol-modified biochar synthesized by a facile ball-milling method for enhanced sorption of inorganic Hg^{2+} and organic CH_3Hg^+. J Hazard Mater 384:121357. https://doi.org/10.1016/j.jhazmat.2019.121357

Lyu P, Wang G, Cao Y, Wang B, Deng N (2021) Phosphorus-modified biochar cross-linked Mg–Al layered double-hydroxide composite for immobilizing uranium in mining contaminated soil. Chemosphere 276:130116. https://doi.org/10.1016/j.chemosphere.2021.130116

Maneechakr P, Mongkollertlop S (2020) Investigation on adsorption behaviors of heavy metal ions (Cd^{2+}, Cr^{3+}, Hg^{2+} and Pb^{2+}) through low-cost/active manganese dioxide-modified magnetic biochar derived from palm kernel cake residue. J Environ Chem Eng 8:104467. https://doi.org/10.1016/j.jece.2020.104467

Nayak A, Bhushan B, Gupta V, Kotnala S (2021) Fabrication of microwave assisted biogenic magnetite-biochar nanocomposite: a green adsorbent from jackfruit peel for removal and recovery of nutrients in water sample. J Ind Eng Chem 100:134–148. https://doi.org/10.1016/j.jiec.2021.05.028

Nyamunda BC, Chivhanga T, Guyo U, Chigondo F (2019) Removal of Zn (II) and Cu (II) ions from industrial wastewaters using magnetic biochar derived from water hyacinth. J Eng 2019:5656983. https://doi.org/10.1155/2019/5656983

Peng G, Jiang S, Wang Y, Zhang Q, Cao Y, Sun Y, Zhang W, Wang L (2020) Synthesis of Mn/Al double oxygen biochar from dewatered sludge for enhancing phosphate removal. J Clean Prod 251:119725. https://doi.org/10.1016/j.jclepro.2019.119725

Priyadarshni N, Nath P, Chanda N (2020) Sustainable removal of arsenate, arsenite and bacterial contamination from water using biochar stabilized iron and copper oxide nanoparticles and associated mechanism of the remediation process. J Water Process Eng 37:101495. https://doi.org/10.1016/j.jwpe.2020.101495

Qian LB, Shang X, Zhang B, Zhang WY, Su AQ, Chen Y, Ouyang D, Han L, Yan JC, Chen MF (2018) Enhanced removal of Cr(VI) by silicon rich biochar-supported nanoscale zero-valent iron. Chemosphere 215:739–745. https://doi.org/10.1016/j.chemosphere.2018.10.030

Qiu B, Duan F (2019) Synthesis of industrial solid wastes/biochar composites and their use for adsorption of phosphate: From surface properties to sorption mechanism. Colloids Surf a: Physicochem Eng Asp 571:86–93. https://doi.org/10.1016/j.colsurfa.2019.03.041

Qiu B, Tao X, Wang H, Li W, Ding X, Chu H (2021) Biochar as a low-cost adsorbent for aqueous heavy metal removal: a review. J Anal Appl Pyrol 155:105081. https://doi.org/10.1016/j.jaap.2021.105081

Qiu Y, Zhang Q, Gao B, Li M, Fan Z, Sang W, Hao H, Wei X (2020) Removal mechanisms of Cr (VI) and Cr (III) by biochar supported nanosized zero-valent iron: synergy of adsorption, reduction and transformation. Environ Pollut 265:115018. https://doi.org/10.1016/j.envpol.2020.115018

Quan G, Sui F, Wang M, Cui L, Wang H, Xiang W, Li G, Yan J (2022) Mechanochemical modification of biochar-attapulgite nanocomposites for cadmium removal: performance and mechanisms. Biochem Eng J 179:108332. https://doi.org/10.1016/j.bej.2022.108332

Reddy DHK, Lee SM (2014) Magnetic biochar composite: facile synthesis, characterization, and application for heavy metal removal. Colloids Surf a: Physicochem Eng Asp 454:96–103. https://doi.org/10.1016/j.colsurfa.2014.03.105

Shaheen SM, El-Naggar A, Wang JX, Hassan NEE, Niazi NK, Wang HL, Tsang DCW, Ok YS, Bolan N, Rinklebe J (2019) Biochar as an (Im)mobilizing agent for the potentially toxic elementsin contaminated soils-science direct. Biochar Biomass Waste 14:255–274. https://doi.org/10.1016/B978-0-12-811729-3.00014-5

Shakya A, Agarwal T (2019) Removal of Cr(VI) from water using pineapple peel derived biochars: adsorption potential and re-usability assessment. J Mol Liq 93:111497. https://doi.org/10.1016/j.molliq.2019.111497

Shen, Z, Zhang J, Hou D, Tsang DCW, Ok YS, Alessi DS (2019) Synthesis of MgO coated corncob biochar and its application in lead stabilization in a soil washing residue. Environ Int 122:357–362. https://doi.org/10.1016/j.envint.2018.11.045

Singh S, Wasewar KL, Kansal SK (2020) Low-cost adsorbents for removal of inorganic impurities from wastewater. In: Inorganic Pollutants in Water. 173–203. https://doi.org/10.1016/B978-0-12-818965-8.00010-X

Sun C, Chen T, Huang Q, Wang J, Lu S, Yan J (2019) Enhanced adsorption for Pb (II) and Cd (II) of magnetic rice husk biochar by $KMnO_4$ modification. Environ Sci Pollut Res 26:8902–8913. https://doi.org/10.1007/s11356-019-04321-z

Tan XF, Liu YG, Gu YL, Xu Y, Zeng GM, Hu XJ, Liu SB, Wang X, Liu SM, Li J (2016) Biochar-based nano-composites for the decontamination of wastewater: a review. Bioresour Technol 212:318–333. https://doi.org/10.1016/j.biortech.2016.04.093

Tho PT, Van HT, Nguyen LH, Hoang TK, Tran TN, Nguyen TT, Nguyen TBH, Nguyen VQ, Le Sy H, Thai VN, Tran QB, Sadeghzadeh SM, Asadpourk R, Thang PQ (2021) Enhanced simultaneous adsorption of As(III), Cd(II), Pb(II) and Cr(VI) ions from aqueous solution using cassava root husk-derived biochar loaded with ZnO nanoparticles. RSC Adv 11:18881. https://doi.org/10.1039/d1ra01599k

Tran DT, Pham TD, Dang VC, Pham TD, Nguyen MV, Dang NM, Ha MN, Nguyen VN, Nghiem LD (2022) A facile technique to prepare MgO-biochar nanocomposites for cationic and anionic nutrient removal. J Water Process 47:102702. https://doi.org/10.1016/j.jwpe.2022.102702

van der Hoek JP, de Fooij H, Struker A (2016) Wastewater as a resource: strategies to recover resources from Amsterdam's wastewater. Resour Conserv Recycl 113:53–64. https://doi.org/10.1016/j.resconrec.2016.05.012

Van Vinh N, Zafar M, Behera SK, Park HS (2015) Arsenic (III) removal from aqueous solution by raw and zinc-loaded pine cone biochar: equilibrium, kinetics, and thermodynamics studies. Int J Environ Sci Technol 12:1283–1294. https://doi.org/10.1007/s13762-014-0507-1

Vishnu D, Dhandapani B, Ramakrishnan SR, Pandian PK, Raguraman T (2020) Fabrication of surface-engineered superparamagnetic nanocomposites (Co/Fe/Mn) with biochar from groundnut waste residues for the elimination of copper and lead metal ions. J Nano Struct Chem 11:215–228. https://doi.org/10.1007/s40097-020-00360-y

Wang H, Wang S, Chen Z, Zhou X, Wang J, Chen Z (2020a) Engineered biochar with anisotropic layered double hydroxide nanosheets to simultaneously and efficiently capture Pb^{2+} and CrO_4^{2-} from electroplating wastewater. Bioresour Technol 306:123118. https://doi.org/10.1016/j.biortech.2020.123118

Wang L, Bolan NS, Tsang DCW, Hou D (2020b) Green immobilization of toxic metals using alkaline enhanced rice husk biochar: effects of pyrolysis temperature and KOH concentration. Sci Total Environ 720:137584. https://doi.org/10.1016/j.scitotenv.2020.137584

Wang L, Wang J, Wang Z, Feng J, Li S, Yan W (2019a) Synthesis of Ce-doped magnetic biochar for effective Sb(V) removal: Performance and mechanism. Powder Technol 345:501–508. https://doi.org/10.1016/j.powtec.2019.01.022

Wang L, Wang J, Wang Z, He C, Lyu W, Yan W, Yang L (2018) Enhanced antimonate (Sb(V)) removal from aqueous solution by La-doped magnetic biochars. Chem Eng J 354:623–632. https://doi.org/10.1016/j.cej.2018.08.074

Wang S, Gao B, Zimmerman AR, Li Y, Ma L, Harris WG, Migliaccio KW (2015) Removal of arsenic by magnetic biochar prepared from pinewood and natural hematite. Bioresour Technol 175:391–395. https://doi.org/10.1016/j.biortech.2014.10.104

Wang S, Zhao M, Zhou M, Li YC, Wang J, Gao B, Sato S, Feng K, Yin W, Igalavithana AD, Oleszczuk P (2019b) Biochar-supported nZVI (nZVI/BC) for contaminant removal from soil and water: a critical review. J Hazard Mater 373:820–834. https://doi.org/10.1016/j.jhazmat.2019.03.080

Wang T, Zhang D, Fang K, Zhu W, Peng Q, Xie Z (2021) Enhanced nitrate removal by physical activation and Mg/Al layered double hydroxide modified biochar derived from wood waste: adsorption characteristics and mechanisms. J Environ Chem Eng 9:105184. https://doi.org/10.1016/j.jece.2021.105184

Wang YY, Lu HH, Liu YX, Yang SM (2016) Removal of phosphate from aqueous solution by SiO_2–biochar nanocomposites prepared by pyrolysis of vermiculite treated algal biomass. RSC Adv 6:83534–83546. https://doi.org/10.1039/C6RA15532D

Wen Z, Xi J, Lu J, Zhang Y, Cheng G, Zhang Y, Chen R (2021) Porous biochar-supported $MnFe_2O_4$ magnetic nanocomposite as an excellent adsorbent for simultaneous and effective removal of organic/inorganic arsenic from water. J Hazard Mater 411:124909. https://doi.org/10.1016/j.jhazmat.2020.124909

Wu J, Li Z, Huang D, Liu X, Tang C, Parikh SJ, Xu J (2020a) A novel calcium-based magnetic biochar is effective in stabilization of arsenic and cadmium co-contamination in aerobic soils. J Hazard Mater 387:122010. https://doi.org/10.1016/j.jhazmat.2019.122010

Wu Z, Chen X, Yuan B, Fu ML (2020b) A facile foaming-polymerization strategy to prepare 3D MnO_2 modified biochar-based porous hydrogels for efficient removal of Cd(II) and Pb(II). Chemosphere 239:124745. https://doi.org/10.1016/j.chemosphere.2019.124745

Xu L, Shu Z, Feng L, Zhou J, Li T, Zhao Z, Wang W (2022) Fresh biomass derived biochar with high-load zero-valent iron prepared in one step for efficient arsenic removal. J Clean Prod 352:131616. https://doi.org/10.1016/j.jclepro.2022.131616

Yan J, Li R (2022) Simple and low-cost production of magnetite/graphene nanocomposites for heavy metal ions adsorption. Sci Total Environ 813:152604. https://doi.org/10.1016/j.scitotenv.2021.152604

Yan L, Dong FX, Lin X, Zhou XH, Kong LJ, Chu W, Diao ZH (2021) Insights into the removal of Cr (VI) by a biochar–iron composite from aqueous solution: reactivity, kinetics and mechanism. Environ Technol Innov 24:102057. https://doi.org/10.1016/j.eti.2021.102057

Yan L, Kong L, Qu Z, Li L, Shen G (2014) Magnetic biochar decorated with ZnS nanocrystals for Pb (II) removal. ACS Sustain Chem Eng 3:125–132. https://doi.org/10.1021/sc500619r

Yang H, Ye S, Zeng Z, Zeng G, Tan X, Xiao R, Wang J, Song B, Du L, Qin M, Yang Y (2020a) Utilization of biochar for resource recovery from water: a review. Chem Eng J 397:125502. https://doi.org/10.1016/j.cej.2020.125502

Yang Y, Sun F, Li J, Chen J, Tang M (2020b) The effects of different factors on the removal mechanism of Pb (II) by biochar-supported carbon nanotube composites. RSC Adv 10:5988–5995. https://doi.org/10.1039/c9ra09470a

Yao Y, Gao B, Chen J, Zhang M, Inyang M, Li Y, Alva A, Yang L (2013) Engineered carbon (biochar) prepared by direct pyrolysis of Mg-accumulated tomato tissues: characterization and phosphate removal potential. Bioresour Technol 138:8–13. https://doi.org/10.1016/j.biortech.2013.03.057

Yao Y, Gao B, Inyang M, Zimmerman AR, Cao X, Pullammanappallil P, Yang L (2011) Removal of phosphate from aqueous solution by biochar derived from anaerobically digested sugar beet tailings. J Hazard Mater 190:501–507. https://doi.org/10.1016/j.jhazmat.2011.03.083

Yin Z, Liu Y, Liu S, Jiang L, Tan X, Zeng G, Li M, Liu S, Tian S, Fang Y (2018) Activated magnetic biochar by one-step synthesis: enhanced adsorption and coadsorption for 17β-estradiol and copper. Sci Total Environ 639:1530–1542. https://doi.org/10.1016/j.scitotenv.2018.05.130

Yu H, Liu J, Shen J, Sun X, Li J, Wang L (2016) Preparation of MnOx-loaded biochar for Pb^{2+} removal: adsorption performance and possible mechanism. J Taiwan Inst Chem Eng 66:313–320. https://doi.org/10.1016/j.jtice.2016.07.010

Yu J, Tang L, Pang Y, Zeng G, Wang J, Deng Y, Liu Y, Feng H, Chen S, Ren X (2019) Magnetic nitrogen-doped sludge-derived biochar catalysts for persulfate activation: internal electron transfer mechanism. Chem Eng J 364:146–159. https://doi.org/10.1016/j.cej.2019.01.163

Zhang A, Li X, Xing J, Xu G (2020) Adsorption of potentially toxic elements in water by modified biochar: a review. J Environ Chem Eng 8:104196. https://doi.org/10.1016/j.jece.2020.104196

Zhang J, Lu W, Zhan S, Qiu J, Wang X, Wu Z, Li H, Qiu Z, Peng H (2021) Adsorption and mechanistic study for humic acid removal by magnetic biochar derived from forestry wastes functionalized with Mg/Al-LDH. Sep Purif Technol 276:119296. https://doi.org/10.1016/j.seppur.2021.119296

Zhang LK, Guo JY, Huang XM, Wamg WD, Sun P, Li YM, Han JH (2019) Functionalized biochar-supported magnetic $MnFe_2O_4$ nanocomposite for the removal of Pb(II) and Cd(II). RSC Adv 9:365–376. https://doi.org/10.1039/C8RA09061K

Zhang M, GaoB YY, Xue Y, Inyang M (2012) Synthesis of porous MgO-biochar nanocomposites for removal of phosphate and nitrate from aqueous solutions. Chem Eng J 210:26–32. https://doi.org/10.1016/j.cej.2012.08.052

Zhang Y, Cao B, Zhao L, Sun L, Gao Y, Li J, Yang F (2018) Biochar-supported reduced graphene oxide composite for adsorption and coadsorption of atrazine and lead ions. Appl Surf Sci 427:147–155. https://doi.org/10.1016/j.apsusc.2017.07.237

Zhao C, Wang B, Theng BK, Wu P, Liu F, Wang S, Lee X, Chen M, Li L, Zhang X (2021) Formation and mechanisms of nano-metal oxide-biochar composites for pollutants removal: a review. Sci Total Environ 767:145305. https://doi.org/10.1016/j.scitotenv.2021.145305

Zheng Y, Wan Y, Chen J, Chen H, Gao B (2020) MgO modified biochar produced through ball milling: a dual-functional adsorbent for removal of different contaminants. Chemosphere 243:125344. https://doi.org/10.1016/j.chemosphere.2019.125344

Zhu D, Chen Y, Yang H, Wang S, Wang X, Zhang S, Chen H (2020a) Synthesis and characterization of magnesium oxide nanoparticle-containing biochar composites for efficient phosphorus removal from aqueous solution. Chemosphere 247:125847. https://doi.org/10.1016/j.chemosphere.2020.125847

Zhu S, Zhao J, Zhao N, Yang X, Chen C, Shang J (2020b) Goethite modified biochar as a multifunctional amendment for cationic Cd (II), anionic As (III), roxarsone, and phosphorus in soil and water. J Clean Prod 247:119579. https://doi.org/10.1016/j.jclepro.2019.119579

Zhu T, Zhang Y, Chen Y, Liu JL, Song XL (2022) Synthesis of novel hydrated ferric oxide biochar nanohybrids for efficient arsenic removal from wastewater. Rare Met 41:1677–1687. https://doi.org/10.1007/s12598-021-01920-z

Zhu Z, Huang CP, Zhu Y, Wei W, Qin H (2018) A hierarchical porous adsorbent of nano-α-Fe_2O_3/Fe_3O_4 on bamboo biochar (HPA-Fe/C-B) for the removal of phosphate from water. J Water Process Eng 25:96–104. https://doi.org/10.1016/j.jwpe.2018.05.010

Biochar-Based Nanocomposites for the Removal of Organic Environmental Contaminants

Arvind Kumar, Tulsi Satyavir Dabodiya, and Duraisamy Ramamoorthy

Abstract

Biochar-based nanocomposites (BNCs) have efficient applications in environmental remediation due to combined catalytic activity and exceptional surface properties. The inclusiveness of nanostructured material with biochar facilitates efficient removal of organic contaminants from water. The synergistically enhanced organic contaminant sorption by biochars microstructure and nanostructure materials' electronic activity envisages high throughput removal of organic impurities. However, the applicability of BNC materials is currently under commencement due to gap between lab scale and field applications with the quality standards. In order to fulfill these knowledge gaps, more studies as well as research should be done to utilize these materials for industrial applications. This chapter will cover the applications of biochar-based nanocomposites for removal of toxic contaminants from environmental media such as antibiotics and dyes.

A. Kumar (✉)
Department of Earth and Atmospheric Sciences, University of Alberta, 126 Earth Sciences Building, Edmonton, Alberta, T6G 2E3, Canada
e-mail: akcimap@gmail.com; biswas3@ualberta.ca

A. Kumar · D. Ramamoorthy (✉)
Department of Ecology and Environmental Sciences, School of Life Sciences, Pondicherry University (A Central University), Dr. R. Vankataraman Nagar, Kalapet, Puducherry, 605014, India
e-mail: d.ramamoorthy01@gmail.com

Tulsi Satyavir Dabodiya
Department of Chemical and Materials Engineering, University of Alberta, Alberta, T6G 1H9, Canada

Centre for Nanoscience and Technology, Madanjeet School of Green Energy Technologies, Pondicherry University (A Central University), Dr. R. Vankataraman Nagar, Kalapet, Puducherry, 605014, India

Keywords

Antibiotics • Adsorption • Catalytic degradation • Dyes • Water treatment

1 Introduction

The progress of global industry and economy causes water pollution as a serious threat to the environment as well as human health (Lu and Astruc 2020). The different kinds of organic pollutants, dye, and antibiotics introduced into the environment due to the discharge of wastewater from the textile, leather, paint and varnish, pulp and paper, and food and pharmaceutical industries cause chemical pollution (Liang et al. 2021). The chemical pollution significantly affects the water quality in different water bodies such as river, lake, and ponds which is a major crisis for global freshwater depletion. Currently, the enormous (about 70 lacks ton annually) use of coloring agent includes cationic as well as anionic dyes, disperse violet 26, methyl red, and crystal violet (Dutta et al. 2021). Another side environmental contamination due to high dosage of antibiotics in animal husbandry is greater than human medicine. The total estimation of antibiotic consumption in 2010 was about (63,151 tons) and could be potentially rise in about 67% in 2030 and estimation suggested that could be two fold times in Brazil, Russia, India, China, and South Africa (BRICS) countries. Improper discharge of antibiotics and active form of antibiotics comes through the feces and urine as parent compound poses promising trouble in wastewater treatment plant (WWTP) and leading to agro-ecosystem contamination. The residual antibiotics into environment cause the adverse impacts on biota of different trophic level as well as human beings through the accumulation of antibiotics resistance bacterial populations in specific niche (Polianciuc et al. 2020). In the last decade, a number of biochar-based nanocomposites has been developed and promising a great material for the

D. Mishra et al. (eds.), *Biochar-Based Nanocomposites for Contaminant Management*, Advances in Science, Technology & Innovation, https://doi.org/10.1007/978-3-031-28873-9_7

removal of organic contaminant from wastewater (Sabzehmeidani et al. 2021). This chapter summarizes the different types of biochar-based nanocomposites on the basis of techniques involved and effects on the remediation of organic pollutants from aqueous media. The number of existing methods for biochar-based nanocomposites synthesis and their potential applications summarized. Furthermore, some uncertainties, knowledge gap, and research need of biochar-based nanocomposites are identified.

2 Biochar-Based Nanocomposites (BNCs)

The raw biochar has certain amount of adsorption capacity which is mostly into high concentration organic pollutants. Besides this, it is hard to segregate from aqueous medium due to smallest particle size (Tan et al. 2016; Sun et al. 2021). To address this, researcher is trying to develop biochar-based composites combining magnetic iron oxide, nano-metal oxides, and functional nanoparticle with different synthetic methods (Ahmed et al. 2017; Santhosh et al. 2020; Chausali et al. 2021; Deng et al. 2021; Zhao et al. 2021). Recently, substantial research work has been done on BNCs for removal of water contaminants (both organic and inorganic), which can fix the problem regarding both water pollution remediation and carbon sequestration (Tan et al. 2015). BNCs can be classified according to the methods of synthesis. The three types of BNCs are classified as (I) complexes of oxide and hydroxide biochar BNCs, (II) ferrimagnetic biochar BNCs, and (III) coated-biochar BNCs with functional nanomaterials. BNCs can be classified based on different kinds of nanomaterials used for functionalization of biochar, for example, metal oxide nanoparticles, hydroxide metal nanoparticles, magnetic metal oxide nanoparticles, and other functional nanoparticles (graphene, chitosan, carbon nanotubes, graphene oxide, ZnS nanomaterials, layered hydroxides, zerovalent iron, graphitic C_3N_4). Each type of composites are produced by specific synthesis approach as discuss below (Fig. 1).

2.1 Complexes of Oxide and Hydroxide BNCs

Oxide and hydroxide BNCs synthesis are carried out using enrichments of target elements by bio-accumulation methods applied to produce nano-metal, pre-treating biomass with metal salt before pyrolysis, and post-treating metal oxide nanoparticles on to the surfaces of biochar after pyrolysis. In target enrichment and metal salts pre-treating methods, the biomass is treated with impregnation of specific metal before pyrolysis. In case of insertion based method, the metallic nano-oxide/hydroxide based as synthesized materials are directly introduced into the biomass/pyrolyzed biomass. For achieving engineered biochar with desired properties, enriched target elements can be implemented via bio-accumulation in biomass. Subsequently, metal precursors in situ converted into metallic nano-oxide/hydroxide after thermal treatment and enriched the biomass with nanomaterials. An innovative strategy has been developed previously via slow pyrolysis of Mg-enriched tomato tissue (Yao et al. 2013; Tan et al. 2016). Similarly, various physiochemical properties have been improved by bio-accumulation of other metal precursor such as Al, Zn, and many others on feedstock biomass to develop nano-metal oxide/hydroxide incorporated biochar. The development of hyperaccumulator plant species biomass technique to produce BNCs provides desired concentration of target elements with potential properties as oxide/hydroxide BNCs. In pre-treating method, chemical reagents are used in biomass before the pyrolysis. The metal ions are absorbed or attached to the surface in the interior of biomass during this procedure. For example, highly porous MgO-BNCs can be prepared by slow pyrolysis of $MgCl_2$-pre-treated biomass (Yao et al. 2013). This result depicts the uniform dispersion of MgO nano-flakes decorated biochar and enhanced catalytic properties of the BNCs. BNCs with metallic oxide nanoparticles can also be synthesized after the biomass pyrolysis by using several common methods such as evaporation, heating and conventional wet impregnation, and direct hydrolyzes. Biochar-iron oxides nanocomposite synthesized by evaporation of $Fe(NO_3)_3{\cdot}9H_2O$. This BNCs composite shown higher surface area than pristine biochar (Cope et al. 2014).

2.2 Magnetic BNCs

On the basis of separation from the aqueous medium, BNCs can be changed into magnetic material using various synthesis methods such as pre-treating of biomass with Fe ions and iron oxide co-precipitation onto biochar surfaces. The magnetic biochar-based composite is separated by an external magnetic field from the solution. Magnetic biochar-based BNCs are influenced by the amount of iron oxide impregnation during synthesis that significantly enhance the removal efficiency toward targeted pollutants from the environmental media. In pre-treatment of biomass with Fe ions method, Fe^{3+} or Fe^{2+} ions are impregnating using co-precipitation with biomass and pyrolyzing under various conditions. Zhang and Gao (2013) improved the ferromagnetic as well as surface properties of biochar by developing of novel magnetic-based nano-sized Fe_2O_3 material hosted into the porous structure of biochar were synthesized via pyrolysis method (Zhang and Gao 2013). Other techniques such as microwave heating method can also be used for preparing magnetic biochar. Furthermore, the hematite (Fe_2O_3) most abundant natural mineral on of

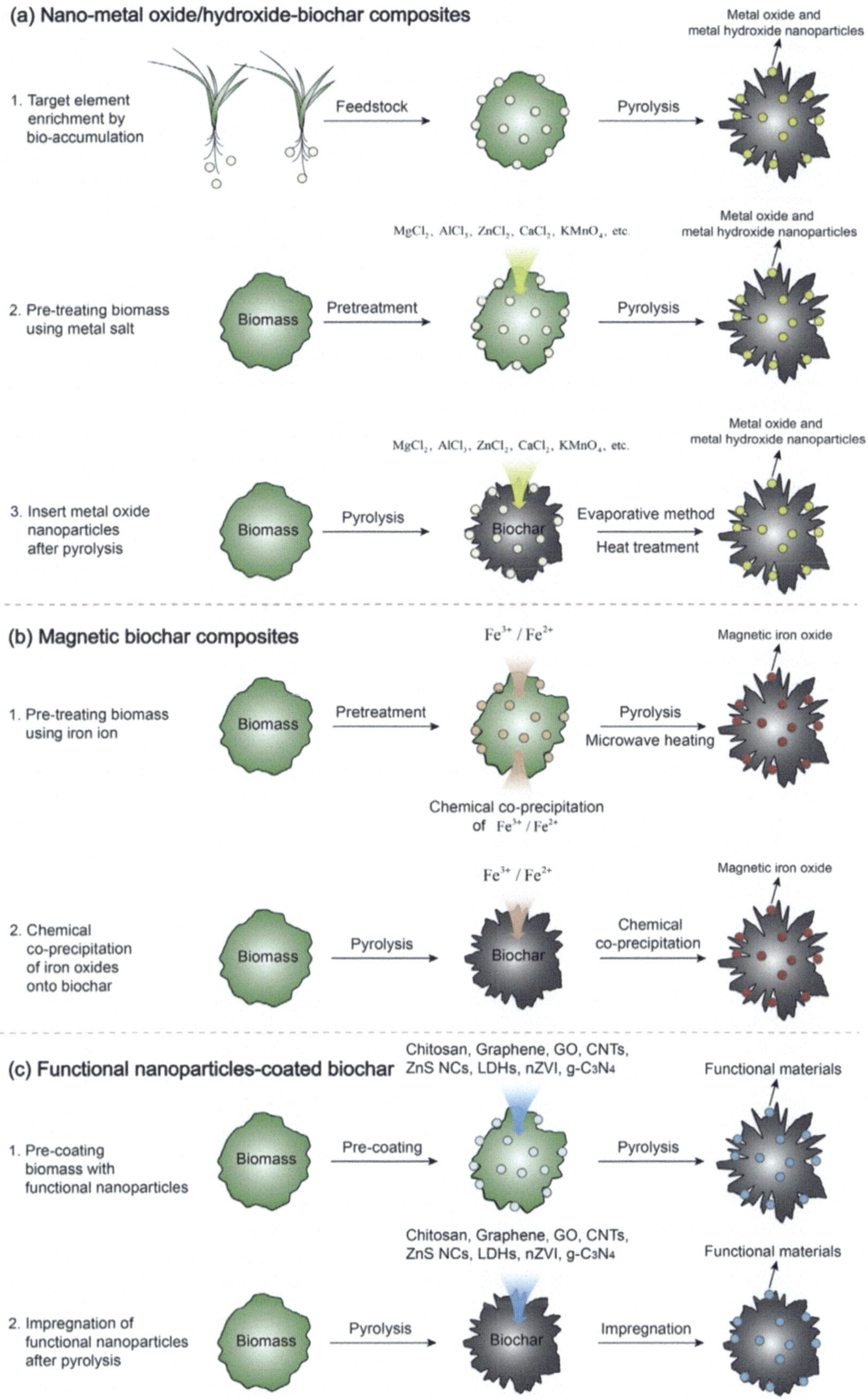

Fig. 1 Schematic repersentation of biochar-based nanocomposites (BNCs) synthesis techniques, **a** oxide/hydroxide type BNC, **b** magnetic BNC, **c** functional nanoparticle BNCs

the earth could be utilized for pre-treatment of biomass to produce magnetic biochar (Wang et al. 2015). However, the hematite treated or blended biochar having potential magnetic property with respect to pristine biochar. The other way of synthesis of magnetic BNCs is direct co-precipitation of Fe ions on biochar obtained from pyrolyzed biomass. In this synthesis approach, the produced biochar by pyrolysis of biomass mixed with aqueous Fe ion solution and followed by aqueous NaOH treatment (Han et al. 2015). Physiochemical properties such as saturation magnetization and thermal stability are mainly dependent on Fe_2O_3 or other magnetic oxide loading content. Comparatively, before the pyrolysis co-precipitating the iron oxide onto biomass consists superior physiochemical properties than after pyrolysis co-precipitation of iron oxides onto biochar.

2.3 Biochar-Coated Functional NCs

Functional nanoparticles (FNPs) such as graphene oxide (GO), carbon nanotubes (CNTs), nano-ZnS, chitosan, nano-zerovalent iron (nZVI), and layered double hydroxide (LDHs) onto surface of biochar can be fulfilled as an effective filler material in the biochar matrix. The involvement of functional nanomaterials increases the potentiality of BNCs with excellent surface area, porosity, and thermal stability which play crucial role into various target applications. In pre-coating process, FNPs are applied to biomass pre-treatment before pyrolysis. For example, biomass can be changed into biochar-based functional nanocomposites by dip-coating method. Here, the FNPs suspensions synthesized by using deionized water, and then, the biomass proper dipped into the suspension and oven dried for later pyrolysis. Similarly, BNCs coated with graphene are synthesized through the treatment of biomass with annealed graphene-pyrene-derivative. Wheat straw biomass was pre-treated with graphene before slow pyrolysis for the synthesis of graphene-based biochar composite. The p–p interactions mechanism developed graphene coating on the surface of biochar with thermal stability and number of functional groups (Tang et al. 2015). Inyang et al. (2014) proposed synthesis of multi-walled carbon nanotube (CNT)-coated biochars for that they dipped biomass into the solution of different concentration of carboxyl-functionalized carbon nanotubes before thermochemical conversion, e.g., slow pyrolysis (Inyang et al. 2014, 2015). The proposed study reveals that addition of carbon nanotubes could improve the thermal stability, surface area, and porosity. Impregnation of FNCs onto freshly prepared biochar via pyrolysis to obtain biochar composites with potentially enriched properties of FNCs. Chitosan, zerovalent iron (ZVI), Hydrogel Mg/Al, and nano-ZnS are generally utilized in this preparation methods. The resulted composite materials have surpassing functions characteristics inherited from both FNCs and biochar. Before introduction of functional nanomaterial, BNCs should be targeted that reduce the undesirable effect on properties of BNCs.

3 Current Emerging Organic Contaminants in the Environment

An increasing industrialization and extensive human activities are directly or indirectly affected water bodies by wastewater effluents, synthetic chemicals, and pesticides as well as other agricultural wastes. Lastly, attention has been given at cosmetic products, and a variety of endocrine disturbing molecules and pharmaceuticals are biologically active and participating as organic contaminants that is needed to be controlled. The number of possible sources for organic contaminants are waste discharges and drugs wastage (hospitals), various kinds of drugs and hormones used to vaccinate to cattle and poultry (animal husbandries), fish fed hormones and antibiotics as well during water processing (aquacultures), and unused and expired medicine into drainage to water body due to leakage in sewers line and septic system (household discharges). All the recent emerging organic contaminants are given below in Table 1.

Table 1 Organic contaminants in the environment (based on/adapted from Pal et al. 2010)

S. no	Compounds
1	Diclofenac, Ibuprofen
2	Diltiazem, acetaminophen, and sulfamethoxazole
3	Mixture of bezafibrate, carbamazepine, cyclophosphamide, ciprofloxacin, furosemide, atenolol, hydrochlorothiazide, ibuprofen, ofloxacin, ranitidine, salbutamol and lincomycin, sulfamethoxazole
4	Ethinylestradiol, zearalonol, 17 β-trenbolone, and melengestrol acetate
5	17 α-ethinylestradiol
6	Chloramphenicol, florfenicol, and thiamphenicol

4 Application of BNCs into Removal of Organic Contaminants

Biochar-based nanocomposites have crucial adsorption characteristic to remediate organic pollutants from environmental media (Arun et al. 2021; Shaikh et al. 2022). The biochar-based nanocomposite interaction with organic contaminants is represented in Table 2. Organic contaminants concerned to environment include Rhodamine B (RhB), Congo red (CR) (Shaikh et al. 2021), Eriochrome Black T (EBT) (Akbarnezhad and Safa 2018), Crystal violet (CV) (Luyen et al. 2020), Acid orange 7 (AO7) (Santhosh et al. 2020), Methylene blue (MB) (Inyang et al. 2014), Methyl orange (MO) (Zhang et al. 2019), Reactive red (RR120) (Wei et al. 2020), Tetracycline hydrochloride (TC) (Ye et al. 2019), Reactive Blue 19 (RB19) (Perveen et al. 2021), Methyl orange (MO) (Chaukura et al. 2017), ofloxacin (Chen et al. 2018), norfloxacin (NOR) (Liang et al. 2022), and tetracycline hydrochloride (TC) (Liu et al. 2019). The adsorption capacity of BNC depends on the biochar substrate, nanomaterials, and targeted organic contaminants. The Langmuir–Freundlich model of adsorption implied for fitting the adsorption data because it is fitted superior than isotherm models. Furthermore, the most common models imitate sorption kinetics such as pseudo-second-order and Elovich. The processing of interaction of biochar-based nanocomposite functional groups with organic pollutants takes place with different mechanisms during the adsorption. The mechanism of interaction with the contaminants includes electrostatic attractions, hydrogen bonding, interaction, ion exchange, and hydrophobic interaction (Amusat et al. 2021). Consequently in same way, nanoparticle coating onto biochar surfaces would increase the number of adsorption sites. The utilization of various functional nanoparticles such as CNTs, chitosan, graphene, NCs, LDHs, and ZnS can increase the number of functional

Table 2 Biochar-based nanocomposites and removal efficiency (%)/sorption capacity (mg/g) of organic contaminants

S. no	Biochar-based nanocomposites	Organic contaminants	Removal efficiency/sorption capacity	References
1	TLB/(nAg–TC)	RhB and CR	RhB-95.89%, CR-94.10%	Shaikh et al. (2022)
2	Oleaster stones BC and composited with iron oxide	EBT	98.11%	Akbarnezhad and Safa (2018)
3	Rice husk BC/IONPs	CV	185.6 mg/g	Luyen et al. (2020)
4	BC-based silver nanocomposite (Ag–nBC)	CR, RhB	90%	Shaikh et al. (2021)
5	Magnetic BC nanocomposites	AO7 dye	110.27 mg/g	Santhosh et al. (2020)
6	BC–BiOCl	Tetracycline	60.3%	Inyang et al. (2014)
7	Hybrid multi-walled carbon nanotube (CNT)-coated biochars	MB	60.3%	Inyang et al. (2014)
8	$CaWO_4$–BC nanocomposites	RhB, MO	More 5% (RhB) and 17% (MO)	Zhang et al. (2019)
9	Copper oxide (CuO)–BC nanocomposites	RR120	LM 1399 mg/g	Wei et al. (2020)
10	Porous graphite biochar (PGBC)–g–MoS_2 Nanosheet	TC	10–50 mg/l	Ye et al. (2019)
11	BC–TiO_2	Phenol	64.1% (UV), 33.6% (Vis)	Lisowski et al. (2017)
12	BC–$ZrFe_2O_5$	RB19	88.82%	Perveen et al. (2021)
13	Fe_2O_3–BC	MO	20.53 mg/g	Chaukura et al. (2017)
14	BC–Co_3O_4	Ofloxacin	90%	Chen et al. (2018)
15	Fe_2O_3/bio-composite	Norfloxacin	38.77 mg/g	Liang et al. (2022)
16	3D hierarchical structure BiO_X(X-Cl, Br)-(CMC)	TC	96.5%	Liu et al. (2019)
17	Magnetic BC/TiO_2	Sulfadiazine	Nearly 88%	Silva et al. (2021)
18	Bi/Bi_2O_3/BC	Estrone	90%	Zhu et al. (2020)
19	$CuWO_4$/BC	Ciprofloxacin	97%	Thiruppathi et al. (2020)
20	BC/Zn/TiO_2	Acetaminophen	92%	Peñas-Garzón et al. (2019)

CR—Congo Red, RhB—Rhodamine B, TLB—Tea Leaf Biochar, BC—Biochar, EBT—Erichrome Black T, CV—Crystal Violet, AO7—Acid Orange, MO—Methyle Blue, MB—Methylene Blue, RR120—Reactive Red, TC—Tetracycline Hydrochloride, RB19—Reactive Blue, LMC—Langumir Maximum Capacity groups, surface area, thermal stability, and removal efficiency of BNCs. The graphene coating onto biochar surfaces improves the potential of interaction mechanism between graphene sheets on carbon surface of biochar and contaminants (Tang et al. 2015). The hydrophobic reactions between organic contaminants and graphitic/hydrophobic sites on CNTs play key role in sorption mechanism for organic pollutants (Pal et al. 2010; Inyang et al. 2014). BC–MnO_2 composite was synthesized using rice husk biochar loaded with MnO_2 for the removal of doxycycline (DC) and of tetracycline hydrochloride (TC). The removal efficiency and physicochemical properties were compared between rice husk biochar (RH–BC) and BC–MnO_2. The adsorption capacity of BC–MnO_2 was found significantly increased than RH-BC with specific surface area, total pore volume, and pore structure (Li et al. 2020). Biochar-based layered double hydroxide (Zn–Co–LDH@BC) nanostructures were prepared for photocatalytic degradation of gemifloxacin (GMF). The nanostructure photocatalyst (Zn–Co–LDH@BC) was degraded 97.70% gemifloxacin (GMF) under UV light irradiation (Gholami et al. 2020). Magnetic nanocomposites (BCmag–TiO_2) were synthesized as a promising photocatalyst for the removal of sulfadiazine (SDZ) and oxolinic acid (OXA) (Silva et al. 2021).

5 Potential Application of BNCs

5.1 Agricultural Applications

Biochar-based nanocomposites enhanced charge density could carried great amount of nutrients that improve seed germination and crop yield (Kumar et al. 2022). The nanocarbons extracted from biochar have promising role as nanofertilizers due to controlled nutrient release and water holding capacity. Biochar and BNCs are highly effective in pesticides sequestration and sorption due to porosity, which facilitates large number of adsorption sites in soil. Similarly, BNCs also reduce pesticides mobility, desorption, leaching, and bio-availability in soil layers which consequently enhance rectification of pesticide-contaminated soils (Silva et al. 2021).

5.2 Energy Applications

Biochar and BNCs can also be used as fuel additives. This procedure provides efficient role in rectifying harmful effects of vehicular emissions while used with fuel. Similarly, nano-bio-additives perform vital role over metallic nano-additives owing to their non-toxic nature. Researcher found that addition of nano-biochar as blend into diesel, biodiesel, and ethanol remarkably affects the performance of engine and emissions. Besides this, bagasse produced biomass [sugarcane nano-biochar (SNB)] provides exceptional performance after substitution with respect to metal fuel additives (Chausali et al. 2021).

5.3 Catalytic Applications

Nano-biochar and BNCs will be used as enzyme supportive reactions and acid catalyst. Nano-biochar-based bio-catalytic systems enhance the controlling efficiency by adjusting loading ability, high surface area, and reducing resistance to mass transfer. BNCs materials provide excellent fixed reinforce material in enzyme systems owing to their porous structure, rich in functional group sites on high surface area to volume ratio, etc. In general, the synthesized BNCs using biochar and functional nanomaterial have a number of physicochemical properties such as functional group, pore structure, active sites on large surface area, and magnetic properties which consequently strengthen the catalytic ability and separations of contaminants. BNCs could be utilized as heterogeneous acid catalysts in the esterification reaction (Chausali et al. 2021).

5.4 Other Potential Applications

Nano-biochar can also be used as sensing and detecting material in different electrochemical devices and techniques. Biochar paper immunosensors are developed for the detection of microcystin-LR toxin in water by decorating conductive particles of BNCs with anti-MCLR antibody on filter paper. Besides this, the waste lignin could be changed into nano-biochar which served as potential candidate for substitute to carbon black in reinforcement of styrene-butadiene rubber matrix. Hence, BNCs and nano-biochar have numerous future applications in different science fields and urgent need of further research efforts needed to be explored.

6 Conclusion and Future Perspective

Biochar-based nanocomposites have been considered as eco-friendly and low-cost material. This book chapter focused on the emerging organic contaminants and its removal using biochar-based nanocomposites. The synthesis process of different BNCs has been discussed such as oxide/hydroxide, magnetic, and functional or coated. Biochar combined nanocomposites have reusable potentiality with high surface area, pore structure, functional group, and multiple adsorption sites promising applications into agriculture, energy, and catalytic activities. Assuring the cost effectiveness and efficiency of BNCs need to explore research on several perspectives elaborated in the following points:

- The selection of biochar feedstock (chemical composition), nanomaterial (metallic oxide/hydroxide and functional material), and synthesis conditions is needed to achieve better performance because these crucially decide the structure, surface area, and functionality of BNCs.
- The treatment of wastewater practically by BNCs via catalytical degradation and adsorption mechanisms is still lacking due to presence of multiple competitive organic contaminants so BNCs need to conduct practical studies in future.
- The use of BNCs for environmental management is increasing, so before utilizing on pilot scale, it is imperative to check its impact on aquatic organism such as bio-toxicity assessment should be looked upon in future.
- The production cost of BNCs at large scale is quite huge, so there is a need to cut through the selection of synthesis methods, feedstocks, and amount of energy consumed.
- The upcoming research should be based on recycling and stability of BNCs into environmental remediation that would improve economic practicability.

References

Ahmed MB, Zhou JL, Ngo HH, Guo W, Johir MA, Sornalingam K, Belhaj D, Kallel M (2017) Nano-Fe^0 immobilized onto functionalized biochar gaining excellent stability during sorption and reduction of chloramphenicol via transforming to reusable magnetic composite. Chem Eng J 322:571–581

Akbarnezhad A, Safa F (2018) Biochar-based magnetic nanocomposite for dye removal from aqueous solutions: response surface modelling and kinetic study. Russ J Appl Chem 91:1856–1866

Amusat SO, Kebede TG, Dube S, Nindi MM (2021) Ball-milling synthesis of biochar and biochar–based nanocomposites and prospects for removal of emerging contaminants: a review. J Water Process Eng 41:101993

Arun J, Nirmala N, Priyadharsini P, Dawn S, Santhosh A, Gopinath K, Govarthanan M (2021) A mini-review on bioderived carbon and its nanocomposites for removal of organic pollutants from wastewater. Mater Lett 2021:131476

Chaukura N, Murimba EC, Gwenzi W (2017) Synthesis, characterisation and methyl orange adsorption capacity of ferric oxide–biochar nano-composites derived from pulp and paper sludge. Appl Water Sci 7:2175–2186

Chausali N, Saxena J, Prasad R (2021) Nanobiochar and biochar-based nanocomposites: advances and applications. J Agric Food Res 5:100191

Chen L, Yang S, Zuo X, Huang Y, Cai T, Ding D (2018) Biochar modification significantly promotes the activity of Co_3O_4 towards heterogeneous activation of peroxymonosulfate. Chem Eng J 354:856–865

Cope CO, Webster DS, Sabatini DA (2014) Arsenate adsorption onto iron oxide amended rice husk char. Sci Total Environ 488:554–561

Deng Y, Li X, Ni F, Liu Q, Yang Y, Wang M, Ao T, Chen W (2021) Synthesis of magnesium modified biochar for removing copper, lead and cadmium in single and binary systems from aqueous solutions: adsorption mechanism. Water 13:599

Dutta S, Gupta B, Srivastava SK, Gupta AK (2021) Recent advances on the removal of dyes from wastewater using various adsorbents: a critical review. Mater Adv 2:4497–4531

Gholami P, Khataee A, Soltani RDC, Dinpazhoh L, Bhatnagar A (2020) Photocatalytic degradation of gemifloxacin antibiotic using Zn-Co-LDH@ biochar nanocomposite. J Hazard Mater 382:121070

Han Z, Sani B, Mrozik W, Obst M, Beckingham B, Karapanagioti HK, Werner D (2015) Magnetite impregnation effects on the sorbent properties of activated carbons and biochars. Water Res 70:394–403

Inyang M, Gao B, Zimmerman A, Zhang M, Chen H (2014) Synthesis, characterization, and dye sorption ability of carbon nanotube–biochar nanocomposites. Chem Eng J 236:39–46

Inyang M, Gao B, Zimmerman A, Zhou Y, Cao X (2015) Sorption and cosorption of lead and sulfapyridine on carbon nanotube-modified biochars. Environ Sci Pollut Res 22:1868–1876

Kumar A, Ramamoorthy D, Kumar N, Verma R, Kumar A, Verma DK, Jayabalan I, Marwein BM, Jaiswal KK (2022) Investigation on the potential of eco-friendly bio-char for amendment in serpentine soils and immobilization of heavy metals contaminants: a review. Biomass Convers Bioref 1–21. https://doi.org/10.1007/s13399-021-02257-4

Li J, Cai X, Liu Y, Gu Y, Wang H, Liu S, Liu S, Yin Y, Liu S (2020) Design and synthesis of a biochar-supported nano manganese dioxide composite for antibiotics removal from aqueous solution. Front Environ Sci 8:62

Liang H, Zhu C, Ji S, Kannan P, Chen F (2022) Magnetic Fe_2O_3/biochar composite prepared in a molten salt medium for antibiotic removal in water. Biochar 4:1–13

Liang L, Xi F, Tan W, Meng X, Hu B, Wang X (2021) Review of organic and inorganic pollutants removal by biochar and biochar-based composites. Biochar 3:255–281

Lisowski P, Colmenares JC, Masek O, Lisowski W, Lisovytskiy D, Kaminska A, Lomot D (2017) Dual functionality of TiO_2/biochar hybrid materials: photocatalytic phenol degradation in the liquid phase and selective oxidation of methanol in the gas phase. ACS Sustain Chem Eng 5:6274–6287

Liu MY, Zheng YF, Song XC (2019) Biomass assisted synthesis of 3D hierarchical structure BiO_X (X Cl, Br)-(CMC) with enhanced photocatalytic activity. J Nanosci Nanotechnol 19:5287–5294

Lu F, Astruc D (2020) Nanocatalysts and other nanomaterials for water remediation from organic pollutants. Coord Chem Rev 408:213180

Luyen NT, Linh HX, Huy TQ (2020) Preparation of rice husk biochar-based magnetic nanocomposite for effective removal of crystal violet. J Electron Mater 49:1142–1149

Pal A, Gin KY-H, Lin AY-C, Reinhard M (2010) Impacts of emerging organic contaminants on freshwater resources: review of recent occurrences, sources, fate and effects. Sci Total Environ 408:6062–6069

Peñas-Garzón M, Gómez-Avilés A, Bedia J, Rodriguez JJ, Belver C (2019) Effect of activating agent on the properties of TiO_2/activated carbon heterostructures for solar photocatalytic degradation of acetaminophen. Materials 12:378

Perveen S, Nadeem R, Ali S, Jamil Y (2021) Biochar caged zirconium ferrite nanocomposites for the adsorptive removal of Reactive Blue 19 dye in a batch and column reactors and conditions optimizaton. Z Phys Chem 235:1721–1745

Polianciuc SI, Gurzău AE, Kiss B, Ştefan MG, Loghin F (2020) Antibiotics in the environment: causes and consequences. Med Pharm Rep 93:231

Sabzehmeidani MM, Mahnaee S, Ghaedi M, Heidari H, Roy VA (2021) Carbon based materials: A review of adsorbents for inorganic and organic compounds. Mater Adv 2:598–627

Santhosh C, Daneshvar E, Tripathi KM, Baltrėnas P, Kim T, Baltrėnaitė E, Bhatnagar A (2020) Synthesis and characterization of magnetic biochar adsorbents for theremoval of Cr (VI) and Acid orange 7 dye from aqueous solution. Environ Sci Pollut Res 27:32874–32887

Shaikh WA, Islam RU, Chakraborty S (2021) Stable silver nanoparticle doped mesoporous biochar-based nanocomposite for efficient removal of toxic dyes. J Environ Chem Eng 9:104982

Shaikh WA, Kumar A, Chakraborty S, Islam RU, Bhattacharya T, Biswas JK (2022) Biochar-based nanocomposite from waste tea leaf for toxic dye removal From facile fabrication to functional fitness. Chemosphere 291:132788

Silva CP, Pereira D, Calisto V, Martins MA, Otero M, Esteves VI, Lima DL (2021) Biochar-TiO_2 magnetic nanocomposites for photocatalytic solar-driven removal of antibiotics from aquaculture effluents. J Environ Manage 294:112937

Sun Y, Wang T, Sun X, Bai L, Han C, Zhang P (2021) The potential of biochar and lignin-based adsorbents for wastewater treatment: comparison, mechanism, and application—a review. Ind Crops Prod 166:113473

Tan X, Liu Y, Zeng G, Wang X, Hu X, Gu Y, Yang Z (2015) Application of biochar for the removal of pollutants from aqueous solutions. Chemosphere 125:70–85

Tan X-F, Liu Y-G, Gu Y-L, Xu Y, Zeng G-M, Hu X-J, Liu S-B, Wang X, Liu S-M, Li J (2016) Biochar-based nano-composites for the decontamination of wastewater: a review. Biores Technol 212:318–333

Tang J, Lv H, Gong Y, Huang Y (2015) Preparation and characterization of a novel graphene/biochar composite for aqueous phenanthrene and mercury removal. Bioresour Technol 196:355–363

Thiruppathi M, Leeladevi K, Ramalingan C, Chen K-C, Nagarajan E (2020) Construction of novel biochar supported copper tungstate nanocomposites: a fruitful divergent catalyst for photocatalysis and electrocatalysis. Mater Sci Semicond Process 106:104766

Wang S, Gao B, Zimmerman AR, Li Y, Ma L, Harris WG, Migliaccio KW (2015) Removal of arsenic by magnetic biochar prepared from pinewood and natural hematite. Bioresour Technol 175:391–395

Wei X, Wang X, Gao B, Zou W, Dong L (2020) Facile ball-milling synthesis of CuO/biochar nanocomposites for efficient removal of reactive red 120. ACS Omega 5:5748–5755

Yao Y, Gao B, Chen J, Yang L (2013) Engineered biochar reclaiming phosphate from aqueous solutions: mechanisms and potential application as a slow-release fertilizer. Environ Sci Technol 47:8700–8708

Ye S, Yan M, Tan X, Liang J, Zeng G, Wu H, Song B, Zhou C, Yang Y, Wang H (2019) Facile assembled biochar-based nanocomposite with improved graphitization for efficient photocatalytic activity driven by visible light. Appl Catal B 250:78–88

Zhang M, Gao B (2013) Removal of arsenic, methylene blue, and phosphate by biochar/AlOOH nanocomposite. Chem Eng J 226:286–292

Zhang Y, Fan R, Zhang Q, Chen Y, Sharifi O, Leszczynska D, Zhang R, Dai Q (2019) Synthesis of $CaWO_4$-biochar nanocomposites for organic dye removal. Mater Res Bull 110:169–173

Zhao Q, Xu T, Song X, Nie S, Choi S-E, Si C (2021) Preparation and application in water treatment of magnetic biochar. Front Bioeng Biotechnol 9:769667. https://doi.org/10.3389/fbioe.2021.769667

Zhu N, Li C, Bu L, Tang C, Wang S, Duan P, Yao L, Tang J, Dionysiou DD, Wu Y (2020) Bismuth impregnated biochar for efficient estrone degradation: the synergistic effect between biochar and Bi/Bi_2O_3 for a high photocatalytic performance. J Hazard Mater 384:121258

Role of Biochar Supported Nano-Photocatalysts for Removal of Dyes

Sudipta Ramola, Diksha Pandey, Sarita Joshi, and Nidhi Rawat

Abstract

Dyes are extensively used in textile, leather accessories, paper, paint, printing, furniture, food, and plastic products. Release of untreated dyes causes adverse effects on environment. Different physical and biological treatments are inefficient to treat dyes. In this regard, photocatalysis using nanoparticles has emerged as an efficient method for degradation of the dyes. TiO_2, Fe_2O_3, and ZnO under UV or sunlight exposure are used as nano-photocatalysts. These nano-photocatalysts have potential to convert toxic pollutants into simpler end products such as CO_2 and H_2O without generation of secondary pollutants. Photocatalytic degradation of dyes is mainly affected by concentration of dye, amount of photocatalyst, intensity of irradiated light, time of irradiation, and effect of dissolved oxygen and pH. Biochar has emerged as an eco-friendly, cost-effective, and multi-functional product prepared from thermo-chemical conversion of different biomass-based wastes. Physico-chemical properties of biochar enable it to support various light-active components with easy isolation and recovery of catalyst. Incorporation of biochar with nano-photocatalyst provides enhanced porosity, specific surface area, number of active sites, chemical stability, charge separation, and photocatalytic efficiency. The present chapter focuses on fundamentals of photocatalytic process, synthesis of biochar supported nano-photocatalysts, their application for removal of dyes, mechanism involved, and future prospects.

Keywords

Biochar • Dyes • Nanoparticles • Nano-photocatalyst • Photocatalysis

1 Introduction

Textile industries are fast growing sectors in developing countries. China is the leading exporter of textile (Ghaly et al. 2014). The global textile dye market is also vast and expected to grow from USD 11.1 billion to 14.0 billion, between year 2022 and 2027, respectively, at a compound annual growth rate of 4.6% (https://www.marketsandmarkets.com/Market-Reports/textile-dye-market-226167405.html). Depending upon the source of origin, the colorant can be a natural or synthetic dye. The natural dyes are purely developed from natural resources (plants, insects, animals, and minerals), whereas synthetic dyes are chemically manufactured. Synthetic dyes are more economical, characterized by their fastness, and can be easily prepared in large quantities in different colors. As per their chemical structure, synthetic dyes are further classified as azo dyes, anthraquinone dyes, sulfur dyes, phthalocyanine dyes, and triarylmethane dyes. Synthetic dyes are also classified as reactive dye, direct dye, disperse dye, basic dye, and vat dye according to their mode of application (Khehra et al. 2006; Popli and Patel 2014; Yaseen and Scholz 2019).

S. Ramola (✉)
Research Group for Advanced Materials and Sustainable Catalysis (AMSC), State Key Laboratory Breeding Base of Green Chemistry-Synthesis Technology, College of Chemical Engineering, Zhejiang University of Technology, Hangzhou, 310014, China
e-mail: sudiptaramola6@gmail.com

D. Pandey
Department of Environmental Science, G.B. Pant, University of Agriculture and Technology, Pantnagar, Uttarakhand 263145, India

S. Joshi
Department of Environmental Science, School of Vocational Studies and Applied Science, Gautam Buddha University, Uttar Pradesh, Greater Noida, 201308, India

N. Rawat (✉)
Department of Agriculture, Doon Business School, Dehradun, Uttarakhand 248001, India
e-mail: nidhirawat05@gmail.com

D. Mishra et al. (eds.), *Biochar-Based Nanocomposites for Contaminant Management*, Advances in Science, Technology & Innovation, https://doi.org/10.1007/978-3-031-28873-9_8

Large volume of water is used in textile industries, for various activities such as dyeing, fixing, and washing. In textile industry, dyeing is an important step where dye along with different chemicals is added to the fabric. After the final product is prepared, all molecules of dyes are not utilized, and a large amount come out with water as effluent. In general, this effluent is a mixture of different pollutants that may include formaldehyde, phenols, chlorophenols, aromatic compounds, heavy metals, phthalates, surfactants, chloride, and sulfate (Hossain et al. 2018; Kishor et al. 2021; Kumar et al. 2021). Release of untreated dye effluent causes aesthetic damage and prevent penetration of solar radiations in water bodies that leads to reduction in photosynthesis and gas solubility in aquatic system. These untreated dye effluents also cause adverse effects on human health (He et al. 2011). Synthetic dyes such as azo (–N=N–) are widely used and are characterized to be toxic and highly persistent in the environment with mutagenic and carcinogenic metabolic products (Xu et al. 2007). Azo dyes are also very harmful for soil microbial communities and growth of plants (Imran et al. 2015; Rehman et al. 2018; Lellis et al. 2019).

Several physical, chemical, and biological methods are used for treatment of dyes (Fig. 1). Major disadvantages of these methods are very high operational costs and less effective removal of dyes from the effluent (Galindo et al. 2001). When treated efficiently, the water can be further used for irrigation purpose or can be reused within the textile factory.

Advanced oxidative processes (AOPs), highlighting the photocatalysis, is an attractive and eco-friendly alternative method for degradation of dyes (Muraro et al. 2020). Photocatalysis is done with the help of photoinduced catalysts. These photocatalysts are semiconductors that change the rate of a chemical reaction on exposure to light. Depending on the physical state of reactants, the photocatalytic reactions can be categorized as homogeneous photocatalysis (semiconductor and reactants in same phase) and heterogeneous photocatalysis (semiconductor and reactants in different phases).

Photocatalysis involves illumination of a semiconductor with a light source having energy more than the gap energy of valence band (VB) and conduction band (CB). This causes the transfer of electron from VB to CB. The electrons produce superoxide anions and hydroxide radicals on reaction with oxygen or water. High oxidizing power of superoxide anions and hydroxide radicals can degrade numerous molecules including industrial dyes. Photocatalysis ensures environmental protection as it completely mineralize the pollutants with no generation of secondary pollutants. Photocatalysis is an economical process as the photocatalysts are self-regenerated and hence can be reused several times (Kumar et al. 2014).

The present chapter covers basic principles and mechanism of photocatalysis, advantages of using nanomaterials for synthesis of photocatalysts, and advances in biochar supported nano-photocatalysts for removal of dyes.

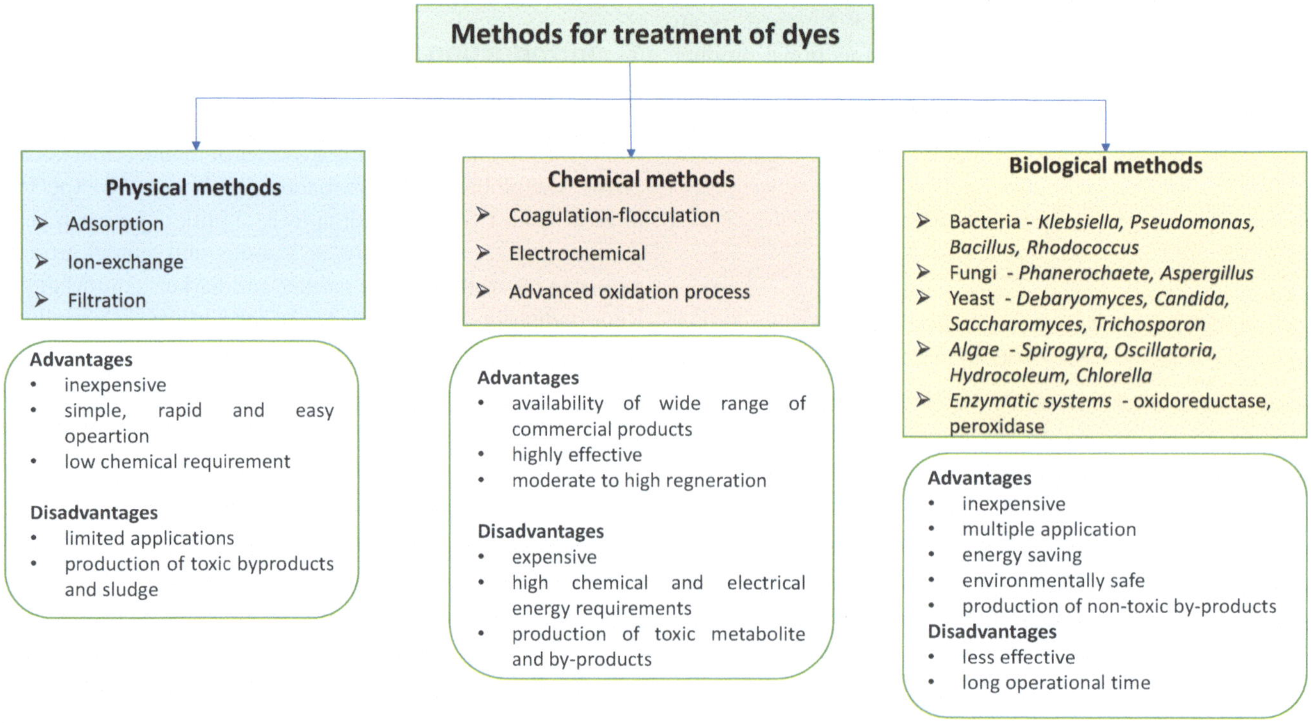

Fig. 1 Different methods for treatment of dyes

2 Mechanism of Photocatalytic Degradation of Dye

Based on band theory principle, there are two mechanisms of photocatalytic degradation, i.e., indirect mechanism and direct mechanism (Fig. 2).

2.1 Indirect Mechanism

It involves generation of electron–hole pair (e^-/h^+) by absorption of light having energy either equal or more than the band gap of the photocatalyst. Illumination of an appropriate wavelength of light on the surface of photocatalyst causes photoexcitation of electrons from VB to CB. This transfer of electrons causes generation of holes in the VB (h_{VB}^+). These e^-/h^+ can also recombine on the surface of semiconductor photocatalyst by releasing the energy as heat or photon. Early recombination of these two can cause lower quantum efficiency of photocatalyst. During the time before recombination, these e^-/h^+ migrate to the surface of the photocatalyst where they undergo charge transfer to initiate redox reactions with contaminant (dyes) to form smaller non-toxic products (Hoffman et al. 1995; Peral et al. 1997). h_{VB}^+ have an important role in photocatalytic degradation of dyes. They have strong oxidation power, and depending upon the semiconductor and pH, the redox potential is between +1.0 and +3.5 V. h_{VB}^+ can react with water that cause ionization of water to produce hydroxyl radical ($OH^{\bullet}$) on the surface of semiconductor. $OH^{\bullet}$ is extremely powerful oxidizing agent that mineralize the contaminants (Saien and Soleymani 2012).

Electrons on the CB must also efficiently react in order to avoid accumulation of excess charges on the surface of semiconductor. Oxygen is commonly employed as a scavenger of electrons, especially for the degradation of dyes. It is cost-effective and has ability to get dissolved in different solutions. The electrons are taken up by the oxygen and generate different oxygen activated species. For example, protonation of $O_2^{-\bullet}$ to form hydroperoxyl radical ($HO_2^{\bullet}$) and then to H_2O_2 which further dissociates into highly reactive $OH^{\bullet}$. These $OH^{\bullet}$ react with dye molecules and convert them to non-toxic products, i.e., CO_2 and H_2O (Zhang et al. 2018).

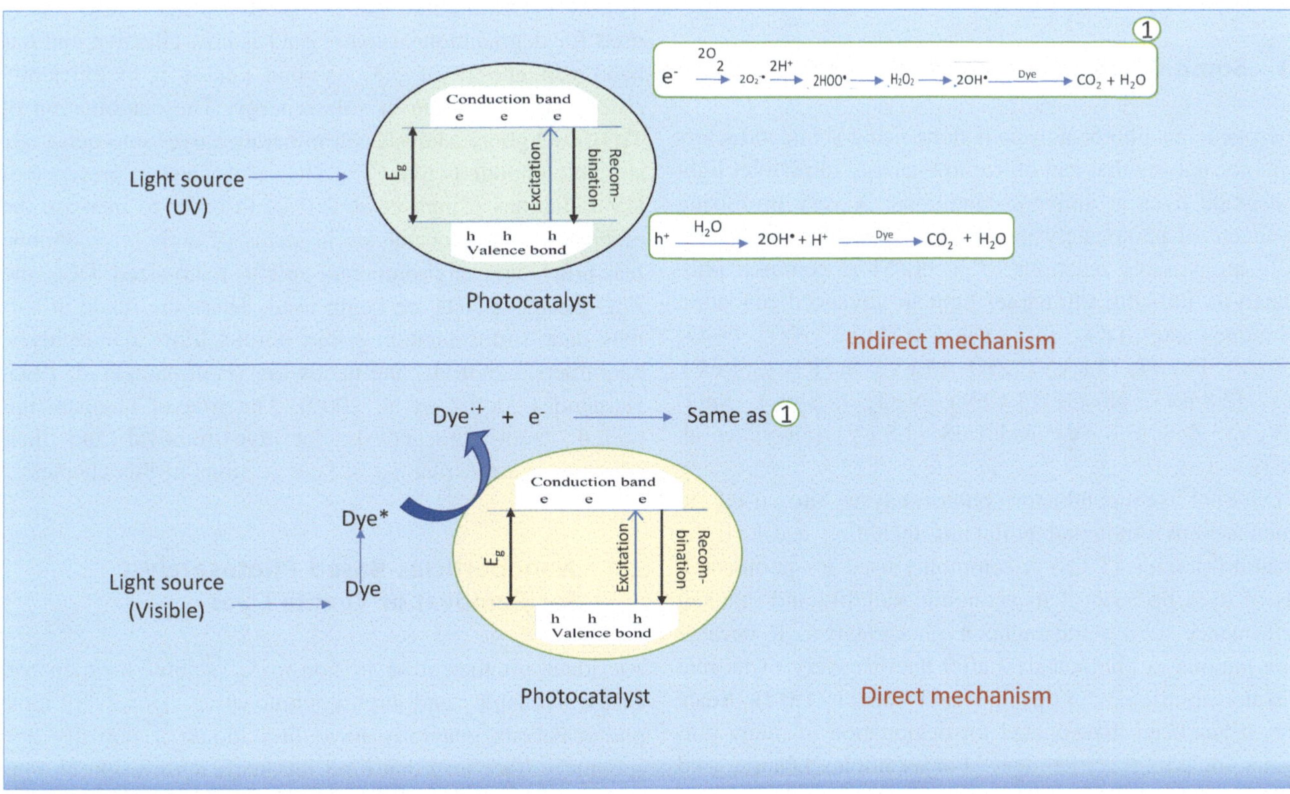

Fig. 2 Different mechanism of photocatalysis of dye. Where [1] is electron mediated generation of hydroxyl radical for photocatalysis of dye; and E_g is band gap

$$\text{Semiconductor} + h\nu \longrightarrow \text{Semiconductor}\,(e^-/h^+)$$
$$H_2O + h^+_{(VB)} \longrightarrow OH^\bullet + H^+$$
$$O_2 + e^- \longrightarrow O_2^{-\bullet}$$
$$O_2^- + H^+ \longrightarrow HO_2^\bullet$$
$$2HO_2^\bullet \longrightarrow H_2O_2 + O_2$$
$$H_2O_2 \longrightarrow 2OH^\bullet$$
$$\text{Dye} + OH^\bullet \longrightarrow CO_2 + H_2O$$

2.2 Direct Mechanism for Dye Degradation

In the direct mechanism of photocatalysis, photoexcitation of the semiconductor generates radicals. Dyes are photoexcited from ground state to excited state under visible light. Further, the photoexcited dye is converted to oxidized radial cation by injecting electrons into CB of photocatalyst. Superoxide radical anions are formed due to reaction between electrons and dissolved oxygen in the system. This in turn form $OH^\cdot$ which oxidize dyes (Ajmal et al. 2014).

2.3 Some Commonly Used Photocatalysts

Heterogeneous photocatalysis is done using semiconductors as photocatalysts that can utilize low-energy ultraviolet light to degrade dyes at ambient conditions. A very promising advantage of photocatalysis is that it allows thermodynamically unfavorable reactions to occur. Most common photocatalysts utilizing ultraviolet light in advanced oxidation techniques are TiO_2 (3.2 eV), ZnO (3.37 eV), Fe_2O_3 (2.2 eV), $FeTiO_3$ (2.8 eV), ZrO_2 (5 eV), $SrTiO_3$ (3.4 eV), WO_3 (2.8 eV), Nb_2O_5 (3.4 eV), V_2O_5 (2.8 eV), SnO_2 (3.5 eV), ZnS (3.6 eV), and CdS (2.5 eV) (Divya et al. 2013).

Different semiconductor photocatalysts are used to remediate environmental pollutants including textile dyes. Titanium dioxide (TiO_2) is commonly used for photocatalysis of dyes because of its economic viability and efficient performance as a semiconductor photocatalyst. It became quite famous as photocatalyst after the discovery of its role in water splitting by Fujishima and Honda (1972). Since then, it has been investigated for degradation of many pollutants including textile dyes. For example, Dhanya and Aparna (2016) have reported the complete degradation of methyl orange dye by photocatalyst chitosan-TiO_2 with 100% efficiency. Vianney et al. (2018) studied the solar-based photocatalytic decolorization of four commercial reactive dyes by using TiO_2–Fe_3O_4 which removes more than 90% color after 3 h of solar irradiation.

TiO_2 exists in two forms—anatase and rutile. Anatase is considered as better nano-photocatalyst for contaminant degradation. It is one of the most efficient photocatalysts, easily available at low cost, toxic free, and chemically stable. Although the band gap of anatase is larger (3.2 eV) than rutile (3.0 eV) but still it is preferred as photocatalyst. The reason for preference of anatase over rutile irrespective of larger band gap is lower charge recombination, high crystalline nature, and higher photocatalytic activity including better adsorption affinity for dyes (Huo et al. 2014; Kim et al. 2014).

Hydrogen peroxide is widely used in combination with TiO_2 and other nano-photocatalysts for degradation of dyes. H_2O_2 gives $OH^\cdot$ that acts as a powerful oxidizing agent. Some of the advantages are that the photocatalysis can be easily done in ambient conditions with no generation of sludge, and oxygen formed helps in aerobic biological decay process (Divya et al. 2013). Several studies have reported successful degradation of dyes using UV/H_2O_2 process. For example, Hoffmann et al. (1995) used fixed bed photoreactor of UV/TiO_2/H_2O_2 to degrade Direct Red 80; Tanaka et al. (2000) used UV/TiO_2 system to degrade seven azo dyes, viz. Orange G, Acid Orange 7, New Coccine, Acid Black 1, Acid Yellow 17, Tartrazine, and Congo Red. ZnO is also widely used for degradation of dyes. ZnO is cost-effective and has band gap energy of 3.37 eV that enable it to efficiently absorb larger fraction of solar energy. The combination of TiO_2/ZnO photocatalysts can mineralize dyes into non-toxic simple compounds like CO_2, HCl, and water in presence of UV radiations (Singh et al. 2013). In order to increase the efficiency of photocatalysis in terms of activity, economic feasibility, and environmental safety, nano-sized TiO_2 and ZnO photocatalysts are being used. These are made in various nano-forms including thin porous films, nanospheres, nanofibers, nanorods, and nanowires (Hoffmann et al. 1995; Hernández-Alonso et al. 2009). The role of biochar supported nano-photocatalysts for dye removal and their advantages are explained in later sections of this chapter.

2.4 Nanoparticles-Based Photocatalyst for Removal of Textile Dyes

An ideal photocatalyst is non-toxic, stable, inexpensive, easily available, and highly photoactive. However, most photocatalysts generally have limitations of porosity and specific surface area that limit intraparticle diffusion of dyes and hence efficiency of photocatalyst decrease. Some modifications are done to the photocatalysts in order to increase their efficiency. In this regard, nano-scale modification of photocatalyst is most promising (Bhadouria et al. 2020). Nano-photocatalysts are ultra-small semiconductors that are few nanometers in size. On the basis of physical and chemical

Table 1 Different photocatalysts systems used for textile dye degradation

Catalyst system	Dyes degraded	Light used	References
ZnO nano-powder	Rhodamine B	UV	Sclafani and Herrmann (1996)
WO_3	Naphthol blue black dye, methylene blue	Halogen lamp, UV–visible	Luo and Hepel (2001)
TiO_2 and activated carbon fibers	Methylene blue	UV	Yuan et al. (2005)
TiO_2, ZnO, SnO_2, ZnS, and CdS	Methyl orange, Rhodamine 6G	UV and sunlight	Kansal et al. (2007)
$GdCoO_3$	Rhodamine B, Rhodamine Blue, Orange G, and Remazol Brilliant Blue R, phenol, 4-methylphenol, 4-nitrophenol, and 4-chlorophenol	UV	Mahata et al. (2007)
Au^{3+}doped SiO_2	Methyl red	Sunlight	Mahmoud et al. (2009)
$CeCrO_3$	Fast green	Visible	Bhati et al. (2010)
TiO_2	Azo dyes (monoazo dye and diazo dye)	UV/Visible	Abo-Farha (2010)
$MnTiO_3$ powder	Methyl orange	Sunlight	He et al. (2010)
CaO	Indigo	Visible and UV	Devarahosahalli Veeranna et al. (2014)
Ag nanoparticles	Methylene blue	Sunlight	Vanaja et al. (2014)
Ag nanoparticles	Methyl violet, Safranin, Methyl orange	Visible	Muthukrishnan et al. (2015)
TiO_2	Methylene blue	Sunlight	Sutisna Rokhmat et al. (2015)
Magnetite with H_2O_2	Methylene blue	UV and sunlight	Reza et al. (2016)
MoS_2	Rhodamine B	Visible	Liu et al. (2016)
SnO_2	Crystal violet, Methyl red, Basic blue	Sunlight	Shinde et al. (2017)
TiO_2–Fe_3O_4	Yellow 22, Red 11, Blue 21	UV	Vianney et al. (2018)
ZnO nanoparticles	Methylene blue and Rhodamine B	UV	Blažeka et al. (2020)
Fe_2O_3 nano-catalyst	Rhodamine B	UV and visible	Muraro et al. (2020)

properties, nanoparticles are broadly classified into different categories. Some of the well-known nanoparticles are ceramic nanoparticles, metal nanoparticles, semiconductor nanoparticles, carbon-based nanoparticles, lipid-based nanoparticles, and polymeric nanoparticles (Tahir et al. 2020).

Metal nanoparticles have high efficiency for the photocatalytic degradation of azo dye (Sarkar et al. 2020). Incorporation of gold nanoparticles (74.32 nm) in biomaterials from fruiting bodies of the Enoki mushroom was done for degradation of methylene blue. This nano-system was able to efficiently degrade about 75% of the dye in 4 h (Rabeea et al. 2020). Gold nanoparticles can efficiently degrade a wide range of organic pollutants due to wide substrate specificity. Similarly, ZnO nano-powder was successful in 95% degradation of Rhodamine B under solar light irradiation by Sclafani and Herrmann (1996) and methylene blue by Taghizadeh et al. (2020). Fe_2O_3 hybridized with silver nano-catalyst (AgNPs–Fe_2O_3) proved to be a promising photocatalyst under ultraviolet (UV) and visible radiation for degradation of Rhodamine B in a slurry reactor. AgNPs–Fe_2O_3 sample showed degradation of Rhodamine B up to 94.1% under UV and 58.36% under visible light. The efficiency of commercial catalyst TiO_2–P25, under the same conditions, was 61.5% and 44.5%, respectively, demonstrating better results in case of nano-structure catalyst (Muraro et al. 2020). Similarly, degradation of methylene blue and rhodamine B was done effectively by nanocomposites-based photocatalyst Al_2O_3–TiO_2 and ZrO_2–TiO_2 under UV-A light, whereas the degradation of methyl orange by same nanocomposites was found to be slow (Karunakaran et al. 2013). An elaborated list of different photocatalysts used for degradation of textile dye is given in Table 1.

2.5 Advantages of Nano-Sized Photocatalyst

Nano-photocatalysts are less toxic, economical, and chemically stable. They have higher specific surface area and porosity that increase their catalytic efficiency with possibility of wider application in heterogeneous photocatalysis. The nano-size-based materials have distinct quantum effects and surface properties as compared to bulk materials that contribute to their enhanced efficiency as photocatalyst. Efficient utilization of electrons is necessary for successful photocatalytic oxidation of organic pollutants. If concurrent oxidation of pollutants and reduction of oxygen do not occur, electrons begin to accumulate in CB of the photocatalyst. This enhances the chances of recombination of electrons and holes. In this regard, nano-sized catalysts have better efficiency as compared to normal photocatalysts (Radhika et al. 2019). The reasons for the same are discussed as following:

(i) ***Quantum size effect***: Nanomaterials can show very different properties compared to macroscale materials. When the particle size is in nanometer range, quantum size effect occurs due to confinement of electrons to small regions. The VB and CB of the semiconductor change into discrete energy levels in nano-photocatalyst. The electric potential of VB changes to more positive and the CB to more negative and thereby increasing the oxidation–reduction potential of the electrons and holes (Colmenares et al. 2009).
(ii) ***High specific surface area***: The efficiency of photocatalysis depends on the time taken by electrons and holes to get transferred to the surface of the catalyst. The surface to volume ratio is higher, and particle diameter is very smaller in nano-photocatalyst that reduces the time for charge carriers to move from inside to the surface in order to start redox reaction (Kim and Kwak 2007).

3 Role of Biochar Supported Nano-Photocatalysts for Degradation of Dyes

3.1 Photocatalysts Derived from Biomass Materials

The biomass waste that is rich in carbon and silica can be utilized as catalyst support. The metallic supported photocatalysts are given less preference than biomass supported photocatalysts derived from biochar as the latter is cheap and eco-friendly and exhibit much chemical stability (Mian and Liu 2018). The other benefits of biomass supported photocatalyst are its easy recoverability and visible light sensitivity.

Biochar (BC) is a carbon-based material prepared from a variety of agriculture waste, forestry waste, animal manure, industrial solid, and urban waste (Singh et al. 2015). The lignin present in cell wall of lignocellulosic plant biomass is a natural cross-linked polymer, composed of three different phenyl propane monomers. It is electronegative and hence has affinity toward electropositive metal ions (Colmenares et al. 2016). Similarly, chitin present in animal-based biomass is the second most abundant natural polysaccharide that has adsorption affinity toward heavy metals and dyes (Reddy and Lee 2013). Biochar used as a supporting material with nanomaterial catalysts forms a novel hybridized nanomaterial known as biochar supported nano-photocatalyst (BSNPs). These BSNPs have excellent physico-chemical and structural characteristic properties, electrical conductivity, chemical and thermal stability, and tunable functional groups. Due to presence of biochar prepared from abundant biomass-based waste, these BSNPs are cost-effective and have improved photodegradation and adsorption properties (Chen et al. 2018).

Different physical, chemical, and thermal methods such as thermal polycondensation, solvothermal method, sol–gel method, and ultrasound sonication are used to prepare BSNPs. These are briefly depicted in Fig. 3. These BSNPs, especially with TiO_2 nanoparticles, can reduce the band gap, prevent early recombination of e^-/h^+ pairs, and perform better visible light photocatalysis that make them highly effective at degrading water pollutants compared to pure nano-photocatalysts (Mian and Liu 2018; Chandra et al. 2021).

3.2 Advantages of Using Biochar Supported Nano-Photocatalysts (Economical, Environmental, and Functional Advantages)

The physico-chemical and functional properties of biochar depend mainly on type of raw materials and pyrolysis conditions. In general, the aromaticity, electrical conductivity, surface area, minerals content, functional groups, and porosity of biochar are more than that of the parent raw materials (Ramola et al. 2020). These changed physico-chemical and structural properties increase the photocatalysis efficiency of BSNPs in comparison with other biomass-based photocatalysts without any pyrolysis. Biochar is an aromatic carbon structure that acts as an excellent platform for uniform dispersion of nanoparticles. Biochar plays important role as reservoir of electron and also in shuttling process of electron. The storage of electron is dependent on presence and oxidation reduction of redox-active moieties on the surface of biochar such as functional groups quinone (main electron acceptors)/hydroquinone (main electron donors), metals (Fe and Mn, as minerals or attached as organometallic moieties), and radicals (aryl radicals, semiquinone, and phenoxyl radicals) (Klüpfel et al. 2014; Chacón et al. 2017). The electron shuttling process is dependent on the dense aromatic structure of biochar. Li et al. (2020) examined different biochar and

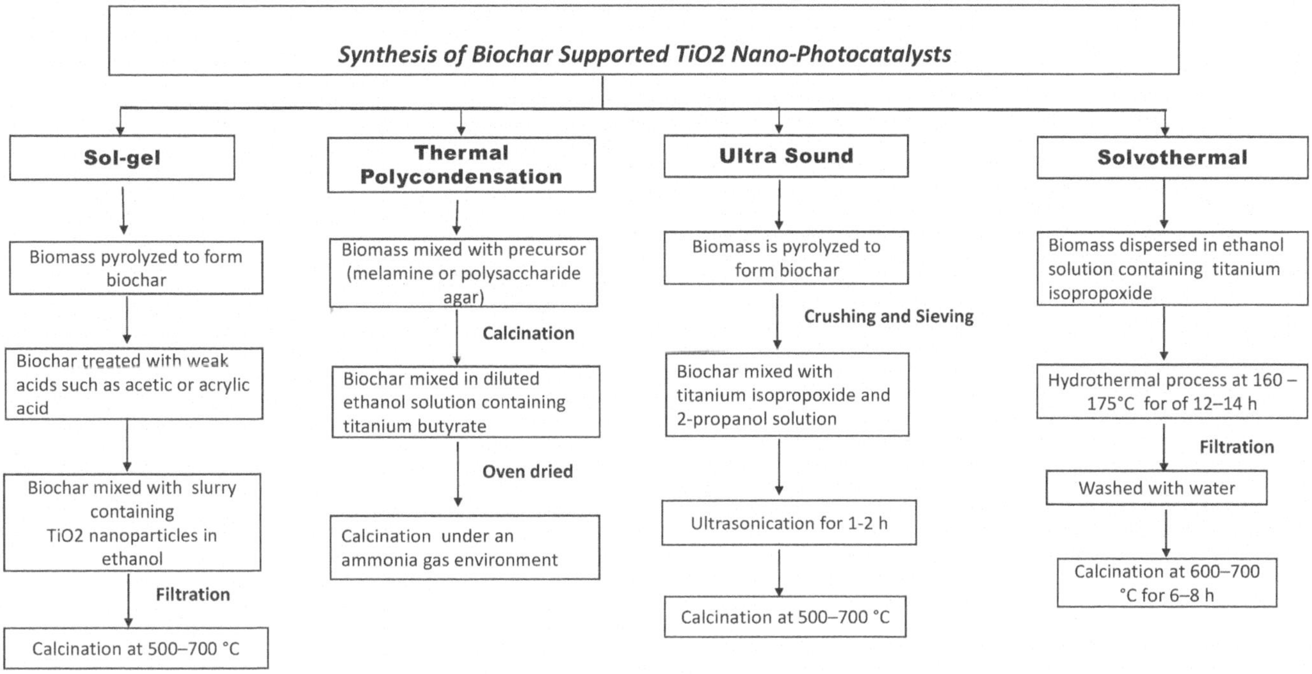

Fig. 3 Different methods to prepare biochar-based TiO_2 supported nano-photocatalysts

concluded that biochar derived from cellulose and lignin exhibited better electron-donating and electron-accepting capacities, respectively. Whereas biochar prepared at 700 °C from starch possessed both high electron-donating and electron-accepting capacities.

The surface area of BSNPs can be more than that of nanoparticles. The well dispersion of nanoparticles and increased surface area would increase the active sites for better light scattering and better photocatalysis. The aromatic structure of biochar has many hydrophilic and hydrophobic functional groups and mineral content such as N, P, K, S, Ca, and Mg that further enhances the chances of removal of pollutants by adsorption (Woolf et al. 2010; Ramola et al. 2014, 2020, 2021).

3.3 Removal of Organic Synthetic Dyes by Biochar Supported Photocatalysts

The photodegradation of dyes by BSNPs is a cost-effective and efficient method(Zangeneh et al. 2015). Organic synthetic dyes are widely photocatalyzed by TiO_2/BC, ZnO/BC, CdS/BC, and Ag/BC (Khan et al. 2014). When BC is coupled with the nanoparticle's catalytic activity, the properties of BC, i.e., porous structure, high conductivity and stability, and optical characteristics result in enhanced dye photodegradation efficiency (Kong et al. 2017; Wang et al. 2017). Some of the applications of BSNPs for dye removal are given in Table 2.

Methylene blue (MB) adsorption by Al_2O_3/BC as a nano-photocatalyst showed that hybrid photocatalyst had a higher adsorption capacity in comparison with pure biochar (Wang et al. 2017). Khataee et al. (2018) used a hydrothermal technique to make cerium nanoparticles composite with biochar (CeO_2–H@BC) for photodegradation of reactive red dye and found that photocatalyst has better photodegradation effectiveness than pure CeO_2–H and BC. Wang et al. (2018b) used the calcination process to produce visibly active bamboo biochar-TiO_2 composite catalysts utilizing tetrabutyl titanate and bamboo powder of various sizes. The TiO_2 nanoparticles were spread over the surface of the charcoal fiber, and the removal efficiency obtained was 99% for MB. In another study, Sharma et al. (2019) used the microwave technique to produce algal biochar-reinforced trimetallic (La/Cu/Zr) nanocomposite catalyst (AlBcTNC). Algal biomass taken from a rivulet was used to make biochar. AlBcTNC was investigated as a potential adsorbent/photocatalyst for the removal of 94% malachite green (MG) by combination of adsorption/photocatalytic approach.

Titanium dioxide-coated biochar (TBCs) was synthesized by pyrolyzing titanium dioxide-treated biomass using the sol–gel method for safranin T (ST) removal (Cai et al. 2017). ST removal capabilities of the TBCs were 7.2 times better than that of the raw biochar with 2 g of the biomass. The biochar and TiO_2 contents of TBC played significant role in removal of ST over a wide pH range. The loading of TiO_2

Table 2 Applications of various biochar supported photocatalysts (BSNPs) for dye removal

S. No	Nano-photocatalysts	Raw material used for biochar	Degradation of dyes	Removal efficiency (%)	References
1	N-TiO_2-C	Waste plum stones	Methylene blue	100	Matos et al. (2013)
2	TiO_2-bamboo	Bamboo	Methylene blue	99	Wu et al. (2015)
3	BC-BiOCl	–	Methyl orange	82	Li et al. (2016)
4	TiO_2-sludge and wheat husk	Sludge and wheat husk	Reactive blue 69	98	Khataee et al. (2017a, b)
5	BC-TiO_2	Coconut shell	Reactive brilliant blue	99.70	Khataee et al. (2017a, b)
6	ZrO_2-sludge and wheat husk	Sludge and wheat husk	Reactive yellow 39	98	Khataee et al. (2017a, b)
7	N–TiO_2–Fe_3O_4-biochar	Agar	Methylene blue	100	Mian and Liu (2018)
8	Bamboo biochar-TiO_2	Bamboo	Methylene blue	99	Wang et al. (2018a, b, c)
9	BC/ZnO	Waste biomass	Orange G	88.50	Vinayagam et al. (2018)
10	Fe impregnated BC	Rice straw, wood chip, and Korean cabbage	Congo red (CR) and Crystal violet (CV)	99.70	Park et al. (2018)
11	BC-TiO_2	*Salvinia molesta*	Acid orange 7	57.60	Silvestri et al. (2019)
12	AlBc-TNC	Algae	Malachite green	94	Sharma et al. (2019)

significantly enhanced the adsorption performance of biochar, and the high specific surface area of the biochar synergistically promoted the photocatalytic activity of TiO_2 with maximum removal capacity of 226.7 mg g^{-1}. Both adsorption and photocatalytic degradation contributed to the removal of ST, with the former mechanism contributing slightly higher. Kulaksiz et al. (2017) produced Fe-modified/BC for photodegradation of malachite green (MG). The surface area and pore volume of the hybrid photocatalysts were higher than that of pure biochar, resulting in faster photocatalytic degradation of MG. Fu et al. (2019) produced $MnFe_2O_4$/biochar composites ($MnFe_2O_4$/MX) (Where X is biochar; CS, CL, and CC represent biochar prepared from corn stems, leaves, and cores, respectively). MX was given to the biochar samples after they were activated and graphitized. The composites were utilized to activate peroxymonosulfate, which was then used to degrade orange (II) dye. Chen et al. (2019) made ZnO-loaded biochar by pyrolyzing biomass and zinc oxide precursor at the same time. Carbonized jute fibers pretreated with zinc acetate were treated at 200 °C and pyrolyzed at 700 °C in a box-type furnace. The removal of MB was shown to be pH dependent, with an optimum removal rate of 99% and a mineralization level of over 93% when exposed to UV light for 30 min at pH 7.0. Wang et al. (2018a, b, c) prepared nanocomposites of ZnO and biochar (nZOR/BC) by adding zinc chloride to a biochar suspension at pH 2 and reducing it with $NaBH_4$. Carboxymethyl cellulose (CMC) was used to synthesize CMC/ZnO/biochar composites (nZORc/BC). The catalysts were employed for MB adsorption and photodegradation, and the saturated sorption of MB anticipated by the Langmuir isotherm for nZORc/BC was 17.01 g kg^{-1}

4 Factor Affecting Rate of Photocatalytic Degradation of Dyes

The rate of photocatalytic degradation of dyes depends on band gap energy, surface area of photocatalysts, concentration of dye, catalyst loading, temperature, light intensity, pH, time of irradiation, and dissolved oxygen (Colón et al. 2006; Mozia et al. 2009; Hussein and Abass 2010; Tahir et al. 2020). The effects of these factors on photodegradation of dyes are discussed below:

i. **Band gap energy**
   Band gap is the energy difference between CB and VB of a photocatalyst. Band gap helps in charge separation and reduces the chances of recombination for better photocatalytic activity. Ideally doping with metal or non-metals is done to reduce the band gap of photocatalysts. This activates the photocatalysts in visible region for enhanced hydrogen production and photocatalytical activity (Tahir et al. 2020).
ii. **Surface area/structure**
   Surface area and crystal structure of photocatalysts have an important role in photocatalysis. Smaller size particles of photocatalyst with larger surface area have more number of active sites as reaction centers. Also, photo-electrons and holes cover shorter distance to

reach the surface of photocatalysts, thus decreasing the recombination rate (Tahir et al. 2020).

iii. **Concentration of dye**: The photodegradation of dye increases as the concentration of dye increases up to a certain concentration. If the dye concentration reaches beyond a critical concentration, degradation starts to decrease. This fall in efficiency occur as more dye molecules compete for limited active sites of photocatalyst. Another factor can be that UV–visible radiations are screened by increased dye molecules and cannot reach photocatalyst, and hence, less degradation occurs (Colón et al. 2006).

iv. **Catalyst loading**: In general, with increase in catalyst loading, better photocatalytic degradation of dyes is observed. This may be attributed to the fact that with higher catalyst loading, more number of active sites are present that contribute to produce more reactive $OH^{\bullet}$. However, after reaching an optimum value of catalyst loading, further increase of catalyst observed no further enhanced efficiency of dye degradation. This may be attributed to the fact that more number of catalyst particles increase the turbidity of the solution and adversely affect the efficiency of photocatalyst to degrade dye. Also, particles of photocatalyst agglomerate at higher concentration of photocatalyst and decrease the available surface area for reaction.

v. **Temperature**: The photodegradation process of dye is dependent on changes in temperature. High temperature generally improves rate of mineralization and affects absorption kinetics. Hussein and Abass (2010) studied that the efficiency of all types of catalyst increased with increase in temperature for decolorization of real textile wastewater. However, too high or too low temperature can alter the percentage of dissolved oxygen and hence can affect the mineralization as well as recombination rates.

vi. **Light intensity**: Dye degradation is dependent on the light intensity used for photocatalysis. As the intensity increases, dye degradation is more since the formation of e^-/h^+ is predominant, and the chances of e^-/h^+ recombination are insignificant (Zhou et al. 2006). Unlike this, dye degradation is very less at low irradiated light intensity. This is due to the fact that at low intensity, separation of e^-/h^+ pair competes for recombination, and free radical formation is very less.

vii. **pH**: The photocatalytic degradation of dye is very sensitive to pH value of solution. pH affects the surface charge, size, and aggregation of photocatalyst. It is observed that most of the dye degradation occur in acidic pH range. For example, TiO_2 has different surface charges at different pH. At acidic pH, the surface of TiO_2 is protonated, and at alkaline pH, it is deprotonated. Due to positive surface charge, TiO_2 behaves as a strong Lewis acid at acidic medium and can easily absorb a strong Lewis base or anionic dyes. However, in alkaline conditions, this process is not favored as there is repulsion between hydroxyl groups and the dye molecule (Mozia et al. 2009).

viii. **Time of irradiation**: In general, with increase in irradiation time, the reaction rate of photocatalysis degradation of dye decreases. This slow kinetics is due to deactivation of active sites of photocatalyst and difficulty in the reaction of short chain aliphatic with $OH^{\bullet}$ radicals (Kumar and Pandey 2017).

ix. **Effect of dissolved oxygen**: Dissolved oxygen is very important factor for efficient working of photocatalyst as it acts like scavenger of electrons and helps in avoiding the recombination of e^-/h^+.

5 Conclusion and Future Perspectives

Untreated dyes released into ambient water bodies have created adverse environmental health. Photocatalysis has emerged as an advanced process to mineralize the dyes to safer end products. It is cost-effective and eco-friendly process with less requirement of energy and easy operational conditions. Also, the end products of this process are generally CO_2 and H_2O that have no further concerns of creating environmental pollution. The limitation of pure metal semiconductors such as low efficiency for degradation is overcome by use of different nanotechnology-based semiconductors. These nano-photocatalysts have quantum size effects and larger surface area that increases the efficiency than bulk material. Factors such as high capital cost of preparation, chances of generation of secondary pollutants, and lower reusability have limited the use of metals and non-metals in synthesizing hybrid composites. Biochar has been used as a secondary material that provide excellent platform for photocatalytic nanoparticles. Biochar enhances photocatalysis process by reducing the quick recombination of electron–hole pairs owing to its electron conductive nature and also helps in immobilization of pollutants owing to its varied surface functional groups and mineral content. The biochar supported nano-photocatalysts also improve the chemical stability, increase surface area and active sites, visible light sensitivity, catalytic degradation of pollutants, and reduce band gap energy. The effectiveness of using biochar supported nano-photocatalyst can be further enhanced by screening of cheap and abundant biomasses/ municipal waste, synthetic process to incorporate catalytic nanoparticles for better synergetic effects, and operational parameters including heating conditions and biomass to nanoparticle ratio.

Some points to consider for future research activity of biochar-based nano-photocatalysts are as follow:

- To find out tunable synthesis, modification process and operational factors for optimized results of biochar-based nano-photocatalysts.
- Screening of hazardous wastes for their safe use as a feedstock for preparation of biochar-based nano-photocatalysts.
- Modification methods for increasing the photoactivity in visible region for making the photocatalyst system green and clean.

References

Abo-Farha S (2010) Photocatalytic degradation of monoazo and diazo dyes in wastewater on nanometer-sized TiO_2. Am J Sci 6(11):130–142

Ajmal A, Majeed I, Malik RN, Idriss H, Nadeem MA (2014) Principles and mechanisms of photocatalytic dye degradation on TiO_2 based photocatalysts: a comparative overview. RSC Adv 4(70):37003–37026. https://doi.org/10.1039/c4ra06658h

Bhadouria R, Mishra D, Singh VK, Singh P, Srivastava P, Tripathi S, Singh R (2020) Nanocatalyst types and their potential impacts in agroecosystems: an overview. In: Nano-materials as photocatalysts for degradation of environmental pollutants, pp 323–344

Bhati I, Punjabi PB, Ameta SC (2010) Photocatalytic degradation of fast green using nanosized $CeCrO_3$. Maced J Chem En 29(2):195. https://doi.org/10.20450/mjcce.2010.166

Blažeka D, Car J, Klobučar N, Jurov A, Zavašnik J, Jagodar A, Kovačević E, Krstulović N (2020) Photodegradation of methylene blue and rhodamine B using laser-synthesized ZnO nanoparticles. Materials 13(19):4357. https://doi.org/10.3390/ma13194357

Cai X, Li J, Liu Y, Yan Z, Tan X, Liu S, Zeng G, Gu Y, Hu X, Jiang L (2017) Titanium dioxide-coated biochar composites as adsorptive and photocatalytic degradation materials for the removal of aqueous organic pollutants. JCTB 93(3):783–791. https://doi.org/10.1002/jctb.5428

Chacón FJ, Cayuela ML, Roig A, Sánchez-Monedero MA (2017) Understanding, measuring and tuning the electrochemical properties of biochar for environmental applications. Rev Environ Sci BioTechnol 16(4):695–715. https://doi.org/10.1007/s11157-017-9450-1

Chandra S, Jagdale P, Medha I, Tiwari AK, Bartoli M, Nino AD, Olivito F (2021) Biochar-supported TiO_2-based nanocomposites for the photocatalytic degradation of sulfamethoxazole in water—a review. Toxics 9(11):313. https://doi.org/10.3390/toxics9110313

Chen M, Bao C, Hu D, Jin X, Huang Q (2019) Facile and low-cost fabrication of ZnO/biochar nanocomposites from jute fibers for efficient and stable photodegradation of methylene blue dye. J Anal and Appl Pyrolysis 139:319–332. https://doi.org/10.1016/j.jaap.2019.03.009

Chen S, Zhou M, Wang H, Wang T, Wang X, Hou H, Song B (2018) Adsorption of reactive brilliant red X-3B in aqueous solutions on clay-biochar composites from bagasse and natural attapulgite. Water 10(6):703. https://doi.org/10.3390/w10060703

Colmenares JC, Luque R, Campelo JM, Colmenares F, Karpiński Z, Romero AA (2009) Nanostructured photocatalysts and their applications in the photocatalytic transformation of lignocellulosic biomass: an overview. Materials 2(4):2228–2258

Colmenares JC, Varma RS, Lisowski P (2016) Sustainable hybrid photocatalysts: titania immobilized on carbon materials derived from renewable and biodegradable resources. Green Chem 18(21):5736–5750

Colón G, Hidalgo MC, Munuera G, Ferino I, Cutrufello MG, Navío JA (2006) Structural and surface approach to the enhanced photocatalytic activity of sulfated TiO_2 photocatalyst. Appl Catal B 63(1–2):45–59. https://doi.org/10.1016/j.apcatb.2005.09.008

Devarahosahalli Veeranna K, Theeta Lakshamaiah M, Thimmasandra Narayan R (2014) Photocatalytic degradation of indigo carmine dye using calcium oxide. Int J Photochem 2014:1–6. https://doi.org/10.1155/2014/530570

Dhanya A, Aparna K (2016) Synthesis and evaluation of TiO_2/Chitosan based hydrogel for the adsorptional photocatalytic degradation of Azo and anthraquinone dye under UV light irradiation. Proced Technol 24:611–618. https://doi.org/10.1016/j.protcy.2016.05.141

Divya N, Bansal A, Jana AK (2013) Nano-photocatalysts in the treatment of colored wastewater-a review. Mater Sci Forum 734:349–363. https://doi.org/10.4028/www.scientific.net/MSF.734.349

Fu H, Ma S, Zhao P, Xu S, Zhan S (2019) Activation of peroxymonosulfate by graphitized hierarchical porous biochar and $MnFe_2O_4$ magnetic nanoarchitecture for organic pollutants degradation: structure dependence and mechanism. Chem Eng J 360:157–170. https://doi.org/10.1016/j.cej.2018.11.207

Fujishima A, Honda K (1972) Electrochemical photolysis of water at a semiconductor electrode. Nature 238(5358):37–38. https://doi.org/10.1038/238037a0

Galindo C, Jacques P, Kalt A (2001) Photooxidation of the phenylazonaphthol AO20 on TiO_2: kinetic and mechanistic investigations. Chemosphere 45(6–7):997–1005. https://doi.org/10.1016/S0045-6535(01)00118-7

Ghaly A, Ananthashankar R, Alhattab M, Ramakrishnan V, Ghaly A (2014) Production, characterization and treatment of textile effluents: a critical review. J Chem Eng Process Technol 05(01). https://doi.org/10.4172/2157-7048.1000182

He HY, Dong WX, Zhang GH (2010) Photodegradation of aqueous methyl orange on $MnTiO_3$ powder at different initial pH. Res Chem Intermed 36(9):995–1001. https://doi.org/10.1007/s11164-010-0213-7

He L, Freeman HS, Lu L, Zhang S (2011) Spectroscopic study of anthraquinone dye/amphiphile systems in binary aqueous/organic solvent mixtures. Dyes Pigm 91(3):389–395. https://doi.org/10.1016/j.dyepig.2011.05.010

Hernández-Alonso MD, Fresno F, Suáreza S, Coronado JM (2009) Development of alternative photocatalysts to TiO_2: challenges and opportunities. Energy Environ Sci 2(12):1231–1257. https://doi.org/10.1039/B907933E

Hoffmann M, Martin S, Choi W, Bahnemann D (1995) Environmental applications of semiconductor photocatalysis. Chem Rev 95(1):69–96. https://doi.org/10.1021/cr00033a004

Hossain L, Sarker SK, Khan MS (2018) Evaluation of present and future wastewater impacts of textile dyeing industries in Bangladesh. Environ Dev 26:23–33. https://doi.org/10.1016/j.envdev.2018.03.005

Huo Y, Chen X, Zhang J, Pan G, Jia J, Li H (2014) Ordered macroporous Bi_2O_3/TiO_2 film coated on a rotating disk with enhanced photocatalytic activity under visible irradiation. Appl Catal B 148–149:550–556. https://doi.org/10.1016/j.apcatb.2013.11.040

Hussein F, Abass T (2010) Photocatalytic treatment of textile industrial wastewater. Int J Chem Sci 8(3):1409–1420

Imran M, Crowley DE, Khalid A, Hussain S, Mumtaz MW, Arshad M (2015) Microbial biotechnology for decolorization of textile wastewaters. Rev Environ Sci Biotechnol 14(1):73–92. https://doi.org/10.1007/s11157-014-9344-4

Kansal SK, Singh M, Sud D (2007) Studies on photodegradation of two commercial dyes in aqueous phase using different photocatalysts. J Hazard Mater 141(3):581–590. https://doi.org/10.1016/j.jhazmat.2006.07.035

Karunakaran C, Magesan P, Gomathisankar P, Vinayagamoorthy P (2013) Photocatalytic degradation of dyes by Al_2O_3-TiO_2 and ZrO_2-TiO_2 nanocomposites. Mater Sci Forum 734:325–333. https://doi.org/10.4028/www.scientific.net/MSF.734.325

Khan M, Lee J, Cho M (2014) Au@TiO_2 nanocomposites for the catalytic degradation of methyl orange and methylene blue: an electron relay effect. J Ind Eng Chem 20(4):1584–1590. https://doi.org/10.1016/j.jiec.2013.08.002

Khataee A, Gholami P, Kalderis D, Pachatouridou E, Konsolakis M (2018) Preparation of novel CeO_2-biochar nanocomposite for sonocatalytic degradation of a textile dye. Ultrason Sonochem 41:503–513. https://doi.org/10.1016/j.ultsonch.2017.10.013

Khataee A, Kayan B, Gholami P, Kalderis D, Akay S (2017a) Sonocatalytic degradation of an anthraquinone dye using TiO_2-biochar nanocomposite. Ultrason Sonochem 39:120–128. https://doi.org/10.1016/j.ultsonch.2017.04.018

Khataee A, Kayan B, Gholami P, Kalderis D, Akay S, Dinpazhoh L (2017b) Sonocatalytic degradation of reactive yellow 39 using synthesized ZrO_2 nanoparticles on biochar. Ultrason Sonochem 39:540–549. https://doi.org/10.1016/j.ultsonch.2017.05.023

Khehra MS, Saini HS, Sharma DK, Chadha BS, Chimni SS (2006) Biodegradation of azo dye CI Acid Red 88 by an anoxic–aerobic sequential bioreactor. Dyes Pigm 70(1):1–7. https://doi.org/10.1016/j.dyepig.2004.12.021

Kim DS, Kwak SY (2007) The hydrothermal synthesis of mesoporous TiO_2 with high crystallinity, thermal stability, large surface area, and enhanced photocatalytic activity. Appl Catal A-Gen 323:110–118. https://doi.org/10.1016/j.apcata.2007.02.010

Kim LJ, Jang JW, Park JW (2014) Nano TiO_2-functionalized magnetic-cored dendrimer as a photocatalyst. Appl Catal 147:973–979. https://doi.org/10.1016/j.apcatb.2013.10.024

Kishor R, Purchase D, Saratale GD, Saratale RG, Ferreira LFR, Bilal M, Chandra R, Bharagava RN (2021) Ecotoxicological and health concerns of persistent coloring pollutants of textile industry wastewater and treatment approaches for environmental safety. J Environ Chem Eng 9(2):105012. https://doi.org/10.1016/j.jece.2020.105012

Klüpfel L, Keiluweit M, Kleber M, Sander M (2014) Redox properties of plant biomass-derived black carbon (biochar). Environ Sci Technol 48:5601–5611. https://doi.org/10.1021/es500906d

Kong R, Zhao Y, Zheng Y, Qu F (2017) Facile synthesis of ZnO/CdS@ZIF-8 core–shell nanocomposites and their applications in photocatalytic degradation of organic dyes. RSC Adv 7 (50):31365–31371. https://doi.org/10.1039/c7ra03918b

Kulaksiz E, Gözmen B, Kayan B, Kalderis D (2017) Adsorption of malachite green on Fe-modified biochar: influencing factors and process optimization. Desalin Water Treat 74:383–394. https://doi.org/10.5004/dwt.2017.20601

Kumar S, Ahlawat W, Bhanjana G, Heydarifard S, Nazhad MM, Dilbaghi N (2014) Nanotechnology-based water treatment strategies. J Nanosci Nanotechnol 14(2):1838–1858. https://doi.org/10.1166/jnn.2014.9050

Kumar A, Pandey G (2017) A review on the factors affecting the photocatalytic degradation of hazardous materials. Int J Eng Sci 1 (3):106–114. https://doi.org/10.15406/mseij.2017.01.00018

Kumar D, Patel Z, Pandit P, Pandit R, Patel A, Joshi M, Joshi C (2021) Textile industry wastewaters from Jetpur, Gujarat, India, are dominated by shewanellaceae, bacteroidaceae, and pseudomonadaceae harboring genes encoding catalytic enzymes for textile dye degradation. Front Environ Sci 9:720707. https://doi.org/10.3389/fenvs.2021.720707

Lellis B, Fávaro-Polonio CZ, Pamphile JA, Polonio JC (2019) Effects of textile dyes on health and the environment and bioremediation potential of living organisms. Biotechnol Res Innov 3(2):275–290. https://doi.org/10.1016/j.biori.2019.09.001

Li M, Huang H, Yu S, Tian N, Dong F, Du X, Zhang Y (2016) Simultaneously promoting charge separation and photoabsorption of BiOX (X = Cl, Br) for efficient visible-light photocatalysis and photosensitization by compositing low-cost biochar. Appl Surf Sci 386:285–295. https://doi.org/10.1016/j.apsusc.2016.05.171

Li S, Shao L, Zhang H, He P, Lü F (2020) Quantifying the contributions of surface area and redox-active moieties to electron exchange capacities of biochar. J Hazard Mater 394:122541. https://doi.org/10.1016/j.jhazmat.2020.122541

Liu P, Liu Y, Ye W, Ma J, Gao D (2016) Flower-like N-doped MOS2for photocatalytic degradation of RHB by visible light irradiation. Nanotechnol 27(22):225403. https://doi.org/10.1088/0957-4484/27/22/225403

Luo J, Hepel M (2001) Photoelectrochemical degradation of naphthol blue black diazo dye on WO_3 film electrode. Electrochim 46 (19):2913–2922. https://doi.org/10.1016/s0013-4686(01)00503-5

Mahata P, Aarthi T, Madras G, Natarajan S (2007) Photocatalytic degradation of dyes and organics with nanosized $GdCoO_3$. J Phys Chem 111(4):1665–1674. https://doi.org/10.1021/jp066302q

Mahmoud MA, Poncheri A, Badr Y, Abd El Waned MG (2009) Photocatalytic degradation of methyl red dye. S Afr J Sci 105(7/8). https://doi.org/10.4102/sajs.v105i7/8.86

Matos J, Hofman M, Pietrzak R (2013) Synergy effect in the photocatalytic degradation of methylene blue on a suspended mixture of TiO_2 and N-containing carbons. Carbon 54:460–471. https://doi.org/10.1016/j.carbon.2012.12.002

Mian M, Liu G (2018) Recent progress in biochar-supported photocatalysts: synthesis, role of biochar, and applications. RSC Adv 8 (26):14237–14248. https://doi.org/10.1039/c8ra02258e

Mozia S, Morawski AW, Toyoda M, Inagaki M (2009) Application of anatase-phase TiO_2 for decomposition of azo dye in a photocatalytic membrane reactor. Desalination 241(1–3):97–105. https://doi.org/10.1016/j.desal.2007.12.048

Muraro PCL, Mortari SR, Vizzotto BS, Chuy G, dos Santos C, Brum LFW, da Silva WL (2020) Iron oxide nanocatalyst with titanium and silver nanoparticles: synthesis, characterization and photocatalytic activity on the degradation of Rhodamine B dye. Sci Rep 10(1):3055. https://doi.org/10.1038/s41598-020-59987-0

Muthukrishnan SBS, Muthukumar MSM, Rao MV, Senthil Kumar T (2015) Catalytic degradation of organic dyes using synthesized silver nanoparticles: a green approach. J Bioremed Biodeg 06 (05):312. https://doi.org/10.4172/2155-6199.1000312

Park J, Wang J, Xiao R, Tafti N, DeLaune R, Seo D (2018) Degradation of Orange G by Fenton-like reaction with Fe-impregnated biochar catalyst. Bioresour Technol 249:368–376. https://doi.org/10.1016/j.biortech.2017.10.030

Peral J, Domènech X, Ollis DF (1997) Heterogeneous photocatalysis for purification, decontamination and deodorization of air. J Chem Technol Biotechnol Int Res Process, Environ Clean Technol 70 (2):117–140

Popli S, Patel UD (2014) Destruction of azo dyes by anaerobic–aerobic sequential biological treatment: a review. IJEST 12(1):405–420. https://doi.org/10.1007/s13762-014-0499-x

Rabeea MA, Owaid MN, Aziz AA, Jameel MS, Dheyab MA (2020) Mycosynthesis of gold nanoparticles using the extract of Flammulina velutipes, Physalacriaceae, and their efficacy for decolorization of methylene blue. J Environ Chem Eng 8(3):103841. https://doi.org/10.1016/j.jece.2020.103841

Radhika NP, Selvin R, Kakkar R, Umar A (2019) Recent advances in nano-photocatalysts for organic synthesis. Arab J Chem 12 (8):4550–4578

Ramola S, Mishra T, Rana G, Srivastava RK (2014) Characterization and pollutant removal efficiency of biochar derived from Baggase, bamboo and tyre. Environ Monit Assess 186(12):9023–9039. https://doi.org/10.1007/s10661-014-4062-5

Ramola S, Belwal T, Li CJ, Liu YX, Wang YY, Yang SM, Zhou CH (2021) Preparation and application of novel rice husk biochar–calcite composites for phosphate removal from aqueous medium. J Clean Prod 299:126802. https://doi.org/10.1016/j.jclepro.2021.126802

Ramola S, Belwal T, Li CJ, Wang YY, Lu HH, Yang SM, Zhou CH (2020) Improved lead removal from aqueous solution using novel porous bentonite - and calcite-biochar composite. Sci Total Environ 709:136171. https://doi.org/10.1016/j.scitotenv.2019.136171

Reddy DH, Lee SM (2013) Application of magnetic chitosan composites for the removal of toxic metal and dyes from aqueous solutions. Adv Colloid Interface Sci 201–202:68–93. https://doi.org/10.1016/j.cis.2013.10.002

Rehman K, Shahzad T, Sahar A, Hussain S, Mahmood F, Siddique MH, Siddique MA, Rashid MI (2018) Effect of reactive black 5 azo dye on soil processes related to C and N cycling. PeerJ 6:e4802. https://doi.org/10.7287/peerj.preprints.26557v1

Reza KM, Kurny A, Gulshan F (2016) Photocatalytic degradation of methylene blue by magnetite+H_2O_2+UV process. IJESD 7(5):325–329. https://doi.org/10.7763/ijesd.2016.v7.793

Saien J, Soleymani AR (2012) Feasibility of using a slurry falling film photo-reactor for individual and hybridized AOPs. J Ind Eng Chem 18(5):1683–1688

Sarkar S, Ponce NT, Banerjee A, Bandopadhyay R, Rajendran S, Lichtfouse E (2020) Green polymeric nanomaterials for the photocatalytic degradation of dyes: a review. Environ Chem Lett 18(5):1569–1580. https://doi.org/10.1007/s10311-020-01021-w

Sclafani A, Herrmann JM (1996) Comparison of the photoelectronic and photocatalytic activities of various anatase and rutile forms of titania in pure liquid organic phases and in aqueous solutions. J Phys Chem A 100(32):13655–13661. https://doi.org/10.1021/jp9533584

Sharma G, Sangeeta B, Vinod KG, Shilpi A, Amit K, Deepak P, Mola GT, Florian JS (2019) Algal biochar reinforced trimetallic nanocomposite as adsorptional/photocatalyst for remediation of malachite green from aqueous medium. J Mol Liq 275:499–509. https://doi.org/10.1016/j.molliq.2018.11.070

Shinde DR, Tambade PS, Chaskar MG, Gadave KM (2017) Photocatalytic degradation of dyes in water by analytical reagent grades ZnO, TiO_2 and SnO_2: a comparative study. Drink Water Eng Sci 10 (2):109–117. https://doi.org/10.5194/dwes-10-109-2017

Silvestri S, Gonçalves M, da Silva Veiga, P, Matos T, Peralta-Zamora P, Mangrich A (2019) TiO_2 supported on Salvinia molesta biochar for heterogeneous photocatalytic degradation of Acid Orange 7 dye. J Environ Chem Eng 7(1):102879. https://doi.org/10.1016/j.jece.2019.102879

Singh R, Babu JN, Kumar R, Srivastava P, Singh P, Raghubanshi AS (2015)Multifaceted application of crop residue biochar as a tool for sustainable agriculture: an ecological perspective.Ecol Eng 77:324–347

Singh P, Mondal K, Sharma A (2013) Reusable electrospun mesoporous ZnO nanofiber mats for photocatalytic degradation of polycyclic aromatic hydrocarbon dyes in wastewater. J Colloid Interface Sci 394:208–215. https://doi.org/10.1016/j.jcis.2012.12.006

Sutisna Rokhmat M, Wibowo E, Murniati R, Khairurrijal, Abdullah M (2015) Application of immobilized titanium dioxide as reusable photocatalyst on photocatalytic degradation of methylene blue. Adv Mat Res 1112:149–153. https://doi.org/10.4028/www.scientific.net/amr

Taghizadeh MT, Siyahi V, Ashassi-Sorkhabi H (2020) Visible-light photocatalytic degradation of textile dye by recyclable and recoverable AgBr–ZnO/chitosan beads. Poly Bullet 8(7):3869–3887. https://doi.org/10.1007/s00289-020-03298-x

Tahir MB, Sohaib M, Sagir M, Rafique M (2020) Role of nanotechnology in photocatalysis. In: Reference module in materials science and materials engineering. https://doi.org/10.1016/B978-0-12-815732-9.00006-1

Tanaka K, Padermpole K, Hisanaga T (2000) Photocatalytic degradation of commercial azo dyes. Water Res 34(1):327–333. https://doi.org/10.1016/S0043-1354(99)00093-7

Vanaja M, Paulkumar K, Baburaja M, Rajeshkumar S, Gnanajobitha G, Malarkodi C, Sivakavinesan M, Annadurai G (2014) Degradation of methylene blue using biologically synthesized silver nanoparticles. Bioinorg Chem Appl 2014:1–8. https://doi.org/10.1155/2014/742346

Vianney YM, Rosalyn I, Angela S (2018) Solar based photocatalytic decolorization of four commercial reactive dyes utilizing bound TiO_2-Fe_3O_4 nanocomposite. Indones J Chem 18(4):621. https://doi.org/10.22146/ijc.27549

Vinayagam M, Ramachandran S, Ramya V, Sivasamy A (2018) Photocatalytic degradation of orange G dye using ZnO/biomass activated carbon nanocomposite. J Environ Chem Eng 6(3):3726–3734. https://doi.org/10.1016/j.jece.2017.06.005

Wang B, Liu B, Ji X, Ma M (2018c) Synthesis, Characterization, and photocatalytic properties of bamboo charcoal/TiO_2 composites using four sizes powder. Materials 11(5):670. https://doi.org/10.3390/ma11050670

Wang Q, Zhang S, He J, Wang L, Shi Y, Zhou W, She H, Wang F (2017) Surface-enhanced palygorskite coated CdS: synthesis, characterization and highly improved photocatalytic degradation efficiency of organic dyes. J Mater Sci Mater Electron 28 (14):10464–10471. https://doi.org/10.1007/s10854-017-6819-4

Wang L, Yan W, He C, Wen H, Cai Z, Wang Z, Chen Z, Liu W (2018a) Microwave-assisted preparation of nitrogen-doped biochars by ammonium acetate activation for adsorption of acid red 18. Appl Surf Sci 433:222–231. https://doi.org/10.1016/j.apsusc.2017.10.031

Wang S, Zhou Y, Han S, Wang N, Yin W, Yin X, Gao B, Wang X, Wang J (2018b) Carboxymethyl cellulose stabilized ZnO/biochar nanocomposites: Enhanced adsorption and inhibited photocatalytic degradation of methylene blue. Chemosphere 197:20–25. https://doi.org/10.1016/j.chemosphere.2018.01.022

Woolf D, Amonette JE, Street-Perrott FA, Lehmann J, Joseph S (2010) Sustainable biochar to mitigate global climate change. Nat Commun 1(1):1–9. https://doi.org/10.1038/ncomms1053

Wu F, Liu W, Qiu J, Li J, Zhou W, Fang Y, Zhang S, Li X (2015) Enhanced photocatalytic degradation and adsorption of methylene blue via TiO_2 nanocrystals supported on graphene-like bamboo charcoal. Appl Surf Sci 358:425–435. https://doi.org/10.1016/j.apsusc.2015.08.161

Xu M, Guo J, Sun G (2007) Biodegradation of textile azo dye by Shewanella decolorationis S12 under microaerophilic conditions. Appl Microbiol Biotechnol 76(3):719–726. https://doi.org/10.1007/s00253-007-1032-7

Yaseen DA, Scholz M (2019) Textile dye wastewater characteristics and constituents of synthetic effluents: a critical review. Int J Environ Sci Technol 16(2):1193–1226. https://doi.org/10.1007/s13762-018-2130-z

Yuan R, Guan R, Shen W, Zheng J (2005) Photocatalytic degradation of methylene blue by a combination of TiO_2 and activated carbon fibers. J Colloid Interface Sci 282:87–91. https://doi.org/10.1016/j.jcis.2004.08.143

Zangeneh H, Zinatizadeh A, Habibi M, Akia M, Hasnain Isa M (2015) Photocatalytic oxidation of organic dyes and pollutants in wastewater using different modified titanium dioxides: a comparative review. J Ind Eng Chem 26:1–36. https://doi.org/10.1016/j.jiec.2014.10.043

Zhang J, Tian B, Wang L, Xing M, Lei J (2018) Mechanism of photocatalysis. In: Photocatalysis. Springer, Singapore, pp. 1–15. https://doi.org/10.1007/978-981-13-2113-9_1

Zhou M, Yu J, Cheng B (2006) Effects of Fe-doping on the photocatalytic activity of mesoporous TiO_2 powders prepared by an ultrasonic method. J Hazard Mater 137(3):1838–1847. https://doi.org/10.1016/j.jhazmat.2006.05.028

Consideration About Regeneration, Reactivity, Toxicity, and Challenges of Biochar-Based Nanocomposites

Ekta Mishra, Shruti Kapse, and Shilpi Jain

Abstract

Innovative nanostructured materials, such as nanobiochar (NB), have provided a sustainable solution to a number of contemporary concerns. Biochar technology and nanobiotechnology could lead to the development of carbon-based nanomaterials, such as biochar-based nanocomposites, which could revolutionise research in this sector. In recent years, NB has arisen as a technique to improve the biochar's physical, chemical, and structural properties and so overcome its limits. Researchers' curiosity about NB's amazing development in recent years has been piqued due to its potential for expansion in a range of applications. As a result, it is a good choice for soil amendment, pollutant remediation, and wastewater treatment. NB's proclivity for polluting soil and water, in addition to its benefits, has been a source of concern. To acquire a comprehensive image of the geochemical transport behaviour of NB and the possible risk of contaminant adsorption onto NB carried in the environment, it is essential to comprehend the reactivity potential of NB. Although NB can successfully adsorb micropollutants, it, like other adsorbents, will get saturated over time, necessitating NB regeneration to maintain system effectiveness and long-term use. The environmental and biological toxic effects of NB should also be thoroughly explored before large-scale industrial or environmental applications. In a nutshell, this chapter covers the most state-of-the-art information on regeneration, reactivity, and toxicity, as well as challenges that need to be addressed in the future research. The main takeaways from this chapter are the necessity of understanding NB's behaviour, transport cycle, and potential risk of pollutant adsorption onto NB being released into the environment. One of the primary challenges to overcome for the successful application of NB in the environment is the absence of characterization methods, poor yields, a lack of ecotoxicological data, and the requirement for cost–benefit analyses and life-cycle assessments for innumerable applications.

Keywords

Biotoxicity • Engineered nanomaterials (ENMs) • Nanobiochar • Reactivity • Recovery

1 Introduction

The research community has been interested in carbon-based nanomaterials because of its diverse uses in the fields of the environment, agriculture, energy, catalysis, and biomedical (Chausali et al. 2021). When biomass is thermochemically decomposed at a high temperature, biochar is created as a solid, carbon-rich by-product. Since its discovery as *terra preta* black soil in the Amazon Basin (De Gisi et al. 2014), biochar has undergone extensive research on all fronts, from its fundamental components and properties to its uses, which have expanded beyond agriculture to include environmental clean-up, climate regulation, and advanced energy materials (Qin et al. 2020). Biochar (BC) has traditionally been associated with low costs in comparison with other carbon-based compounds, such as fullerenes, graphene, and carbon nanotubes, due to the abundance of raw materials available, including crop remains, wood biomass, animal litter, and solid wastes (Xiao et al. 2018; Qin et al. 2020).

Depending on the pyrolysis process, BC particles can range in size from micrometres to centimetres (Lehmann and Joseph 2015). Better adsorption capacity was achieved by reducing particle size in the micro range (10–600 μm) by increasing the number of adsorption sites available (Lonappan et al. 2016; Taheran et al. 2016; Naghdi et al. 2019).

E. Mishra · S. Kapse · S. Jain (✉)
Department of Environmental Studies, The Maharaja Sayajirao University of Baroda, Vadodara, 390002, Gujarat, India
e-mail: shilpijain55@gmail.com

D. Mishra et al. (eds.), *Biochar-Based Nanocomposites for Contaminant Management*, Advances in Science, Technology & Innovation, https://doi.org/10.1007/978-3-031-28873-9_9

Further reducing BC particle size to the nanoscale, down to 100 nm or less, improved its characteristics, increasing surface energy, adsorption potential, and ultimately biological effectiveness (Chausali et al. 2021). As a result, NB is popular right now since it combines BC technology with the benefits of nanotechnology.

There are many benefits to employing BC as the substrate material for the development of nanocomposite (NC) materials as opposed to other NC materials. First off, the feedstocks for BC production are cheap and plentiful, primarily coming from solid waste and agricultural biomass (Lehmann and Joseph 2012; Sohi 2012). Additionally, BC manufacturing, which is typically produced at very moderate temperatures (700 °C), is inexpensive and requires little energy (Xiao et al. 2018). Additionally, the thermochemical processing of biomass can produce syngas and biofuels in addition to the generation of biochar (Lehmann and Joseph 2012). As a result, the synthesis of BC-based NC may meet four integrated goals, including the elimination of pollutants, the reduction of waste, the sequestration of carbon, and the generation of energy (Fig. 1) (Sohi 2012; Tan et al. 2015).

This potential of nanobiochar (NB) to expand in a variety of applications has sparked researchers' interest in recent years. Additionally, since carbonaceous NC is transported with other substances, there may be possible environmental hazards. Despite the fact that BC has a repute for being a sustainable soil additive, it is use in wastewater treatment and pollution clean-up, NB's penchant for polluting soil and water has been a subject of concern. Selection of the wrong feedstock material, preparation conditions, and preparation techniques, the dangerous components [polycyclic aromatic hydrocarbons (PAHs), heavy metals, ecologically persistent free radicals (EPFRs), perfluorochemicals (PFCs) and dioxins] might be created (Xiang et al. 2021). Due to BC's potentially toxic components and its assorted interactions with the environment, recent studies have focussed on the adverse impacts of BC on the ecosystem (El-Naggar et al. 2020; Cui et al. 2021).

Furthermore, carbonaceous NC can operate as an active carrier, and there may be environmental issues associated with co-transportation (Song et al. 2019). Transporting biochar may cause the onset of a soil water gas cycle (Chen et al. 2018b). NCs may migrate from terrestrial to aquatic ecosystems because of relocation and leaching, from aquatic to terrestrial because of runoff, from the atmosphere to the soil or water because of free settlement and precipitation, and lastly from the atmosphere to the terrestrial or aquatic systems because of wind erosion and weathering (Wang et al. 2013a,b; Xiang et al. 2021). Similar to the engineered carbonaceous nanoparticles, considerable NBs dispersion, and associated contaminants in natural environments, however, may expose aquatic life to danger (Ema et al. 2016; Freixa et al. 2018). Thus, understanding the reactivity potential of NB is critical for obtaining a thorough picture of the

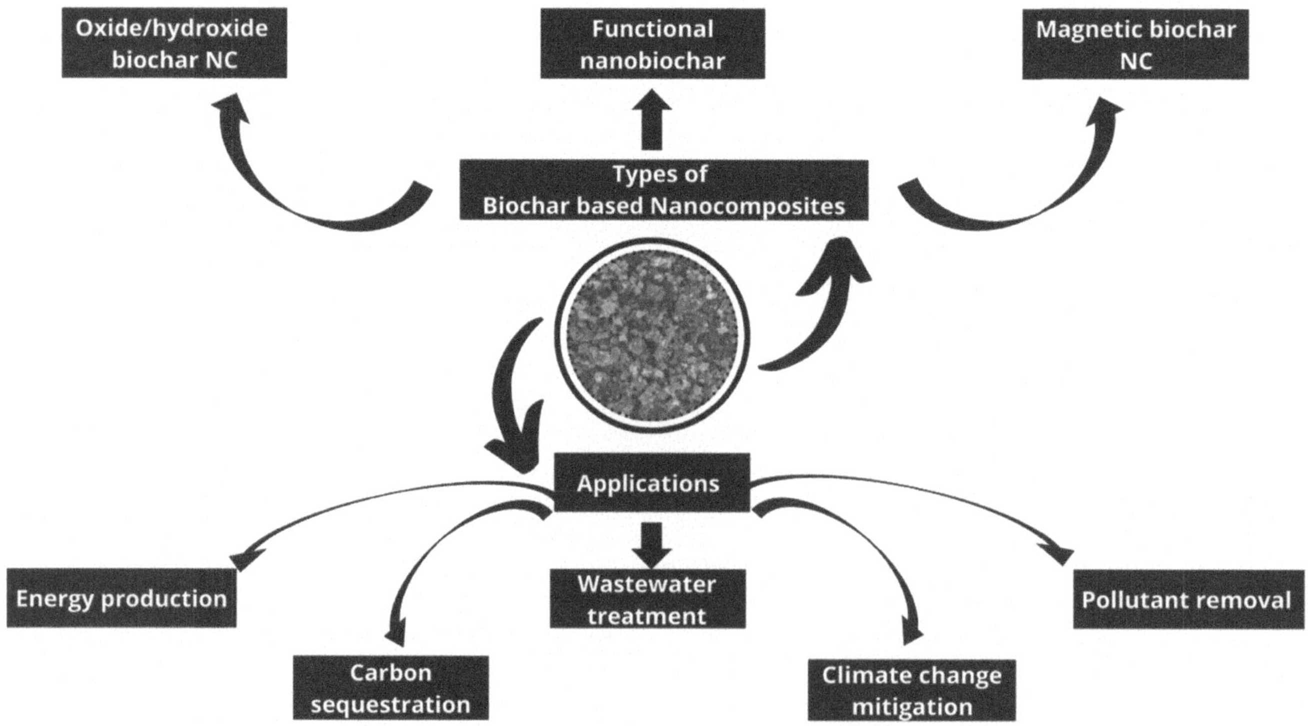

Fig. 1 Schematic representation of types and applications of biochar-based nanocomposites (BNC)

transport cycle and behaviour of NB and the potential danger of contaminant adsorption onto NB conveyed in the environment.

Although NB can effectively adsorb micropollutants, it, like other adsorbents, will get saturated over time, requiring NB regeneration to ensure system performance and long-term use. The sustainable management of used adsorbents represents one of the toughest logistical problems with regard to resource recovery (Hossain et al. 2020). The sustainable management of used adsorbents represents one of the biggest logistical challenges in the context of resource recovery (Hossain et al. 2020). To redevelop and recycle the used adsorbents, several methods involving photodegradation, desorption, and biodegradation of sorbed pollutants have been investigated (Lata et al. 2015; Vakili et al. 2019). Several processes, such as precipitation, filtering, centrifuge, and magnetic extraction methods, are utilised during wastewater treatment to separate and recover spent adsorbents (Vakili et al. 2019; Hassan et al. 2020). Used adsorbents are afterwards recycled or safely disposed of by landfilling and incineration (Kozyatnyk et al. 2020). Before large-scale industrial or environmental uses, the environmental and biological hazard consequences of NB should be thoroughly investigated. In a summary, this chapter contains the most up-to-date knowledge on regeneration, reactivity, and toxicity, as well as research issues that must be addressed in the future.

2 Environmental Risks

Nanobiochar (NB) has received a lot of attention lately due to its abilities as a soil amendment for the removal of heavy metals as well as other environmental benefits. As a result, it is critical to fully comprehend the environmental danger posed by unanticipated NB's on plants and humans before pursuing large-scale industrial or agricultural uses (Fig. 2) (Ramanayaka et al. 2020).

The risk associated with these nanoscale particles will enter into the environment, and contaminate groundwater or the drinking water supply is a big worry because it could impair flora, animals, and human health (Corsi et al. 2018; Zhang et al. 2019; Lowry et al. 2019). Still, many significant issues remain unresolved. There are not many reports on NB's ecotoxicity (Zhang et al. 2022).

The pollution caused by free radicals (FRs), pollutant-accumulating hydrocarbons (PAHs), and HMs in the biochar medium and the flotation of very fine biochar dust could speed the environmental effects (Liu et al. 2022). Though the presence and discharge of engineered nanomaterials (ENMs) might be beneficial occasionally to soil inhabitants, they are harmful to soil, microbes, plants, and ecosystem when concentrations and safety limits are exceeded (Shrivastava et al. 2019).

2.1 Potential Phyto/Microbial Toxicity

It has been previously stated that one of the main mechanisms underlying NB cytotoxicity, which rises with decreasing particle size, is particle-induced oxidative stress. The majority of the particles that are environmentally persistent free radicals (EPFR) have an aerodynamic diameter of less than 1 m (Pan et al. 2019; Jia et al. 2021). Process used to prepare BC has an impact on its toxicity. The effects of the six varieties of NB on rice, tomato, and reed seedling germination and growth were also investigated (Zhang et al. 2020a, b). According to the research, NB derived from lignin-rich raw material inhibited seeds and drastically decreased biomass and bud length. This occurrence resulted from phenolic compounds depositing on the NB during BC pyrolysis. These compounds have an adverse impact on plants because they are cytotoxic to fibroblasts (Sigmund et al. 2017; Oliveira et al. 2019). Graphene and carbon nanotubes are not as hazardous to *Streptomyces* as NB, according to Liu et al. (2019). With just a 68% survival rate, they found that *Streptomyces* cells were severely damaged at the moderately low treatment dose of 10 mg/l.

Whether or not NB-sorbed contaminants, particular heavy metals, will be released after adsorption and what effects this would have on plants are still not known. Zhang et al. (2020a, b) have shown that NB can be engrossed by and travel throughout plants. It is important to note that no experimental research has been done on the mechanism(s) through which NB migrates in plants (Zhang et al. 2022). According to two studies, the entry of graphene oxide and fullerene into plants occurs via endocytosis, interacting with carrier proteins, ion channels, and water molecular channels (Chen et al. 2010; Gao and Song 2019). According to Gao and Song (2019), after being absorbed by root tissues, graphene oxide was subsequently transferred to mesophyll cells, where it caused damage to cellular structures. According to the information now available, both the potential advantages of NB for use in diverse environmental applications and any potential risks to the environment must be taken into account. NB prepared from ball milling displays more biotoxicity than NB made by other procedures because its spherical shape makes it more convenient to touch and strike with cells in comparison with the original BC or another nanomaterial like tubes and sheets (Lyu et al. 2018; Liu et al. 2019). For example, BC that has been ball-milled can enter cells and cause the release of harmful reactive oxygen species (ROS). The ROS generated can lessen the number of starch granules required to keep the

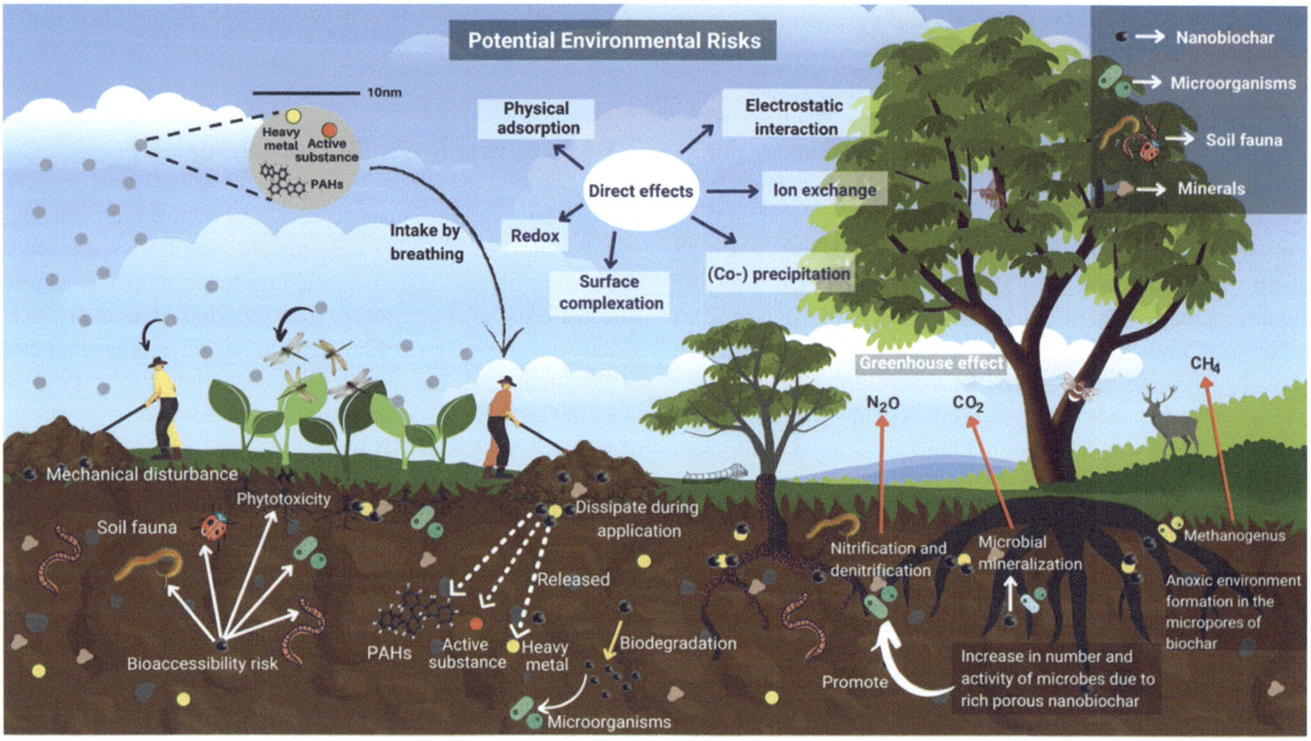

Fig. 2 Illustrative diagram showing possible environmental risks from using biochar-based nanocomposites for environmental remediation

cell's osmotic pressure, which may further worsen damage to the cell's internal structure. Cell mortality ultimately rises as a result of this (Lin et al. 2012). As a result, the process used to prepare NB can determine its biotoxicity.

2.2 Potential Risks in Terrestrial Ecosystems

It is important to consider the environmental risks that could arise from NB. Recent research has demonstrated that the stabilisation and agglomeration of NB may be influenced by the internal physical and chemical features as well as the synergy with natural soil colloidal particles (Saleh et al. 2008; Wang et al. 2013a; Chen et al. 2017). For instance, according to Yang et al. (2019b), more negatively charged NB surfaces are produced by the existence of O-containing functional groups, which in turn upsurges the electrostatic repulsion amongst the particles and formulates them more steady in aqueous solutions. BC minerals are capable of dissolving and releasing cations into aqueous solutions, where the cationic bridging effect lowers the repelling energy barrier amongst colloidal particles and accelerates the formation of NB (Liu et al. 2018a). However, negatively charged inorganic soil colloidal particles, as an example kaolin, can enhance NB steadiness in the ternary system of NB-soil colloids to increase their propensity to move in the environment. The movement of NB, however, can be restricted by positively charged soil inorganic colloids by charge neutralisation. Humic acid (HA), a kind of natural OM, affects NB aggregates function by ligand exchange, energy, hydrophobic forces, and van der Waals forces. As an illustration, when HA is adsorbed onto the NB's surface, the zeta potential is changed, and the electrostatic attraction amongst the NB particles is intensified (Gui et al. 2021). This rise in electrostatic repulsion upsurges the likelihood of NB movement, which also has an impact on the efficiency of adsorption, the toxicity of the environment, and the relocation of NB and pollutants. Though, one study discovered that higher HA concentrations (about 5 mg) and high concentrations of divalent cations to promote cation bridging increased the accumulation of biochar colloids in soil (Yang et al. 2019a). Additionally, low-temperature pyrolyzed NB with functional surface groups is frequently less likely to concentrate in the soil solution, leading to enhanced fluidity in the soil (Cely et al. 2015; Yang et al. 2019a).

In summation, soil solutions, particularly in acid soils with low base saturation, can dissolve a stable suspension of biochar particles. OM that has been dissolved can keep NB even more stable, increasing the possibility that NB would not be carried by soil water. In order to determine the transit and fate of NB whilst utilising BC for agricultural or environmental remediation, it is necessary to simultaneously consider the source material, the formulating temperature, and the configuration of the soil colloids.

2.3 Potential Risks from Air Exposure

Due to the existence of semiquinone and phenoxyl radicals, NB may enter into the air when the wind blows and induce ecological toxicity in organisms through respiration (Odinga et al. 2020). An emerging environmental concern is the toxicity of BC-induced dust emissions (Gelardi et al. 2019). Dust, as previously mentioned, may increase the chances of harmful compounds in biochar spreading to the environment. Dust created during the manufacture and biochar's ground applications might be poses a health risk because it is particulate in nature and has a size distribution linked to cytotoxic effects (Sigmund et al. 2017). It is important to take into account the inherent dust exposure of relevant personnel and proposes using safety measures (Liu et al. 2022).

3 Reactivity

Particles of nanobiochar (NB) are typically less than 100 nm. The sources of NB in the environment could beseparated into two groups: (i) foremost NB, which is accidently formulated during the formulation or purposely manufactured in the laboratory using grinding, ultrasonic, and other methods and (ii) secondary NB, which is developed after application by the interface of bulk BC with the environment (Zhang et al. 2020a, b). The O content of NB created through ultrasonic processing was 19.2–31.8% greater than that of the unique structure in terms of structural properties (Xiang et al. 2021). Although NB demonstrates higher dispersion in water, it exhibits reduced C stability due to its less aromatic composition (Liu et al. 2018b).

3.1 Reactivity with Heavy Metals and Nutrients in the Soil

When applied to soil, NB helps to release heavy metal ions into the soil. According to Kim et al. (2018), biochar particles size of less than 0.45 μm may enhance the diffusion and migration of arsenic (As) in soil. Additionally, the feedstock has an impact on the capacity of biochar to co-migrate with heavy metals. The contaminant co-migration characteristics of biochar formulated from nine distinct types of biomasses were studied by Song et al. (2019), who found that NB formed from plants had more fused aromatic rings and functional groups than NB formulated from urban locations. Contaminants (such as Cd^{+2}) have a greater chance of co-transporting with NB produced from plants. By varying the retaining and relocation of P, NB actually indorses the losing of P in basic soil, the contrary of its favourable effect on sustaining soil fertility. This results in a decline in soil fertility (Xiang et al. 2021). Arai and Livi (2013) claim that P can create P-Fe/Al soil colloids by soil ligand adsorption and electrostatic attraction, which indorses the liberation of P-Fe/Al soil colloids and their movement to the groundwater system. Furthermore, NB has a high potential for co-transporting bound P to groundwater when used as a hauler for P movement in acid or base soils (Chen et al. 2018a; Liu et al. 2018b). As a result, applying biochar to soils may cause nutrient leaching and environmental degradation.

3.2 Geochemical Transport Behaviour

Due to its minor size and adsorption capacity, increased soil sponginess, interactions with soil particles, and interactions with organic matter, engineered nanomaterials (ENMs) may have an impact on the soil's geophysical and chemical properties in addition to any potential effects on the soil's local flora, fauna, and human population (Ben-Moshe et al. 2013; Bayat et al. 2019). Ben-Moshe et al. (2013) employed three-dimensional fluorescence spectroscopy (such as 3D-FTIR) to compare the differences amongst soils with and without amended metal oxide NMs (CuO and Fe_3O_4) and discovered that ENMs modified the humic components in soils rather than the total organic content (Qian et al. 2020). In contrary, once amended in the soil, BC can move longitudinally by surface run or upwards to groundwater systems, eventually entering the ocean through environmental channels like streams or the atmosphere (Novak et al. 2009). For instance, upon the application of biochar, NB readily splits from the BC matrix and travels with the soil solution (Wang et al. 2013a, b), which is brought by either physical (such as water erosion and abrasion) or biological processes (e.g. biodegradation) (Xiang et al. 2021). Additionally, NB is very mobile during the ascendent and downhill movement in soil and aquifers (Qu et al. 2016). As like biochar, NB is a rich in mineral and O content, alkalinity, and dynamic stability (Xiang et al. 2021).

3.2.1 Behaviour in Aquatic Environment

The reactivity of NB in aquatic environments is high (Song et al. 2019). Water pollution and aquatic organism biotoxicity are both exacerbated by NB due to its greater dispersibility in water and greater co-migration influence on pollutants in aquatic ecosystems (Wang et al. 2013b; Liu et al. 2018b). Additionally, in wind-driven conditions, NB may enter into the atmosphere, and due to the existence of semiquinone and phenoxyl radicals, produce ecological toxicity in species through respiration (Xiang et al. 2021).

4 Lack in Long-Term Field Studies

It is important to note that there are currently few field-scale and long-term investigations reported on the effects of environmental exposure of biochar-based NC (Qian et al. 2020). To assess the long-lasting efficacy of the aforesaid ENMs and nano-enabled remedial techniques, it is important to underline the importance of conducting field studies in actual contaminated sites. Through a three-year field experiment, Wang et al. (2020b) recently shown that mercapto-functionalized nano-silica (MPTS/nano-silica) may show a long-lasting stability of Cd in polluted soil. Further technical information is mandatory to determine whether the pollutants that have been sequestered may someday be released back into the ecosystem.

5 Ageing Issue

The physicochemical properties of NB will alter as biochar ages after being utilised in the field due to a variety of environmental factors, including temperature variations, precipitation, and soil characteristics. NB erosion, fragmentation, oxidation, and disagglomeration/ disaggregation may occur as a result of ageing (Wang et al. 2020a). The kind of minerals in soil systems will affect the chemical ageing of NB, which could, for example, lead to the formation of more O-containing surface functional groups (Ren et al. 2018). Additionally, biological activities linked to soil microorganisms can produce additional surface functional groups on NB that contain oxygen (Hale et al. 2011). As a result of improved surface complexation amongst metal cations/complexes and negative O-containing functional groups, NB typically immobilise more heavy metals (Chang et al. 2019; Hua et al. 2020). Contrarily, several research have revealed that the ageing process might cause a decrease in the ability of BC to adsorb contaminants. For instance, Zhang et al. (2016) showed that when BC ages, soil-dissolved organic debris blocks the pores, limiting its capacity to sorb contaminants.

Future research is required to determine how NB impacts the retention and adsorption of heavy metals throughout time because the ageing process is complex (Zhang et al. 2022). It is also important to note that soil characteristics influence speciation of heavy metal and therefore bioavailability. Numerous reports have revealed that soil-dissolved organic matter and BC can adsorb heavy metals together to improve the capacity of BC to do so in soils with a high organic matter concentration (Lian et al. 2021). Long-term direct monitoring research has not done an adequate job of analysing the time-based progression of changes to NB in the environment. It would be impossible to refute the fact that thorough studies of nano-physicochemical BC's characteristics and how they mark its fate and transit are necessary to comprehend the environmental risks associated with this material.

6 Recovery and Regeneration

Despite the endless potential of BC-based products, using them involves a number of practical concerns. Specifically, comprehensive analyses of key environmental issues, economic efficiency, ageing and disintegration, and recovery and regeneration, as lately emphasised (Cui et al. 2019). The concentration of the de-sorbent and its impact on regeneration are crucial. Studies have shown that materials made of BC have outstanding regeneration and recovery capabilities (Table 1).

In order to evaluate the cost-benefits of used BC-based goods, future research should develop experimental formulas relating the raw material preparation price, making, transport, revival, and post-treatment by repeated testing. BC regeneration and cost reduction can be done by using different methods to handle different types of BC (Liu et al. 2020). BC costs are determined by the availability of local raw materials, pyrolysis conditions, process flow, reactor availability, and recovery and service life concerns. BC, on the other hand, might be a by-product of various operations (Mohan et al. 2014). There will be substantial diversity in NB prices in the future, since the potential value of NB has yet to be established and the BC market is not yet developed enough, which will aid key individuals in obtaining complete BC price and cost data. The economic benefits of NB are secondary to its application value; the environmental benefits of NB are far more significant (Dai et al. 2019).

On the other hand, renewal lowers secondary pollution whilst recycle in other applications increases cost effectiveness and source retrieval, according to Baskar et al. (2022). Some of the techniques used to recover and regenerate used adsorbents include magnetic separation, filtering, thermal desorption, solvent regeneration, microwave irradiation, supercritical fluid regeneration, enhanced oxidation process, and microbial-assisted adsorbent regeneration. A sorbent is heated at a particular temperature in order to dissolve the chemical and physical linkages that bind the sorbent to the sorbate is known as thermal regeneration (Shahadat and Isamil 2018). The carbon framework and any volatile compound of biochar are destroyed when heated in the existence of air at temperatures lower than 500 °C. By using microbial degradation of organic contaminants as an adsorbent, a spent adsorbent can be revived with the help of microbes. This procedure is frequently carried out using a single microbial culture or microbial consortium, such as bacteria, fungi, and

Table 1 Summary of few more studies that studied regeneration of biochar-based nanocomposites (BNC)

Sr. No	Type of biochar-based NC used	Type of feedstock	Application	Separation and regeneration technique	Regeneration condition	Highlights of the study	References
1	SiO_2-biochar nanocomposites	Algal biomass	Removal of phosphate from aqueous solution	NaOH solution	25 °C at 120 rpm for 24 h	The SiO_2-biochar nanocomposites developed in this study are cost-effective and beneficial in waste reusing, and they may be used in the future as multifunctional and highly direct phosphate adsorbents. The adsorbed SiO_2-biochar nanocomposites were efficiently regenerated by NaOH solution	Wang et al. (2016)
2	ZnO-biochar nanocomposites	Jute fibers	Photodegradation of methylene blue dye	–	110 °C for 8 h	During seven cycles of use and regeneration, the ZnO nanorod/jute fiber (JF) composite was stable and demonstrated a high efficiency of over 80%. These composite materials were predicted to be used in environmental engineering and manufacturing for the elimination of organic dyes because of their high photocatalytic efficiency, ease of recycling, and long-term stability	Chen et al. (2019)
3	Mg-biochar NC	Rice husk and corn cob	Cationic and anionic nutrient removal	NaOH solution	–	The desorption and recyclability of the RMgN500 and CMgN500 were exceptionally successful by utilising 0.5 M NaOH arrangement for PO_4^{3-} and weakened HCl arrangements for NH_4^+	Tran et al. (2022)
4	Fe_3O_4-graphene-biochar composite	Rice straw	Adsorption of crystal violet	Recovery via magnetic separation and regeneration via chemical treatment	For 2 h at 30 °C	The recovered composite's crystal violet (CV) absorbability was 157.31 mg/g, which was little less than pristine (199 mg/g). These results demonstrated how the used sorbent may be recovered and reused	Du et al. (2020)
5	Green biochar/iron oxide	Banana peel	Removal of methylene blue	Chemical treatment	At 80 °C	After five cycles, the adsorption effectiveness of the biochar/iron oxide composite showed a slight decline, but it was still well within the acceptable range	Zhang et al. (2020a, b)
6	MgAl-LDH (layered double hydroxide)/biochar composites	Bovine bone	Adsorption of methylene blue	Recovery via filtration and regeneration via chemical treatment	For 2 h at 60 °C	After six cycles, the composite's elimination capacity fell from 65–70 mg/g to 40–45 mg/g. The adsorptive capacity was more stable due to the biochar's presence	Meili et al. (2019)
7	Magnesium hydroxide-coated pyrolytic biochar	Wheat straw	Adsorption of anionic dye	Microwave irradiation	At 320, 480, 640 W for 5 min	When used for 5 min at 320 W, 480 W, or 640 W, respectively, microwave irradiation of used magnesium hydroxide-coated pyrolytic biochar produced regeneration yields of 98.5%, 89.0%, and 85.5%	Zhang et al. (2014)

algae. The adsorbent surface must not be poisonous to the active microorganisms and support microbial habitat and development in order to be used in microorganism-assisted regeneration. In batch bio-regeneration of used adsorbents, microorganisms and their sources of carbon and nutrition are combined with adsorbents rich in organic material. Through microbial action, the contaminants are mineralized, and the adsorbents are renewed (Baskar et al. 2022).

In conclusion, recycling products made in BC have major positive effects on both the economy and the environment. When utilised in a soil system, BC particles, for instance, might be put into porous sponges, filter papers, or another large-volume hauler to encourage separation, regaining, and renewal (Cui et al. 2019). In recent investigations (Kumar et al. 2020), BC was employed as a catalyst support for advanced oxidation processes (AOPs), or the AOPs were used to regenerate used biochar.

7 Challenges

Due to its distinct and varied features, NB has recently attracted a lot of attention amongst tailored biochars. In addition to having a significantly larger surface area to mass ratio than macrobiochar, the nanoscale transformation of biochar made it possible for it to be used in a variety of different ways than as an adsorbent, such as a sensor, capacitor, and photocatalytic material. In relation to the advancement of NB research and its applications, this study revealed research gaps and obstacles (Fig. 3).

The main issues are as follows: (1) characterization of NB alone is scarce because of most technical studies concentrated on biochar produced by ball-mill, which contains equally colloidal and nano-sized fractions; (2) few large-scale studies because of low yields, necessitating the requirement for improved preparation methods for high yields; (3) there is a missing information on the ecotoxicological data on the destiny and transportation of NB; (4) fully comprehend chemical changes during the production and disposal stages not reported yet; and (5) to advance the use of NB, a cost–benefit analysis and life-cycle evaluation for various applications are urgently required. According to Zhou et al. (2021), nanomaterials can facilitate an uncontrolled interaction between NPs and soil factors when used as single agrochemical carriers, which can cause drastic changes in the activity of soil microbiota and, in some cases, negatively affect soil fertility and consequently crop productivity. More specifically, the introduction of several different types of NPs into soil causes a variety of reactions in the enzymatic activities of advantageous bacteria. The thorough analysis of NP synthesis, utilisation, and mechanism of action on a variety of soil resource components should provide crucial information for improved management of NPs discharged into the environment (Rajput et al. 2021). Although some of the nanoparticles included in the matrix might leak out into handled water if

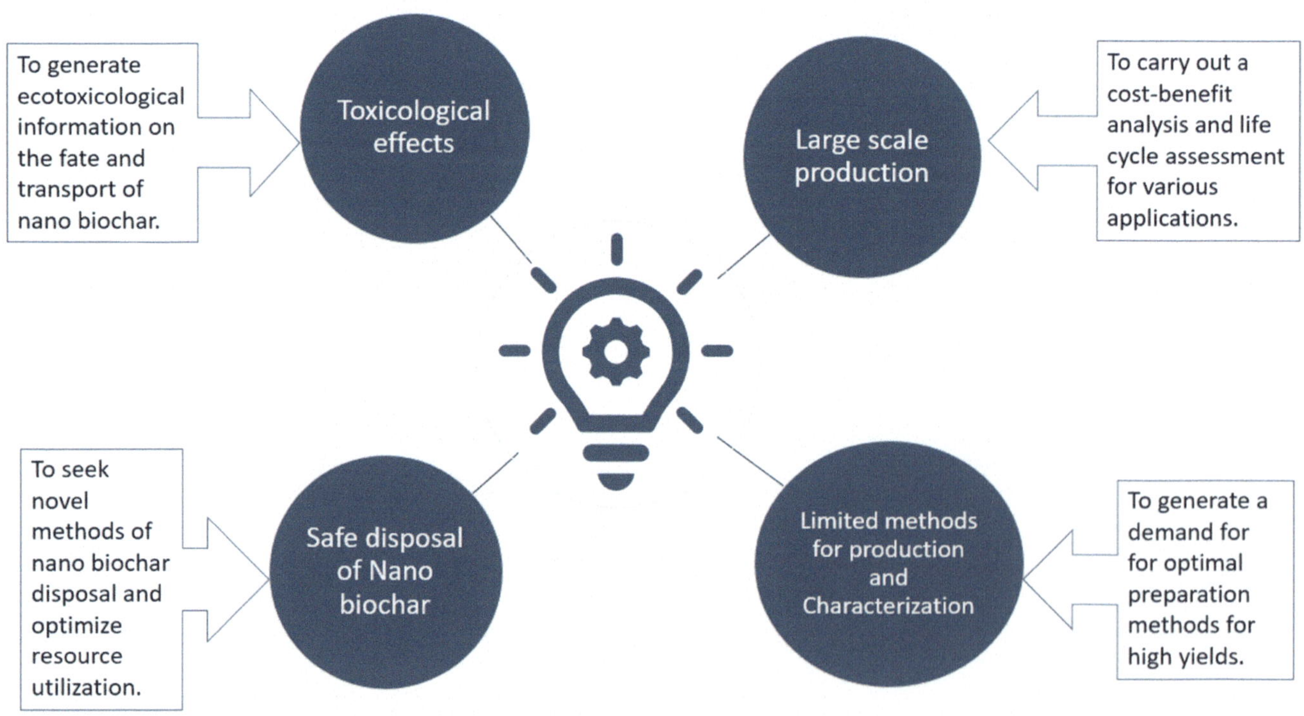

Fig. 3 Major challenges related to nanobiochar production and commercialization

incorrectly anchored, there have been questions raised about the stability of these biochar-based nanocomposites after use (Amusat et al. 2021). Due to its increased effort with pollution, the growth and usage of nano-biochar produced using ball milling technology are still difficult in the field of environmental restoration (Zhou et al. 2021).

Biochar can be used to improve the qualities of soil since it has the right cationic exchange ability. However, there are still issues with the separation, retrieval, and dumping of consumed adsorbent sludge at both the pilot and full industrial scales (Yang et al. 2020). Therefore, reaching the BC's scalability and commercialization levels is yet in its initial phases of growth (Premarathna et al. 2019). The practicability of BC, however, has not been fully utilised, and its benefits are anticipated to last past the existing service. Due to their moderate sensitivity, chemical steadiness, ecological non-toxicity, abundance, and low cost, semiconductor materials like TiO_2 nanoparticles make suitable photocatalysts. Such a semiconductor's function is to have a large band gap and quick electron–hole pair recombination. Although hybridising with BC is a potential way to close such gaps, doping of transition metals remains costly and impractical (Mian and Liu 2018). The benefits of metal doping are decreasing the band gap and improving visible light absorption, whereas its drawbacks contain causing secondary pollution, photo-corrosion, and an increase in the rate at which electron gap pairs recombine when they are present in excess (Mian and Liu 2018). Before use on either a pilot or an industrial scale, the nanocomposite must be stabilised in order to reduce its aspect effects (Fito et al. 2022). For the past ten years, it has been difficult to scale up dust carbon adsorbents, for example, activated carbon and BC for the application of NB and biochar materials in wastewater treatment systems. The principal disadvantages were the post-treatment slurry separation, less dense adsorbent to settle (sedimentation), clarifying, filtration (centrifugation), reuse, and regeneration (Son et al. 2018; Guo et al. 2020). The performance, time, and financial costs of using technology advancements are increased by these obstacles. The BC amendment was formerly a subject of required research that the scientific community had to look into in order to address these difficulties in particular. In keeping with this, the BC-derived spinel ferrites (SPF) nanocomposite is currently emerging as a useful nanomaterial to be used in a variety of water and wastewater treatment applications. The BC-based SPF nanocomposites had good catalytic performance, which was due to their magnetic properties, oxygenated functional groups, graphite structure, and persistent free radicals (Feng et al. 2021). Whilst SPFs are responsible for photodegradation, BC typically enhances pollutant adsorption. The practical project in the water purification industry is the focus of this paradigm transition from biochar to BC-based SPF nanocomposites. After being treated with an external magnetic field, the BC-SPF nanocomposites will be eliminated (Feng et al. 2021). However, another significant task that must be addressed for the scale-up challenge is the industrial application of external magnetic fields (Fito et al. 2022).

8 Future Perspectives

Additional study should focus on: (1) seeking out more inventive solid wastes to use as raw materials for the production of biochar (BC) to scavenge potentially toxic substances; (2) discover more high-quality and environmentally friendly techniques for BC preparation; (3) at some point during BC pyrolysis, production techniques should be adopted to minimise or eliminate the creation of polycyclic aromatic hydrocarbons (PAHs), which could be discharged into the environment and affect people and other species; (4) the current focus should be on the toxicological effects on organisms as a result of widespread application, such as when metals or other environmentally hazardous substances may have a negative effect on living things in the environment; (5) to prevent the build-up of BC substances in the environment with large-scale utility and the re-release of PTEs into the environment, deem adequate ways to safely dispose of the BC materials after the adsorption of PTEs (Deng et al. 2020).

9 Conclusion

Although the majority of studies show both nanobiochar's benefits and drawbacks, there is still much to learn about it. It has proven effective as a sorbent for a variety of pollutants, including heavy metals, dyes, and pharmaceuticals. Additionally, NB can be used to improve soil, fertilise plants, and lessen the harmful effects of pesticides. Understanding NB's reactivity potential is crucial for comprehending its behaviour, transport cycle, and possible risk of pollutant adsorption onto NB that results in environmental release. The requirement for cost–benefit analyses and life-cycle assessments for various applications, a dearth of NB characterization methods, poor yields, and a lack of ecotoxicological data are some of the major obstacles to the use of NB. Future research will, therefore, focus on finding ways to increase NB yield and production whilst reducing agglomeration.

References

Amusat SO, Kebede TG, Dube S, Nindi MM (2021) Ball-milling synthesis of biochar and biochar–based nanocomposites and prospects for removal of emerging contaminants: a review. J Water Proc Eng 41:101993. https://doi.org/10.1016/j.jwpe.2021.101993

Arai Y, Livi KJ (2013) Underassessed phosphorus fixation mechanisms in soil sand fraction. Geoderma 192:422–429. https://doi.org/10.1016/j.geoderma.2012.06.021

Baskar AV, Bolan N, Hoang SA, Sooriyakumar P, Kumar M, Singh L, Jasemizad T, Padhye LP, Singh G, Vinu A, Sarkar B, Kirkham MB, Rinklebe J, Wang S, Wang H, Balasubramanian R, Siddique KH (2022) Recovery, regeneration and sustainable management of spent adsorbents from wastewater treatment streams: a review. Sci Tot Environ 822:153555. https://doi.org/10.1016/j.scitotenv.2022.153555

Bayat H, Kolahchi Z, Valaey S, Rastgou M, Mahdavi S (2019) Iron and magnesium nano-oxide effects on some physical and mechanical properties of a loamy Hypocalcic Cambisol. Geoderma 33:57–68. https://doi.org/10.1016/j.geoderma.2018.08.007

Ben-Moshe T, Frenk S, Dror I, Minz D, Berkowitz B (2013) Effects of metal oxide nanoparticles on soil properties. Chemosphere 90:640–646. https://doi.org/10.1016/j.chemosphere.2012.09.018

Cely P, Gascó G, Paz-Ferreiro J, Méndez A (2015) Agronomic properties of biochars from different manure wastes. J Anal Appl Pyrol 111:173–182. https://doi.org/10.1016/j.jaap.2014.11.014

Chang R, Sohi SP, Jing F, Liu Y, Chen J (2019) A comparative study on biochar properties and Cd adsorption behavior under effects of ageing processes of leaching, acidification and oxidation. Environ Pol 254:113123. https://doi.org/10.1016/j.envpol.2019.113123

Chausali N, Saxena J, Prasad R (2021) Nanobiochar and biochar based nanocomposites: advances and applications. J Agric Food Res 5:100191. https://doi.org/10.1016/j.jafr.2021.100191

Chen M, Alim N, Zhang Y, Xu N, Cao X (2018a) Contrasting effects of biochar nanoparticles on the retention and transport of phosphorus in acidic and alkaline soils. Environ Pollut 239:562–570. https://doi.org/10.1016/j.envpol.2018.04.050

Chen H, Xie A, You S (2018b) A review: advances on absorption of heavy metals in the waste water by biochar. In: IOP conference series: materials science and engineering, vol 301, no 1. IOP Publishing, p 012160. https://doi.org/10.1088/1757-899X/301/1/012160

Chen M, Bao C, Hu D, Jin X, Huang Q (2019) Facile and low-cost fabrication of ZnO/biochar nanocomposites from jute fibers for efficient and stable photodegradation of methylene blue dye. J Anal Appl Pyrol 139:319–332. https://doi.org/10.1016/j.jaap.2019.03.009

Chen M, Wang D, Yang F, Xu X, Xu N, Cao X (2017) Transport and retention of biochar nanoparticles in a paddy soil under environmentally-relevant solution chemistry conditions. Environ Pollut 230:540–549. https://doi.org/10.1016/j.envpol.2017.06.101

Chen R, Ratnikova TA, Stone MB, Lin S, Lard M, Huang G, Hudson JS, Ke PC (2010) Differential uptake of carbon nanoparticles by plant and mammalian cells. Small 6:612–617. https://doi.org/10.1002/smll.200901911

Corsi I, Winther-Nielsen M, SethiI R, Punta C, Della Torre C, Libralato G, Lofrano G, Sabatini L, Aiello M, Fiordi L, Cinuzzi F, Caneschi A, Pellegrini D, Buttino I (2018) Ecofriendly nanotechnologies and nanomaterials for environmental applications: key issue and consensus recommendations for sustainable and ecosafe nanoremediation. Ecotoxicol Environ Saf 154:237–244. https://doi.org/10.1016/j.ecoenv.2018.02.037

Cui H, Li D, Liu X, Fan Y, Zhang X, Zhang S, Zhou J, Fang G, Zhou J (2021) Dry-wet and freeze-thaw aging activate endogenous copper and cadmium in biochar. J Clean Prod 288:125605. https://doi.org/10.1016/j.jclepro.2020.125605

Cui J, Jin Q, Li Y, Li F (2019) Oxidation and removal of As(iii) from soil using novel magnetic nanocomposite derived from biomass waste. Environ Sci Nano 6:478–488. https://doi.org/10.1039/C8EN01257A

Dai Y, Zhang N, Xing C, Cui Q, Sun Q (2019) The adsorption, regeneration and engineering applications of biochar for removal organic pollutants: a review. Chemosphere 223:12–27. https://doi.org/10.1016/j.chemosphere.2019.01.161

De Gisi S, Petta L, Wendland C (2014) History and technology of Terra Preta sanitation. Sustainability 6:1328–1345

Deng R, Huang D, Wan J, Xue W, Wen X, Liu X, Chen S, Lei L, Zhang Q, Zhang Q (2020) Recent advances of biochar materials for typical potentially toxic elements management in aquatic environments: a review. J Clean Prod 255:119523. https://doi.org/10.1016/j.jclepro.2019.119523

Du C, Song Y, Shi S, Jiang B, Yang J, Xiao S (2020) Preparation and characterization of a novel Fe_3O_4-graphene-biochar composite for crystal violet adsorption. Sci Total Environ 711:134662. https://doi.org/10.1016/j.scitotenv.2019.134662

El-Naggar A, Lee MH, Hur J, Lee YH, Igalavithana AD, Shaheen SM, Ryu C, Rinklebe J, Tsang DCW, Ok YS (2020) Biochar-induced metal immobilization and soil biogeochemical process: an integrated mechanistic approach. Sci Total Environ 698:134112. https://doi.org/10.1016/j.scitotenv.2019.134112

Ema M, Hougaard KS, Kishimoto A, Honda K (2016) Reproductive and developmental toxicity of carbon-based nanomaterials: a literature review. Nanotoxicology 10:391–412. https://doi.org/10.3109/17435390.2015.1073811

Feng Z, Yuan R, Wang F, Chen Z, Zhou B, Chen H (2021) Preparation of magnetic biochar and its application in catalytic degradation of organic pollutants: a review. Sci Total Environ 765:142673. https://doi.org/10.1016/j.scitotenv.2020.142673

Fito J, Kefeni KK, Nkambule TT (2022) The potential of biochar-photocatalytic nanocomposites for removal of organic micropollutants from wastewater. Sci Total Environ 829:154648. https://doi.org/10.1016/j.scitotenv.2022.154648

Freixa A, Acuña V, Sanchís J, Farré M, Barceló D, Sabater S (2018) Ecotoxicological effects of carbon based nanomaterials in aquatic organisms. Sci Total Environ 619:328–337. https://doi.org/10.1016/j.scitotenv.2017.11.095

Gao M, Song Z (2019) Toxicity of cadmium to wheat seedling roots in the presence of graphene oxide. Chemosphere 233:9–16. https://doi.org/10.1016/j.chemosphere.2019.05.127

Gelardi DL, Li C, Parikh SJ (2019) An emerging environmental concern: biochar-induced dust emissions and their potentially toxic properties. Sci Total Environ 678:813–820. https://doi.org/10.1016/j.scitotenv.2019.05.007

Gui X, Song B, Chen M, Xu X, Ren Z, Li X, Cao X (2021) Soil colloids affect the aggregation and stability of biochar colloids. Sci Total Environ 771:145414. https://doi.org/10.1016/j.scitotenv.2021.145414

Guo Z, Chen R, Yang R, Yang F, Chen J, Li Y, Zhou R, Xu J (2020) Synthesis of amino-functionalized biochar/spinel ferrite magnetic composites for low-cost and efficient elimination of Ni(II) from wastewater. Sci Total Environ 722:137822. https://doi.org/10.1016/j.scitotenv.2020.137822

Hale S, Hanley K, Lehmann J, Zimmerman A, Cornelissen G (2011) Effects of chemical, biological, and physical aging as well as soil addition on the sorption of pyrene to activated carbon and biochar. Environ Sci Technol 45:10445–10453. https://doi.org/10.1021/es202970x

Hassan M, Liu Y, Naidu R, Parikh SJ, Du J, Qi F, Willett IR (2020) Influences of feedstock sources and pyrolysis temperature on the properties of biochar and functionality as adsorbents: a meta-analysis. Sci Total Environ 744:140714. https://doi.org/10.1016/j.scitotenv.2020.140714

Hossain N, Bhuiyan MA, Pramanik BK, Nizamuddin S, Griffin G (2020) Waste materials for wastewater treatment and waste adsorbents for biofuel and cement supplement applications: a critical review. J Clean Prod 255:120261. https://doi.org/10.1016/j.jclepro.2020.120261

Hua Y, Zheng X, Xue L, Han L, He S, Mishra T, Feng Y, Yang L, Xing B (2020) Microbial aging of hydrochar as a way to increase cadmium ion adsorption capacity: process and mechanism. Biores Technol 300:122708. https://doi.org/10.1016/j.biortech.2019.122708

Jia C, Luo J, Fan J, Clark JH, Zhang S, Zhu X (2021) Urgently reveal longly hidden toxicant in a familiar fabrication process of biomass-derived environment carbon material. J Environ Sci 100:250–256. https://doi.org/10.1016/j.jes.2020.08.001

Kim HB, Kim SH, Jeon EK, Kim DH, Tsang DC, Alessi DS, Kwon EE, Baek K (2018) Effect of dissolved organic carbon from sludge, Rice straw and spent coffee ground biochar on the mobility of arsenic in soil. Sci Total Environ 636:1241–1248. https://doi.org/10.1016/j.scitotenv.2018.04.406

Kozyatnyk I, Yacout DM, Van Caneghem J, Jansson S (2020) Comparative environmental assessment of end-of-life carbonaceous water treatment adsorbents. Bioresour Technol 302:122866. https://doi.org/10.1016/j.biortech.2020.122866

Kumar M, Xiong X, Wan Z, Sun Y, Tsang DC, Gupta J, Ok YS et al (2020) Ball milling as a mechanochemical technology for fabrication of novel biochar nanomaterials. Bioresour Technol 312:123613. https://doi.org/10.1016/j.biortech.2020.123613

Lata S, Singh PK, Samadder SR (2015) Regeneration of adsorbents and recovery of heavy metals: a review. Int J Environ Sci Technol 12:1461–1478. https://doi.org/10.1007/s13762-014-0714-9

Lehmann J, Joseph S (2012) Biochar systems. In: Biochar for environmental management. Routledge, pp 179–200. https://doi.org/10.4324/9781849770552

Lehmann J, Joseph S (eds) (2015) Biochar for environmental management: science, technology and implementation. Routledge

Lian F, Wang Z, Xing B (2021) Nano-black carbon (biochar) released from pyrogenic carbonaceous matter as a super suspending agent in water/soil environments. Biochar 3(1):1–3. https://doi.org/10.1007/s42773-020-00075-x

Lin Y, Munroe P, Joseph S, Kimber S, Van Zwieten L (2012) Nanoscale organo-mineral reactions of biochars in ferrosol: an investigation using microscopy. Plant Soil 357:369–380. https://doi.org/10.1007/s11104-012-1169-8

Liu G, Zheng H, Jiang Z, Wang Z (2018a) Effects of biochar input on the properties of soil nanoparticles and dispersion/sedimentation of natural mineral nanoparticles in aqueous phase. Sci Total Environ 634:595–605. https://doi.org/10.1016/j.scitotenv.2018.04.019

Liu G, Zheng H, Jiang Z, Zhao J, Wang Z, Pan B, Xing B (2018b) Formation and physicochemical characteristics of nano biochar: insight into chemical and colloidal stability. Environ Sci Technol 52:10369–10379. https://doi.org/10.1021/acs.est.8b01481

Liu X, Tang J, Wang L, Liu Q, Liu R (2019) A comparative analysis of ball-milled biochar, graphene oxide, and multi-walled carbon nanotubes with respect to toxicity induction in Streptomyces. J Environ Manage 243:308–317. https://doi.org/10.1016/j.jenvman.2019.05.034

Liu W, Li Y, Feng Y, Qiao J, Zhao H, Xie J, Liang S et al (2020) The effectiveness of nanobiochar for reducing phytotoxicity and improving soil remediation in cadmium-contaminated soil. Sci Rep 10(1):858. https://doi.org/10.1038/s41598-020-57954-3

Liu M, Almatrafi E, Zhang Y, Xu P, Song B, Zhou C, Zeng G, Zhu Y (2022) A critical review of biochar-based materials for the remediation of heavy metal contaminated environment: applications and practical evaluations. Sci Total Environ 806:150531. https://doi.org/10.1016/j.scitotenv.2021.150531

Lonappan L, Rouissi T, Das RK, Brar SK, Ramirez AA, Verma M, Surampalli RY, Valero JR (2016) Adsorption of methylene blue on biochar microparticles derived from different waste materials. Waste Manage 49:537–544. https://doi.org/10.1016/j.wasman.2016.01.015

Lowry GV, Avellan A, Gilbertson LM (2019) Opportunities and challenges for nanotechnology in the agri-tech revolution. Nat Nanotechnol 14:517–522. https://doi.org/10.1038/s41565-019-0461-7

Lyu H, Gao B, He F, Zimmerman AR, Ding C, Huang H, Tang J (2018) Effects of ball milling on the physicochemical and sorptive properties of biochar: experimental observations and governing mechanisms. Environ Pollut 233:54–63. https://doi.org/10.1016/j.envpol.2017.10.037

Meili L, Lins PV, Zanta CLPS, Soletti JI, Ribeiro LMO, Dornelas CB, Silva TL, Vieira MGA (2019) MgAl-LDH/biochar composites for methylene blue removal by adsorption. Appl Clay Sci 168:11–20. https://doi.org/10.1016/j.clay.2018.10.012

Mian MM, Liu G (2018) Recent progress in biochar-supported photocatalysts: synthesis, role of biochar, and applications. RSC Adv 8:14237–14248. https://doi.org/10.1039/C8RA02258E

Mohan D, Sarswat A, Ok YS, Pittman CU Jr (2014) Organic and inorganic contaminants removal from water with biochar, a renewable, low cost and sustainable adsorbent–a critical review. Biores Technol 160:191–202. https://doi.org/10.1016/j.biortech.2014.01.120

Naghdi M, Taheran M, Pulicharla R, Rouissi T, Brar SK, Verma M, Surampalli RY (2019) Pine-wood derived nanobiochar for removal of carbamazepine from aqueous media: adsorption behavior and influential parameters. Arab J Chem 12:5292–5301. https://doi.org/10.1016/j.arabjc.2016.12.025

Novak JM, Busscher WJ, Laird DL, Ahmedna M, Watts DW, Niandou MA (2009) Impact of biochar amendment on fertility of a southeastern coastal plain soil. Soil Sci 174:105–112. https://doi.org/10.1097/SS.0b013e3181981d9a

Odinga ES, Waigi MG, Gudda FO, Wang J, Yang B, Hu X, Li S, Gao Y (2020) Occurrence, formation, environmental fate and risks of environmentally persistent free radicals in biochars. Environ Int 134:105172. https://doi.org/10.1016/j.envint.2019.105172

Oliveira DF, Costa VA, Terra WC, Campos VP, Paula PM, Martins SJ (2019) Impact of phenolic compounds on Meloidogyne incognita in vitro and in tomato plants. Exp Parasitol 199:17–23. https://doi.org/10.1016/j.exppara.2019.02.009

Pan B, Li H, Lang D, Xing B (2019) Environmentally persistent free radicals: occurrence, formation mechanisms and implications. Environ Pollut 248:320–331. https://doi.org/10.1016/j.envpol.2019.02.032

Premarathna KSD, Rajapaksha AU, Sarkar B, Kwon EE, Bhatnagar A, Ok YS, Vithanage M (2019) Biochar-based engineered composites for sorptive 823 decontamination of water: a review. Chem Eng J 372:536–550. https://doi.org/10.1016/j.envpol.2019.02.032

Qian Y, Qin C, Chen M, Lin S (2020) Nanotechnology in soil remediation−applications vs. implications. Ecotoxicol Environ Saf 201:110815. https://doi.org/10.1016/j.ecoenv.2020.110815

Qin C, Wang H, Yuan X, Xiong T, Zhang J, Zhang J (2020) Understanding structure-performance correlation of biochar materials in environmental remediation and electrochemical devices. Chem Eng J 382:122977. https://doi.org/10.1016/j.cej.2019.122977

Qu X, Fu H, Mao J, Ran Y, Zhang D, Zhu D (2016) Chemical and structural properties of dissolved black carbon released from biochars. Carbon 96:759–767. https://doi.org/10.1016/j.carbon.2015.09.106

Rajput VD, Singh A, Singh VK, Minkina TM, Sushkova S (2021) Impact of nanoparticles on soil resource. In: Nanomaterials for soil remediation. Elsevier, pp 65–85. https://doi.org/10.1016/B978-0-12-822891-3.00004-9

Ramanayaka S, Vithanage M, Alessi DS, Liu WJ, Jayasundera AC, Ok YS (2020) Nanobiochar: production, properties, and multifunctional applications. Environ Sci Nano 7:3279–3302. https://doi.org/10.1039/D0EN00486C

Ren X, Sun H, Wang F, Zhang P, Zhu H (2018) Effect of aging in field soil on biochar's properties and its sorption capacity. Environ Pollut 242:1880–1886. https://doi.org/10.1016/j.envpol.2018.07.078

Saleh NB, Pfefferle LD, Elimelech M (2008) Aggregation kinetics of multiwalled carbon nanotubes in aquatic systems: measurements and environmental implications. Environ Sci Technol 42:7963–7969. https://doi.org/10.1021/es801251c

Shahadat M, Isamil S (2018) Regeneration performance of clay-based adsorbents for the removal of industrial dyes: a review. RSC Adv 8 (43):24571–24587. https://doi.org/10.1039/C8RA04290J

Shrivastava M, Srivastav A, Gandhi S, Rao S, Roychoudhury A, Kumar A, Singhal RK, Jha SK, Singh SD (2019) Monitoring of engineered nanoparticles in soil-plant system: a review. Environ Nanotechnol Monitor Manag 11:100218. https://doi.org/10.1016/j.enmm.2019.100218

Sigmund G, Huber D, Bucheli TD, Baumann M, Borth N, Guebitz GM, Hofmann T (2017) Cytotoxicity of biochar: a workplace safety concern? Environ Sci Technol Lett 4:362–366. https://doi.org/10.1021/acs.estlett.7b00267

Sohi SP (2012) Carbon storage with benefits. Science 338:1034–1035

Son EB, Poo KM, Chang JS, Chae KJ (2018) Heavy metal removal from aqueous solutions using engineered magnetic biochars derived from waste marine macro-algal biomass. Sci Total Environ 615:161–168. https://doi.org/10.1016/j.scitotenv.2017.09.171

Song B, Chen M, Zhao L, Qiu H, Cao X (2019) Physicochemical property and colloidal stability of micron-and nano-particle biochar derived from a variety of feedstock sources. Sci Total Environ 661:685–695. https://doi.org/10.1016/j.scitotenv.2019.01.193

Taheran M, Naghdi M, Brar SK, Knystautas EJ, Verma M, Ramirez AA, Surampalli RY, Valéro JR (2016) Adsorption study of environmentally relevant concentrations of chlortetracycline on pinewood biochar. Sci Total Environ 571:772–777. https://doi.org/10.1016/j.scitotenv.2016.07.050

Tan X, Liu Y, Zeng G, Wang X, Hu X, Gu Y, Yang Z (2015) Application of biochar for the removal of pollutants from aqueous solutions. Chemosphere 125:70–85. https://doi.org/10.1016/j.chemosphere.2014.12.058

Tran DT, Pham TD, Dang VC, Pham TD, Nguyen MV, Dang NM, Ha MN, Nguyen VN, Nghiem LD (2022) A facile technique to prepare MgO-biochar nanocomposites for cationic and anionic nutrient removal. J Water Process Eng 47:102702. https://doi.org/10.1016/j.jwpe.2022.102702

Vakili M, Deng S, Cagnetta G, Wang W, Meng P, Liu D, Yu G (2019) Regeneration of chitosan-based adsorbents used in heavy metal adsorption: a review. Sep Purif Technol 224:373–387. https://doi.org/10.1016/j.seppur.2019.05.040

Wang D, Zhang W, Zhou D (2013a) Antagonistic effects of humic acid and iron oxyhydroxide grain-coating on biochar nanoparticle transport in saturated sand. Environ Sci Technol 47:5154–5161. https://doi.org/10.1021/es305337r

Wang D, Zhang W, Hao X, Zhou D (2013b) Transport of biochar particles in saturated granular media: effects of pyrolysis temperature and particle size. Environ Sci Technol 47:821–828. https://doi.org/10.1021/es303794d

Wang H, Wang H, Wang Z, Tang L, Zeng G, Xu P, Chen M, Xiong T, Zhou C, Li X, Huang D, Zhu Y, Wang Z, Tang J (2020a) Covalent organic framework photocatalysts: structures and applications. Chem Soc Rev 49:4135–4165. https://doi.org/10.1039/D0CS00278J

Wang YY, Lu HH, Liu YX, Yang SM (2016) Removal of phosphate from aqueous solution by SiO_2–biochar nanocomposites prepared by pyrolysis of vermiculite treated algal biomass. RSC Adv 6:83534–83546. https://doi.org/10.1039/C6RA15532D

Wang Y, Liu Y, Zhan W, Zheng K, Lian M, Zhang C, Ruan X, Li T (2020b) Long-term stabilization of Cd in agricultural soil using mercapto-functionalized nano-silica (MPTS/nano-silica): a three-year field study. Ecotoxicol Environ Saf 197:110600. https://doi.org/10.1016/j.ecoenv.2020.110600

Xiang L, Liu S, Ye S, Yang H, Song B, Qin F, Shen M, Tan C, Zeng G, Tan X (2021) Potential hazards of biochar: the negative environmental impacts of biochar applications. J Hazard Mater 420:126611. https://doi.org/10.1016/j.jhazmat.2021.126611

Xiao X, Chen B, Chen Z, Zhu L, Schnoor JL (2018) Insight into multiple and multilevel structures of biochars and their potential environmental applications: a critical review. Environ Sci Technol 52:5027–5047. https://doi.org/10.1021/acs.est.7b06487

Yang W, Shang J, Sharma P, Li B, Liu K, Flury M (2019a) Colloidal stability and aggregation kinetics of biochar colloids: effects of pyrolysis temperature, cation type, and humic acid concentrations. Sci Total Environ 658:1306–1315. https://doi.org/10.1016/j.scitotenv.2018.12.269

Yang X, Ng W, Wong BSE, Baeg GH, Wang CH, Ok YS (2019b) Characterization and ecotoxicological investigation of biochar produced via slow pyrolysis: effect of feedstock composition and pyrolysis conditions. J Hazard Mater 365:178–185. https://doi.org/10.1016/j.jhazmat.2018.10.047

Yang H, Ye S, Zeng Z, Zeng G, Tan X, Xiao R, Xu F et al (2020) Utilization of biochar for resource recovery from water: a review. Chem Eng J 397:125502. https://doi.org/10.1016/j.cej.2020.125502

Zhang K, Wang Y, Mao J, Chen B (2020a) Effects of biochar nanoparticles on seed germination and seedling growth. Environ Pollut 256:113409. https://doi.org/10.1016/j.envpol.2019.113409

Zhang P, O'Connor D, Wang Y, Jiang L, Xia T, Wang L, Tsang DCW, Ok YS, Hou D (2020b) A green biochar/iron oxide composite for methylene blue removal. J Hazard Mater 384:121286. https://doi.org/10.1016/j.jhazmat.2019.121286

Zhang T, Lowry GV, Capiro NL, Chen J, Chen W, Chen Y, Alvarez PJ et al (2019) In situ remediation of subsurface contamination: opportunities and challenges for nanotechnology and advanced materials. Environ Sci Nano 6:1283–1302. https://doi.org/10.1039/C9EN00143C

Zhang XN, Mao GY, Jiao YB, Shang Y, Han RP (2014) Adsorption of anionic dye on magnesium hydroxide-coated pyrolytic bio-char and reuse by microwave irradiation. Int J Environ Sci Technol 11:1439–1448. https://doi.org/10.1007/s13762-013-0338-5

Zhang X, Sarmah AK, Bolan NS, He L, Lin X, Che L, Tang C, Wang H (2016) Effect of aging process on adsorption of diethyl phthalate in soils amended with bamboo biochar. Chemosphere 142:28–34. https://doi.org/10.1016/j.chemosphere.2015.05.037

Zhang X, Wells M, Niazi N, Bolan N, Shaheeng S, Hou D, Gao B, Wang H, Rinklebe J, Wang Z (2022) Nanobiochar-rhizosphere interactions: implications for the remediation of heavy-metal contaminated soils. Environ Pollut 299:118810. https://doi.org/10.1016/j.envpol.2022.118810

Zhou S, Ni X, Zhou H, Meng X, Sun H, Wang J, Yin X (2021) Effect of nZVI/biochar nanocomposites on Cd transport in clay mineral-coated quartz sand: facilitation and rerelease. Ecotoxicol Environ Saf 228:112971. https://doi.org/10.1016/j.ecoenv.2021.112971

Engineered Biochar-Based Nanocomposites: A Sustainable Solution for Smart Agriculture

Mayank Singh

Abstract

Soil degradation and contamination have now become a serious global issue for agricultural production. The huge application of pesticides and chemicals for more and more agricultural production deteriorates the soil quality for crop production. For the last few years, biochar has been identified as a promising tool for the remediation of contaminants from agricultural land. By the several kinds of research, many engineered or enriched nanocomposite-based biochar were produced to decontaminate the soil through different methodologies. These opted pathways for their physico-chemical modifications help in increasing their sorption capacity, enhancing the soil bacterial abundance, decreases the mobility of xenobiotics, and increasing their bioavailability in soil. Apart from this, they also impart an important role for carbon sequestration, soil fertility, decreased phytotoxicity, increased nutrient availability, higher biomass, decreased bioaccumulation of contaminants, and overall improved crop performance. This book chapter summarized the recent methods, types, mechanisms, and efficacy of engineered biochar-based nanocomposites in the sustainability of agriculture and environment.

Keywords

Carbon sequestration • Nanocomposites • Phytotoxicity • Pyrolysis

M. Singh (✉)
Environmental Chemistry Laboratory, Crop Production and Protection Division, CSIR-Central Institute of Medicinal and Aromatic Plants, Lucknow, 226015, India
e-mail: mayanksingh1699@gmail.com

Academy of Scientific and Innovative Research (AcSIR), Ghaziabad, India

1 Introduction

Nowadays, climate change is the biggest issue, which aggregates many other problems such as global warming, environmental problems like eutrophication, runoff, etc., and irrational use of resources such as natural gas, crude oil, and petroleum that are not refillable. Due to all these problems, agriculture is negatively affected, and due to the vigorous increase in population, these problems are increasing rapidly on a day-to-day basis (Rajput et al. 2022). Carbon-based nanomaterials are attracting more attention from researchers due to their multidimensional application in agriculture, environment, energy, and biomedicine. In 1985 first time for plant growth enhancement, carbon nanotube which was made of fullerene was used, which ameliorates fertilizer and pesticide performance. Previous studies reveal that it enhances total crop production and increases the number of flowers and fruits in tomato plants (Khodakovskaya et al. 2012). The fabrication of carbon-based nanotubes is easy because biochar naturally contains nanomaterials, and due to the low cost of biochar, it can be a sustainable precursor for carbon nanotubes (CNTs) and its diverse applications. Although the main raw material for biochar production comes from plants, it has much better potential for the remediation of various pollutants, including inorganic and organic contaminants and heavy metals present in polluted water by using plants.

Biochar is a carbon-rich product derived from biomass raw materials, which includes agricultural residues and some others such as lignocellulosic substances, through pyrolysis or gasification processes under the controlled process of thermal decomposition (Singh et al. 2015). Biochar can also be made from solid wastes such as litter and animal residues by various thermochemical conversions of biomass. Biochar technology has evolved a major research area for researchers as it can reduce the risk of climate change, detoxify contaminants, and enhances plant growth and fertility of soil (Tripathi et al. 2011; Mohamed et al. 2018). In addition,

D. Mishra et al. (eds.), *Biochar-Based Nanocomposites for Contaminant Management*, Advances in Science, Technology & Innovation, https://doi.org/10.1007/978-3-031-28873-9_10

biochar production is very cost-effective, environmentally friendly, and waste efficient as it requires much less energy. Furthermore, biochar, biofuels, and *syn*-gas were also obtained as by-products from the thermochemical treatment of feedstock (Lehmann et al. 2006; Tan et al. 2015). Comparison to combustion, it is more economical for biomass use. However, the restricted functionalities limit the water treatment, depending on the feedstock, reaction conditions, and pyrolysis method (Tan et al. 2016a, b).

When raw biochar is used as an adsorbent, it causes difficulties in separating it from the aqueous solution due to its smaller size and limited adsorption efficiency (Chen et al. 2011; Yao et al. 2013). On the other hand, these limitations can be overcome with the advent of nano-biochar, which improves its physical, structural, and chemical properties. For example, increasing the surface area improves mechanical strength and thermal stability due to their nano-size and the presence of many functional groups. Furthermore, improved adsorption ability compared to their macrobiochar counterparts may be due to a smaller hydrodynamic radius, zeta potentials are negative, and a larger number of functional groups with oxygen and carbon defects that can generate reactive organic species (Ramanayaka et al. 2020a, b). Physicochemical properties are strongly influenced by the defect of carbon present in carbon nanomaterials, which are mainly due to topological or edge defects. The edges of the material are mainly filled with a larger number of unpaired electrons, which can reduce the formation energy and increase the rate of electron transfer. In addition, carbon defects have a significant role in the catalytic activity in the oxygen reduction process (Greco et al. 2019). Conventional fertilizers have high productivity, which eventually leads to significant economic and agronomic losses due to problems such as leaching (DeSA 2013; Ditta and Arshad 2016). In order to overcome these problems, there is a need to prepare eco-friendly and inexpensive materials such as uses of fertilizers, which will enhance nutrient uptake to plants, enhance the crop yield, reduce nutrient loss, and has less to no adverse effect on soil agriculture and the economy (Liu and Lal 2015). Bearing this in mind, various types of nano-fertilizers are made, e.g., biochar-based, laminated with a polymer, or encapsulated materials and zeolite-based.

Biochar-based and zeolitic materials are most preferred as they are environmentally friendly and derived from natural sources (Lateef et al. 2019). Nano-biochar demonstrates the combined benefits of nanotechnology with biochar technology (Chausali et al. 2021). Engineered biochar is mainly derived from unmodified biochar, and modification is done by physical, chemical, and biological means for increasing the specific surface area, number of pore structures, mobility of cation, surface functional group, pH, adsorption capacity, and other properties (Wang et al. 2017). Biochar-based nanocomposite improves plant growth by accumulating nutrients over a longer period of time, improves soil fertility, and mitigates environmental risks such as global warming and land contamination caused by nutrient leaching and volatilization from overuse of fertilizers (Lateef et al. 2019). Nano-biochar positively affects soil properties such as soil fertility and water storage capacity, which directly increases crop production, which is mainly influenced by various human activities such as industrialization, overgrazing, and deforestation (Chausali et al. 2021). Some observations such as slow-release column result, water absorption efficiency, salinity index, and retention studies showed that biochar-based nanocomposite has significant potential to be used as nano-fertilizer and that it is an eco-friendly, economical, and viable material for sustainable agriculture (Lateef et al. 2019). In this chapter, a brief insight has been given on methodologies for the production of engineered biochar-nanocomposites, a comparative account on pristine biochar and biochar-nanocomposites, and the application of biochar-nanocomposites in the agriculture sector.

2 Methodologies for the Production of Engineered Biochar-Nanocomposites

Biochar is produced from various types of feedstocks, such as grasses, manure, agricultural waste, municipal solid waste, and wood, through various processes, including carbonization or slow and fast pyrolysis, gasification, and flash carbonization (Gaunt and Lehmann 2008; Singh et al. 2015; Ramanayaka et al. 2020a). Slow pyrolysis is the process performed without oxygen, resulting in increased biochar yield and production of bio-oil in small quantities (Ramanayaka et al. 2020a). As the pyrolysis temperature increases, it enhances the particle size, which leads to the formation of larger particles, and it also converts the disordered low-density carbon to a nano-size particle easily by formation of dense fractal mass structures (Zhou et al. 2017; Nath et al. 2019; García-Sánchez et al. 2020). The amount of biochar produced from wood is higher in comparison with the amount of biochar produced from municipal solid waste (Ashiq et al. 2019).

Biochar can be converted into nano-biochar by various physical and chemical processes depicted in Fig. 1, which includes:

(i) Mechanical grinding: It generates nanosized particles. The higher content of carbon, hydrogen, and nitrogen was observed in coconut nano-biochar prepared by ultrafine grinding as compared to the biochar (Yang et al. 2020).
(ii) Flash heating technique: It results in graphitic nano-sheets.

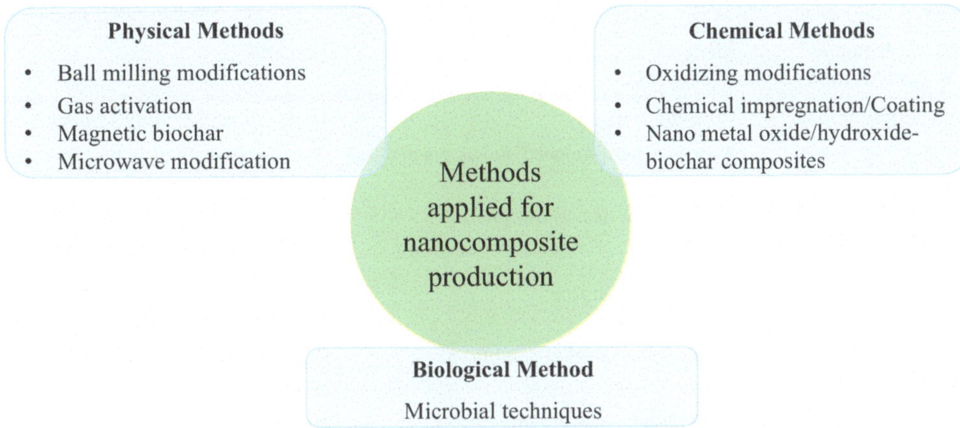

Fig. 1 Methods for the production of nanocomposites

(iii) Ultrasonic vibrator: It is used to physically disperse the biochar and enhance its adsorption capacity.
(iv) Sonication: It is used to produce nano-size particles (Chausali et al. 2021). It peels off the carbon structure and opens the clogged pores by shockwaves; this leads to improved adsorption capacity as the minerals are easily leached into the pores (Sajjadi et al. 2020), and properties of the nano-biochar can be changed with the difference in the amplitude, probe size, temperature, and power. The ash content production is less by mineral separation and increased content of oxygen by the formation of functional groups (Liu et al. 2018).
(v) Carbonization: It is used in combination with after or before treatment (ball milling or chemical treatment) to improve the size and the surface property of nano-biochar; the reason behind the formation of nano-biochar is the pores collapsing and thickening of the carbon matrix (Khare 2021).
(vi) Centrifugation: In this process, the gravitational force, buoyancy force, density gradient of molecules, and interactions between the analyte help in the separation at the time ranging from (2–30 min) and speed (3500–1000 rpm) (Xu and Cölfen 2021).

The method of synthesizing nanoparticles is important to the nature of the nanoparticle since the method mainly depends on the surface area. Disk milling shows the minimal range of nano-biochar composite values compared to the available literature (5–370 m^2/g); this could be because of the presence of nanopores and graphitic nature (Naghdi et al. 2017, 2019a; Lyu et al. 2018; Weber and Quicker 2018; Ramanayaka et al. 2020a).

Ball milling, double-disk milling, and vibration disk milling are some techniques where vibratory disk milling obtains an enormous amount of nano-biochar with more uniformity of size and shape compared with other techniques used. The ball mill process and double-disk milling have not been primarily used because of their higher operating cost. The vibrating disk milling method shows high results by shear and abrasion stress compared to ball milling (Bayram and Öner 2007; Karinkanta et al. 2018; Ramanayaka et al. 2020a; Chausali et al. 2021). Hydrothermal reaction is done using agricultural residues as feedstock, and by-products are used to produce nano-biochar (Chausali et al. 2021). The ball milling process is the most effective technique to produce nano-biochar (Fan et al. 2016). It includes the speed of rotation, the time properly set, and the sphere-to-force mass ratio, which affects the particle size and their surface energy (Shah et al. 2015). This process resulted in the production of various types of carbon nanoparticles, like nanotubes and nanofibers at fix rotational speed and at different times. The formation of the thin graphite plate lies in the time of the process and can also be destroyed if the process takes too long to complete (Awasthi et al. 2002). A process that takes a long time leads to high levels of iron contamination (Naghdi et al. 2017). When the milling time increases from 120 to 240 min it, leads to a minimization of the particle size from 315 to 260 nm, and increasing the milling time has also been reported to increase the particle size of nano-biochar (Wang et al. 2018b). Ball milling avoids the use of harmful chemicals, has high atom economy, zero waste generation, and uses the biodegradable material and renewable sources (Naghdi et al. 2015, 2017; Richard et al. 2016; Wang et al. 2018b; Kumar et al. 2020; Li et al. 2020a). The preparation can be done by two methods—(i) Wet method: It gives smaller particles with more functional groups and more dispersive nature, and the process requires less labor, and (ii) Dry method (Khare 2021).

During high-energy ball milling, some amount (up to 673 mg g^{-1}) of organic matter losses due to minerals which are not tightly bound to the carbon. It can be seen when the collision occurs between metallic balls and biochar for reducing the size of the biomass and also generating a lot of energy resulting the temperature increases. This elevated temperature induces the collision of some amorphous

Table. 1 Advantages and disadvantages of various methods used for engineered biochar-based nanocomposite production

S. no	Method of production	Merits	Demerits
1	Ball milling	Cost-effective	Deprived applicability to large-scale production
		Nano-biochar produced shows greater pore properties	On a large scale, it is possibly an energy-intensive process
		No chemical or harmful process is involved	Less purities and homogeneity
			Environmental impacts and financial viability are not recorded
2	Sonication	Shorter reaction time	Disproportionate ultrasound irradiation can have a negative impact, such as environmental and health hazards issues
		Controllable reaction conditions	Because of its higher production cost, it is not cost-effective at a larger scale
		High purities and uniform shapes	High energy-consuming process
3	Carbonization	Nano-sized particles are produced	High temperatures are required
		Surface modification can be done quickly	Toxic gases are released
			Chemical treatment or ball milling processes are also required to get nano-size particles
4	Centrifugation	Easy to handle	Hazardous solvents are used
		Produced highly stable nano-biochar	Lower yield of nano-biochar is obtained
			Environmental impacts and financial viability are not recorded
5	Chemical treatments	Produced highly improved surface characteristics of nano-biochar	Expensive method
		Tailored-made nano-biochar can be prepared	Hazardous chemicals are used
			By treating with acid, corrosivity occurs in the wall of nano-biochar micropores
			Due to alkaline treatment, unevenness appears on the surface of nano-biochar
			Some metals can cause toxicity if they are used in higher amounts

carbon, which becomes volatile and starts to leave behind some inorganic components, which in turn leads to low carbon and ash content, which is higher in the nano-biochar (Wang et al. 2018b). Techniques, mainly Fourier transform infrared spectroscopy (FTIR), scanning electron microscopy (SEM), X-ray photoelectron spectroscopy, and Brunauer–Emmett–Teller (BET) surface area, were primarily used for characterization of phosphorus and nitrogen-enriched biochar-nanocomposites, which shows the graphitic and flaky nature of the nano-biochar particles with a surface area of 28 m^2/g and in the presence of various functional groups on its surface such as –OH, C=O, –NH, and CH_3 (Ramanayaka et al. 2020a). The chemical property of nano-biochar can be determined by analyzing pH, electrical conductivity (EC), oxidation–reduction potential, ash content, fixed carbon content, zeta potential, volatile matter, elemental analysis, PAH content and categorized by heavy metal content. The physical property can be categorized as surface area, particle size, and specific gravity. Hydraulic properties can be categorized as water holding capacity and moisture content (Naghdi et al. 2015). The nano-biochar produced using manure has a maximum ash content as compared to the plant originated waste. The nano-biochar can be produced by high or low temperatures and has differences in its bulk density, carbon content, and number of extractable cations. The carbon content of nano-biochar varies depending on the raw material, as does the ash content, e.g., coconut shells have a higher carbon content compared to that of nano-biochar derived from sewage (Wang et al. 2013). Merit and demerits of all the above methods are shown in Table 1.

2.1 Effect on Pore Size of Engineered Biochar-Nanocomposites Prepared from Different Processes

The pore size of nano-biochar produced by ball milling process has 7.4 times more which provides high pore volume (Khare 2021). The size of nano-biochar obtained from various biochar raw materials is as follows: (i) Ball milling methods: corn straw—100–600 nm, rice husk—100–600 nm, pinewood—60 nm (Ma et al. 2019); (ii) Sonication method: wheat straw—100 nm, pine needle—100 nm

(Naghdi et al. 2017; Lian et al. 2018); and (iii) Centrifugation—agricultural wastes, animal manures, herbaceous residues, woody biomass, and sewage sludge—100 nm (Song et al. 2019). The average size of nano-biochar was around 70–80 nm by waste product of gasification (Goswami et al. 2020; Khare 2021).

3 Comparison Between Biochar and Engineered Biochar-Based Nanocomposites

Besides improving soil health, biochar has many uses (wastewater treatment, photocatalyst, and enzyme immobilization) due to its pore properties and surface functionality (Singh et al. 2021; Zhao et al. 2021). Biochar is mainly used as an adsorbent to detoxify organic and inorganic pollutants from water, like as toxic heavy metals, antibiotics, and dyes (Ahmad et al. 2014; Rajapaksha et al. 2014; Tan et al. 2015; Inyang et al. 2016; Singh et al. 2022). However, the multifunctionality of biochar applications other than agricultural uses has been limited due to limited surface functionalities (Ahmad et al. 2014; Tan et al. 2016a; Ramanayaka et al. 2020b). Biochar also enhances soil fertility and crop yield, increases nutrient levels and water retention capacity, and minimizes emissions of other greenhouse gases. Nanotechnology is seen as a tempting approach to improve the material's surface properties such as surface area, porosity, and surface functional groups. Therefore, the nanoscaling of materials has been tried in many areas, such as the production of nano-fertilizers, nano-pesticides, and renewable energy sources (Wang and Nguyen 2018; Pascoli et al. 2019). Likewise, various processes have been carried out to convert biochar into engineered biochar-based nanocomposite with desirable properties and a wide range of applications in environmental, agricultural, and other fields. A significant disadvantage of these processes was the use of toxic chemicals, high cost, and harsh conditions (Jiang et al. 2020). At the same time, biochar-based metal oxide-based composites (SiO_2, ZnO, MnO, MgO, CaO, and AlOOH nanoparticles) were also fabricated for enhanced adsorption of organic and inorganic toxins (Tu et al. 2020). However, the main disadvantage of these composites was the low surface area due to the filling of micropores with metal oxide particles (Michálková-Richveisová et al. 2017).

In general, pure biochar has a low adsorption capacity than modified biochar to remove the pollutants present in aqueous solutions, particularly those present in higher concentrations in wastewater. It is due to its low density and small particle size; pristine biochar cannot be easily separated from water, which severely confines its application (Tan et al. 2016b). The properties of engineered biochar-based nanocomposite mainly depend on the pyrolysis process, raw material, and modification methods (Sun et al. 2014; Wang et al. 2015d). Various feedstocks from forest areas and crop residues have been used to produce engineered biochar-based nanocomposites. The adsorption properties of engineered biochar-based nanocomposites are primarily determined by the pore structure properties, surface functional groups, and specific surface area; biochar-based nanocomposites are rich in these properties compared to the ordinary biochar (Tan et al. 2016b). Consequently, the research and development of various engineered biochar-based nanocomposites are done according to the need, which can enhance their application in the environment and agriculture, which has become the current objective of nowadays research. Properties of raw materials, manufacturing methods, and process parameters can alter the adsorption efficiency of biochar-based nanocomposites (Tripathi et al. 2016).

4 Application of Engineered Biochar-Based Nanocomposites in Agriculture

Nano-biochar helps to provide a better solid substrate for enzyme immobilization due to their comparatively larger specific surface area, microporosity, and hydrophobicity, along with the presence of higher functional groups attached to the surface. The particles of nano-biochar depict the wider range of absorbency to the pollutants like polycyclic aromatic hydrocarbons (PAHs), polychlorinated biphenyls (PCBs), herbicides, and heavy metals present in the environment. They also help in counteracting soil acidity (Li et al. 2020b) because of their higher alkaline nature, as shown in Table 2. Nanocomposites allow enhancement in pore properties and functional groups which in turn improves destructive catalytic ability and active sites. The separation also proved to be very easy and quick (Chausali et al. 2021). The size of nano-biochar is extremely smaller, resulting in an increased surface area that shows enhanced potential for environmental remediation and improved soil quality and crop production (Rajput et al. 2022) and other applications which are shown in Fig. 2 and Table 3. Nano-biochar increases the germination of seed and plant biomass by enhancing microbial diversity as well as its availability to plants. Several groups of bacteria, such as Actinobacteria and Bacteroidetes, are boosted through the application of nano-biochar, which are reported to immediately play a beneficial role in remediating contaminated soil (Chausali et al. 2021). This makes it a valuable and inexpensive material in terms of environmental remediation along with sustainable agriculture (Khodakovskaya et al. 2009; Liu et al. 2020; Rajput et al. 2022). Nano-biochar shows tremendous implementations to environmental issues such as phytoremediation, treatment of wastewater,

Table 2 Engineered biochar-based nanocomposites produced from different feedstock and their applications in pollutant removal

Feedstock	Composite type	Pyrolysis temperature (°C)	Contaminants removed	Removal/adsorption capacity	References
Artichoke leaves	Nano-biochar coated with sodium hydroxide	–	Metformin hydrochloride (MFH)	Removal of MFH drug (10 mg/L)—tap water (87.0%), sewage (97.0%), seawater (92.0%)	Mahmoud et al. (2020)
Bagasse	Zinc nitrate-pretreated	450	Cr (VI)	102.66 mg g^{-1}	Gan et al. (2015)
	Montmorillonite-implanted	600	Methylene blue	11.26 mg g^{-1}	Yao et al. (2014)
Bamboo	Modified using Chitosan	600	Pb (II)	14.3 mg g^{-1}	Zhou et al. (2013)
Coconut husk	KOH	700	Removal of methylene blue	434.78 mg g^{-1}	Tan et al. (2008)
Corn straw	MBC (Manganese oxide-modified biochar)	–	As	14.36 mg/g	Yu et al. (2015)
	$KMnO_4$ modification	600	Cu (II)	160.3 mg g^{-1}	Song et al. (2014)
	Iron (Fe)-impregnated biochar (FBC)	–	As (V)	Pristine biochar (0.017 mg/g), FBC (6.80 mg/g), (400-fold increase)	He et al. (2018)
Hickory wood	$KMnO_4$-treated	600	Cd (II)	28.104 mg g^{-1}	Wang et al. (2015a)
Maize straw	Ce-MSB (Maize straw biochar functionalized with nanocerium oxide)	–	PO_4^{3-}	78 mg g^{-1}	Feng et al. (2017)
Paper mill sludge	Ni–ZVI magnetic adsorbents	700	Pentachlorophenol	50 mg g^{-1}	Devi and Saroha (2015)
Pinewood	Manganese oxide–modified biochar (MBC) —$MnCl_2 \cdot 4H_2O$ (MPB), birnessite (BPB)	600	As (V) and Pb (II)	Pristine biochar (0.20 and 2.35 g/kg), MPB (0.59 and 4.91 g/kg), BPB (0.91 and 47.05 g/kg)	Wang et al. (2015c)
	Nano-biochar derived from Pinewood	–	Carbamazepine (CBZ)	95%	Naghdi et al. (2019b)
	Biochar composite/manganese oxide (BC/Mn)	600	Pb (II)	Removal efficiency of lead (II) increased from 6.4 to 98.9%	Wang et al. (2015b)
	Modified biochar with Hematite	–	As (V)	PB (Pinewood biochar) (265 mg/kg), Hematite modified (429 mg/kg)	Wang et al. (2015e)
	Birnessite-modified	600	As (V)	0.91 mg g^{-1}	Wang et al. (2015c)
Rice straw	HAP-BC (Hydroxyapatite–biochar-nanocomposite)	–	Pb (II), Cu (II) and Zn (II)	Pb (II), Cu (II), and Zn (II)–BC 561.80, 826.45, and 925.93 mg/g; HAP-BC 1000, 833.33, and 935.93 mg/g	Wang et al. (2018c)
	Clay composites/functionalized biochar	–	As and Cd	Attapulgite/biochar 1 (treated with 2.5 g of $ZnCl_2$), Attapulgite/biochar 2 (treated with 5 g of $ZnCl_2$)—79 and 82% reduction (As), and the 36 and 44% (Cd)	Wang et al. (2019)
	Rice straw nano-biochar	–	Genes with antibiotic resistance *amp* C, *erm* B	Adsorption of eDNA-Nano 700 (60%), Nano 400 (31.3%)	Lian et al. (2020)
	Calcium-based magnetic biochar	–	As and Cd	CEC was increased in soil by—7.1–11.5 cmolc/kg and pH was increased 0.3–0.6 units to 0.3–0.9 units (at 32nd day)	Wu et al. (2020)
Rice husk	Iron oxide amended	550	As (V)	1.15 mg g^{-1}	Cope et al. (2014)
	H_2SO_4	450–550	Tetracycline	58.82 mg g^{-1}	Liu et al. (2012)
	KOH	450–550	Tetracycline	23.26 mg g^{-1}	Liu et al. (2012)
	Pristine	450–550	Tetracycline	16.95 mg g^{-1}	Liu et al. (2012)
	Methanol modified	450–550	Tetracycline	18.53 mg g^{-1}	Jing et al. (2014)

(continued)

Table 2 (continued)

Feedstock	Composite type	Pyrolysis temperature (°C)	Contaminants removed	Removal/adsorption capacity	References
Sugarcane	MgO decorated magnetic	550	Phosphate	121.25 mg g^{-1}	Li et al. (2016b)
Safflower seed	$ZnCl_2$	900	Methylene blue	128.21 mg g^{-1}	Angın et al. (2013)
Wheat straw	Magnetic nano-biochar	–	Tetracycline and mercury	268.3 and 127.4 mg/g	Li et al. (2020a)
	Wheat straw nano-biochar	–	Cd	Untreated soil groups possess higher cd content which was lowered after the treatment of soil groups ($P < 0.05$)	Liu et al. (2020)
	$FeCl_3{\cdot}6H_2O$	450	Phosphate	16.58 mg g^{-1}	Li et al. (2016a)
	Graphene pretreated	600	Phenanthrene	25.3 mg g^{-1}	Tang et al. (2015)

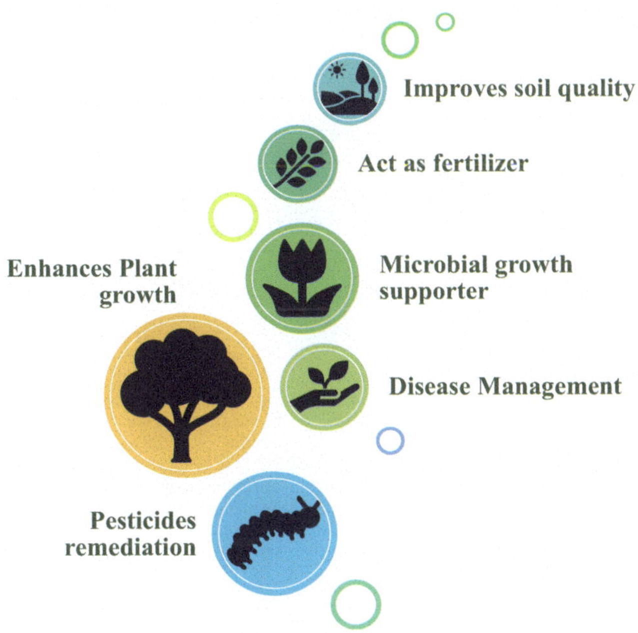

Fig. 2 Application of biochar-based nanocomposites in agriculture

management of waste, carbon sequestration, and energy production (Fig. 3).

Biochar-derived nanocarbons have attracted a great deal of interest by virtue of possessing the ability to hold water and nutrients; therefore, they can be easily used as nano-fertilizers (Chausali et al. 2021; Mukherjee et al. 2022; Sarma et al. 2022). Nano-scale biochar fertilizer helps to reduce and manage the waste by turning them to a value-added product, simultaneously preventing the pollution. It also prevents the leaching problems of traditional fertilizers and supplies nutrients to plants, and thus, improving the overall yield of crop (Lateef et al. 2019). The different types of feedstocks, surface functionalities, and pore structures characterize the different application of nano-biochar. For example, municipal solid waste gives a less amount of nano-biochar and a higher amount of carbonate sulfate and aluminosilicate found in it, which leads to the co-precipitation of heavy metals (Song et al. 2019). The plant-derived nano-biochar with large aromatic cluster sizes and more organic-containing surface functional groups shows greater affinity to organic contaminants and causes the elimination of contamination by binding (Khare 2021). Nano-biochar increases the water retention capacity and improves the capillary force of the soil (Tan et al. 2016a). Soil enriched with nano-biochar (10% w/w) was found to be effective in reducing the rate of evaporation as well as cracking propagation (Khare 2021). This cracking destruction in nano-biochar is due to their porous structure, which provides them with a higher water retention capacity (Bordoloi et al. 2018a, b). Lignin-derived nano-biochar is implemented as a renewable reinforcing filler for styrene-butadiene rubber (Jiang et al. 2020). Biochar does not only have higher carbon content depending upon the raw material; it contains larger concentrations of macronutrients such as nitrogen (N), phosphorous (P), potassium (K), sulfur (S), magnesium (Mg), and calcium (Ca), as well as micronutrients like iron (Fe), copper (Cu), zinc (Zn), manganese (Mn), and ash, making it an efficient organic fertilizer (Hossain et al. 2011). Whereas, when it is produced from animal sources like sewage sludge and broiler litter, it provides comparatively higher concentrations of P and N than the one produced from green wastes and woods, originating from plant biomass. Biochar, when gets combined with other fertilizers like urea, depicts a higher rate of production, but the nitrogen input was found to be decreased sufficiently (Qian et al. 2014). Added minerals and organic matters to biochar are also evaluated to be helpful in providing other fertilizer-related benefits to it. Nanocarbons derived from biochar have gained more attention as nano-fertilizers because of their water and nutrient storage capacity (Tiwari et al. 2014). The implementation of biochar carbon

Table 3 Engineered biochar-based nanocomposite as fertilizer for enhancement of plant growth

Nano-biochar fertilizers	Nutrient	Plants	Growth enhancements	References
Biochar-derived nanocarbon fertilizers (BDNC)	–	Tomato, Rice, Gram	Increases yield of crop	Lin et al. (2009, Sonkar et al. (2012), Khodakovskaya et al. (2013)
		Tobacco, Gram	Growth stimulation	Tripathi et al. (2011), Khodakovskaya et al. (2012), Giraldo et al. (2014)
		Tomato	Seed germination	Alimohammadi et al. (2011), Khodakovskaya et al. (2011)
		Tomato, Tobacco	Protein expressions	Khodakovskaya et al. (2011, 2012)
		Wheat	Root elongation	Wang et al. (2012)
		Gram	Enhanced water uptake	Tripathi et al. (2011), Sonkar et al. (2012)
Calcite, $CaCO_3$	Macro-nutrient Ca	Peanut seedlings	Mixing 1 g L^{-1} of Ca–NPs and humic acid increases the yield	Xiumei et al. (2005)
CuO and Cu_2O	Micro-nutrient Cu	*Elodea densa* planch	At all concentrations $CuSO_4$ act as inhibitor	Nekrasova et al. (2011)
Metallic Cu		Lettuce seeds	Major effect can be found when Cu–NPs added to the soil before 15-day of seed showing	Shah and Belozerova (2009)
Metallic Mn	Micro-nutrient Mn	Mung bean plant	$MnSO_4$ showed phytotoxic, but no phytotoxic were observed in Mn–NPs	Pradhan et al. (2013)
Mg–NPs	Macro-nutrient Mg	Black-eyed pea	Mixing 500 mg L^{-1} Mg–NPs and Fe results in maximum yield	Delfani et al. (2014)
Modified biochar with MgO	–	Rice	In saline-alkaline soil, phosphate retention and yields were increased	Wu et al. (2019)
Magnetic biochar decorated with MgO		Ryegrass	Enhances plant growth and phosphate sorption	Li et al. (2016b)
Mo	Micro-nutrient Mo	Chickpea	Microbial treatment further enhanced the effects	Taran et al. (2014)
Multi-wall carbon nanotubes (MWCNTs)	–	*Brassica napus* L., *Helianthus annus* L., *Cannabis sativa* L	Root is affected by the salts of Cd and Pb and increases shoot growth	Oloumi et al. (2018)
Metallic Zn	Micro-nutrient Zn	Ryegrass seed	Optimum phytotoxicity was observed when Zn–NPs were at a higher concentration	Lin and Xing (2007)
Nanocarbon as water holding fertilizer	–	Mustard tubers	Increases yield, quality and fertilizer uptake efficiency	Wang et al. (2018a)
Wheat straw BCF	–	Rice	Increase yield, pH, and nutrient content (N (40%), P (46%)), plant biomass (67%)	Chew et al. (2020)
Water-soluble carbon nanoparticles CNPs (ws-CNPs)		Wheat	Better seedling growth and shoot and root length increased three times with 50 mg $L^{\text{"}-1}$ treatment	Saxena et al. (2014)
Liquid nanocarbon		Chinese cabbage	Enhances growth and yield	Oloumi et al. (2018)
ZnO	Micro-nutrient Zn	Mung bean and chickpea seedlings	Optimum phytotoxicity was observed when Zn–NPs were at a higher concentration	Mahajan et al. (2011)
		Rape seeds	Optimum phytotoxicity was observed when Zn–NPs were at a higher concentration	Lin and Xing (2007)
		Cucumber	No negative effects found	Zhao et al. (2013)

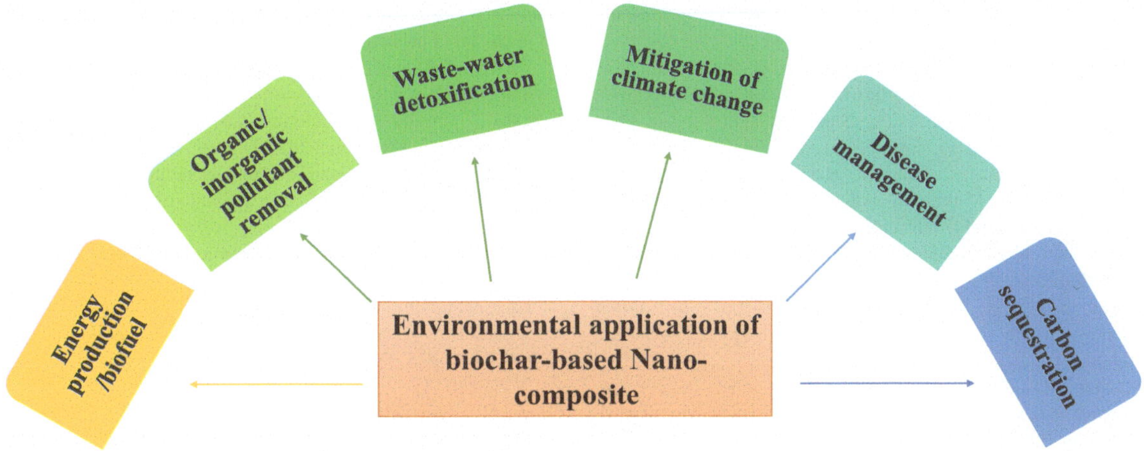

Fig. 3 Environmental application of biochar-based nanocomposites

nanoparticles is significantly important to promote the development and productivity of the wheat crop. However, the effectiveness of biochar-based fertilizers to simultaneously immobilize heavy metals and release nitrogen is required to be further evaluated (Ramadan and Abd-Elsalam 2020).

5 Conclusion and Future Prospective

Applying biochar-based nanocomposites can be a good option in smart agriculture according to need. They can also be used for remediation purposes of different toxicants present in soil and water. These new types of applications create a better basis for the technical use of engineered biochar-based nanocomposites in the future. Although a lot of research has been done for the modification of biochar into nanocomposites, major part of the research work is limited to theoretical labs and field phases. Few large-scale applications of engineered biochar-based nanocomposites and its performance are reported. Large-scale industrial production of biochar-based nanocomposites and their specific application need to be identified, and more technical support needs to be provided. Reducing the modification cost of biochar-based nanocomposites can play a crucial role in the promotion and application of engineered biochar-based nanocomposites. Modification using chemicals has greater utility compared to physical modification methods due to its larger surface area, the presence of porous structure, and large number of oxygen-containing functional groups present on its surface, which can be easily modified using the chemical modification method instead of any other method. Green approaches such as using organic acids such as citric, acetic, and tartaric acids have been used as modifiers for biochar. The effect of engineered biochar-based nanocomposites on ecosystem systems such as soil biota and aquatic ecosystem is not well explored, therefore needs further research. Because of the very low density and nano-size of biochar composites, they may enter into water bodies with the adsorbed organic or inorganic toxins, potentially threatening the whole ecosystem. Therefore, further investigations are needed on the collection, reclamation, and regeneration of spent biochar-based composites under different conditions.

Future Prospective

- The use of biochar-based nanocomposites needs to be improved for agricultural purposes.
- The modification needs to be inexpensive, and green chemical reagents are to be used for modification in biochar-based nanocomposites.
- The major risk analysis of biochar for the environment and ecosystems must be explored prior to large-scale production of biochar-based nanocomposites.
- Application of biochar to sequester carbon present in ecosystem, without proper upfront planning and management, can be detrimental to ecosystem health.
- The reuse and recovery of engineered biochar-based nanocomposites and their effect on the environment and human health will be the main aim of upcoming research.

References

Ahmad M, Rajapaksha AU, Lim JE, Zhang M, Bolan N, Mohan D, Vithanage M, Lee SS, Ok YS (2014) Biochar as a sorbent for contaminant management in soil and water: a review. Chemosphere 99:19–33. https://doi.org/10.1016/j.chemosphere.2013.10.071

Alimohammadi M, Xu Y, Wang D, Biris AS, Khodakovskaya MV (2011) Physiological responses induced in tomato plants by a two-component nanostructural system composed of carbon nanotubes conjugated with quantum dots and its in vivo multimodal detection. Nano-Technology 22:295101

Angın D, Altintig E, Köse TE (2013) Influence of process parameters on the surface and chemical properties of activated carbon obtained from biochar by chemical activation. Bioresour Technol 148:542–549. https://doi.org/10.1016/j.biortech.2013.08.164

Ashiq A, Adassooriya NM, Sarkar B, Rajapaksha AU, Ok YS, Vithanage M (2019) Municipal solid waste biochar-bentonite composite for the removal of antibiotic ciprofloxacin from aqueous media. J Environ Manage 236:428–435. https://doi.org/10.1016/j.jenvman.2019.02.006

Awasthi K, Kamalakaran R, Singh A, Srivastava O (2002) Ball-milled carbon and hydrogen storage. Int J Hydrog Energy 27:425–432. https://doi.org/10.1016/S0360-3199(01)00134-3

Bayram M, Öner MD (2007) Bulgur milling using roller, double disc and vertical disc mills. J Food Eng 79:181–187. https://doi.org/10.1016/j.jfoodeng.2006.01.042

Chausali N, Saxena J, Prasad R (2021) Nanobiochar and biochar based nano-composites: advances and applications. J Agric Res 5:100191. https://doi.org/10.1016/j.jafr.2021.100191

Chen B, Chen Z, Lv S (2011) A novel magnetic biochar efficiently sorbs organicpollutants and phosphate. Bioresour Technol 102:716–723. https://doi.org/10.1016/j.biortech.2010.08.067

Chew J, Zhu L, Nielsen S, Graber E, Mitchell DR, Horvat J, Mohammed M, Liu M, van Zwieten L, Donne S (2020) Biochar-based fertilizer: supercharging root membrane potential and biomass yield of rice. Sci Total Environ 713:136431. https://doi.org/10.1016/j.scitotenv.2019.136431

Cope CO, Webster DS, Sabatini DA (2014) Arsenate adsorption onto iron oxide amended rice husk char. Sci Total Environ 488:554–561. https://doi.org/10.1016/j.scitotenv.2013.12.120

Delfani M, Baradarn Firouzabadi M, Farrokhi N, Makarian H (2014) Some physiological responses of black-eyed pea to iron and magnesium nanofertilizers. Commun Soil Sci Plant Anal 45:530–540. https://doi.org/10.1080/00103624.2013.863911

DeSA U (2013) World population prospects: the 2012 revision population division of the department of economic and social affairs of the United Nations Secretariat, New York 18

Devi P, Saroha AK (2015) Simultaneous adsorption and dechlorination of pentachlorophenol from effluent by Ni–ZVI magnetic biochar composites synthesized from paper mill sludge. Chem Eng J 271:195–203. https://doi.org/10.1016/j.cej.2015.02.087

Ditta A, Arshad M (2016) Applications and perspectives of using nanomaterials for sustainable plant nutrition. Nanotechnol Rev 5:209–229. https://doi.org/10.1515/ntrev-2015-0060

Fan X, Chang DW, Chen X, Baek J-B, Dai L (2016) Functionalized graphene nanoplatelets from ball milling for energy applications. Curr Opin Chem Eng 11:52–58. https://doi.org/10.1016/j.coche.2016.01.003

Feng Y, Lu H, Liu Y, Xue L, Dionysiou DD, Yang L, Xing B (2017) Nano-cerium oxide functionalized biochar for phosphate retention: preparation, optimization and rice paddy application. Chemosphere 185:816–825. https://doi.org/10.1016/j.chemosphere.2017.07.107

Gan C, Liu Y, Tan X, Wang S, Zeng G, Zheng B, Li T, Jiang Z, Liu W (2015) Effect of porous zinc–biochar nano-composites on Cr (VI) adsorption from aqueous solution. RSC Adv 5:35107–35115. https://doi.org/10.1039/C5RA04416B

García-Sánchez F, Simón-Grao S, Martínez-Nicolás JJ, Alfosea-Simón M, Liu C, Chatzissavvidis C, Pérez-Pérez JG, Cámara-Zapata JM (2020) Multiple stresses occurring with boron toxicity and deficiency in plants. J Hazard Mater 397:122713. https://doi.org/10.1016/j.jhazmat.2020.122713

Gaunt JL, Lehmann J (2008) Energy balance and emissions associated with biochar sequestration and pyrolysis bioenergy production. Environ Sci Technol 42:4152–4158. https://doi.org/10.1021/es071361i

Giraldo JP, Landry MP, Faltermeier SM, McNicholas TP, Iverson NM, Boghossian AA, Reuel NF, Hilmer AJ, Sen F, Brew JA (2014) Plant nanobionics approach to augment photosynthesis and biochemical sensing. Nat Mater 13:400–408. https://doi.org/10.1038/nmat3890

Goswami L, Pakshirajan K, Pugazhenthi G (2020) Biological treatment of biomass gasification wastewater using hydrocarbonoclastic bacterium Rhodococcus opacus in an up-flow packed bed bioreactor with a novel waste-derived nano-biochar based bio-support material. J Clean Prod 256:120253. https://doi.org/10.1016/j.jclepro.2020.120253

Greco C, Cosentino U, Pitea D, Moro G, Santangelo S, Patanè S, D'Arienzo M, Fiore M, Morazzoni F, Ruffo R (2019) Role of the carbon defects in the catalytic oxygen reduction by graphite nano-particles: a spectromagnetic, electrochemical and computational integrated approach. Phys Chem Phys 21:6021–6032. https://doi.org/10.1039/C8CP07023G

He R, Peng Z, Lyu H, Huang H, Nan Q, Tang J (2018) Synthesis and characterization of an iron-impregnated biochar for aqueous arsenic removal. Sci Total Environ 612:1177–1186. https://doi.org/10.1016/j.scitotenv.2017.09.016

Hossain MK, Strezov V, Chan KY, Ziolkowski A, Nelson PF (2011) Influence of pyrolysis temperature on production and nutrient properties of wastewater sludge biochar. J Environ Manage 92:223–228. https://doi.org/10.1016/j.jenvman.2010.09.008

Inyang MI, Gao B, Yao Y, Xue Y, Zimmerman A, Mosa A, Pullammanappallil P, Ok YS, Cao X (2016) A review of biochar as a low-cost adsorbent for aqueous heavy metal removal. Crit Rev Environ Sci Technol 46:406–433. https://doi.org/10.1080/10643389.2015.1096880

Jiang C, Bo J, Xiao X, Zhang S, Wang Z, Yan G, Wu Y, Wong C, He H (2020) Converting waste lignin into nano-biochar as a renewable substitute of carbon black for reinforcing styrene-butadiene rubber. J Waste Manag 102:732–742. https://doi.org/10.1016/j.wasman.2019.11.019

Jing X-R, Wang Y-Y, Liu W-J, Wang Y-K, Jiang H (2014) Enhanced adsorption performance of tetracycline in aqueous solutions by methanol-modified biochar. Chem Eng J 248:168–174. https://doi.org/10.1016/j.cej.2014.03.006

Karinkanta P, Ämmälä A, Illikainen M, Niinimäki J (2018) Fine grinding of wood–overview from wood breakage to applications. Biomass Bioenergy 113:31–44. https://doi.org/10.1016/j.biombioe.2018.03.007

Khare P (2021) A comprehensive evaluation of inherent properties and applications of nano-biochar prepared from different methods and feedstocks. J Clean Prod 320:128759. https://doi.org/10.1016/j.jclepro.2021.128759

Khodakovskaya M, Dervishi E, Mahmood M, Xu Y, Li Z, Watanabe F, Biris AS (2009) Carbon nanotubes are able to penetrate plant seed coat and dramatically affect seed germination and plant growth. ACS Nano 3:3221–3227. https://doi.org/10.1021/nn900887m

Khodakovskaya MV, De Silva K, Biris AS, Dervishi E, Villagarcia H (2012) Carbon nanotubes induce growth enhancement of tobacco cells. ACS Nano 6:2128–2135. https://doi.org/10.1021/nn204643g

Khodakovskaya MV, de Silva K, Nedosekin DA, Dervishi E, Biris AS, Shashkov EV, Galanzha EI, Zharov VP (2011) Complex genetic, photothermal, and photoacoustic analysis of nanoparticle-plant interactions. Proc Natl Acad Sci 108:1028–1033. https://doi.org/10.1073/pnas.1008856108

Khodakovskaya MV, Kim BS, Kim JN, Alimohammadi M, Dervishi E, Mustafa T, Cernigla CE (2013) Carbon nanotubes as plant growth regulators: effects on tomato growth, reproductive system, and soil microbial community. Small 9:115–123. https://doi.org/10.1002/smll.201201225

Kumar M, Xiong X, Wan Z, Sun Y, Tsang DC, Gupta J, Gao B, Cao X, Tang J, Ok YS (2020) Ball milling as a mechanochemical technology for fabrication of novel biochar nanomaterials. Bioresour Technol 312:123613. https://doi.org/10.1016/j.biortech.2020.123613

Lateef A, Nazir R, Jamil N, Alam S, Shah R, Khan MN, Saleem M (2019) Synthesis and characterization of environmental friendly corncob biochar based nano-composite–a potential slow release nano-fertilizer for sustainable agriculture. Environ Nanotechnol Monit Manag 11:100212. https://doi.org/10.1016/j.enmm.2019.100212

Lehmann J, Gaunt J, Rondon M (2006) Bio-char sequestration in terrestrial ecosystems–a review. Mitig Adapt Strat Glob Change 11:403–427. https://doi.org/10.1007/s11027-005-9006-5

Li J-h, Lv G-h, Bai W-b, Liu Q, Zhang Y-c, Song J-q (2016a) Modification and use of biochar from wheat straw (Triticum aestivum L.) for nitrate and phosphate removal from water. Desalin Water Treat 57:4681–4693. https://doi.org/10.1080/19443994.2014.994104

Li R, Wang JJ, Zhou B, Awasthi MK, Ali A, Zhang Z, Lahori AH, Mahar A (2016b) Recovery of phosphate from aqueous solution by magnesium oxide decorated magnetic biochar and its potential as phosphate-based fertilizer substitute. Bioresour Technol 215:209–214. https://doi.org/10.1016/j.biortech.2016.02.125

Li R, Zhang Y, Deng H, Zhang Z, Wang JJ, Shaheen SM, Xiao R, Rinklebe J, Xi B, He X (2020a) Removing tetracycline and Hg (II) with ball-milled magnetic nanobiochar and its potential on polluted irrigation water reclamation. J Hazard Mater 384:121095. https://doi.org/10.1016/j.jhazmat.2019.121095

Li Y, Xing B, Ding Y, Han X, Wang S (2020b) A critical review of the production and advanced utilization of biochar via selective pyrolysis of lignocellulosic biomass. Bioresour Technol 312:123614. https://doi.org/10.1016/j.biortech.2020.123614

Lian F, Yu W, Wang Z, Xing B (2018) New insights into black carbon nanoparticle-induced dispersibility of goethite colloids and configuration-dependent sorption for phenanthrene. Environ Sci Technol 53:661–670. https://doi.org/10.1021/acs.est.8b05066

Lian F, Yu W, Zhou Q, Gu S, Wang Z, Xing B (2020) Size matters: nano-biochar triggers decomposition and transformation inhibition of antibiotic resistance genes in aqueous environments. Environ Sci Technol 54:8821–8829. https://doi.org/10.1021/acs.est.0c02227

Lin D, Xing B (2007) Phytotoxicity of nano-particles: inhibition of seed germination and root growth. Environ Pollut 150:243–250. https://doi.org/10.1016/j.envpol.2007.01.016

Lin S, Reppert J, Hu Q, Hudson JS, Reid ML, Ratnikova TA, Rao AM, Luo H, Ke PC (2009) Uptake, translocation, and transmission of carbon nanomaterials in rice plants. Small 5:1128–1132. https://doi.org/10.1002/smll.200801556

Liu G, Zheng H, Jiang Z, Zhao J, Wang Z, Pan B, Xing B (2018) Formation and physicochemical characteristics of nano biochar: insight into chemical and colloidal stability. Environ Sci Technol 52:10369–10379. https://doi.org/10.1021/acs.est.8b01481

Liu P, Liu W-J, Jiang H, Chen J-J, Li W-W, Yu H-Q (2012) Modification of bio-char derived from fast pyrolysis of biomass and its application in removal of tetracycline from aqueous solution. Bioresour Technol 121:235–240. https://doi.org/10.1016/j.biortech.2012.06.085

Liu R, Lal R (2015) Potentials of engineered nano-particles as fertilizers for increasing agronomic productions. Sci Total Environ 514:131–139. https://doi.org/10.1016/j.scitotenv.2015.01.104

Liu W, Li Y, Feng Y, Qiao J, Zhao H, Xie J, Fang Y, Shen S, Liang S (2020) The effectiveness of nanobiochar for reducing phytotoxicity and improving soil remediation in cadmium-contaminated soil. Sci Rep 10:1–10. https://doi.org/10.1038/s41598-020-57954-3

Lyu H, Gao B, He F, Zimmerman AR, Ding C, Huang H, Tang J (2018) Effects of ball milling on the physicochemical and sorptive properties of biochar: experimental observations and governing mechanisms. Environ Pollut 233:54–63. https://doi.org/10.1016/j.envpol.2017.10.037

Ma S, Jing F, Sohi SP, Chen J (2019) New insights into contrasting mechanisms for PAE adsorption on millimeter, micron-and nano-scale biochar. Environ Sci Pollut Res 26:18636–18650. https://doi.org/10.1007/s11356-019-05181-3

Mahajan P, Dhoke S, Khanna A (2011) Effect of nano-ZnO particle suspension on growth of mung (Vigna radiata) and gram (Cicer arietinum) seedlings using plant agar method. J Nanotechnol 2011. https://doi.org/10.1155/2011/696535

Mahmoud ME, El-Ghanam AM, Saad SR, Mohamed RHA (2020) Promoted removal of metformin hydrochloride anti-diabetic drug from water by fabricated and modified nanobiochar from artichoke leaves. Sustain Chem Pharm 18:100336. https://doi.org/10.1016/j.scp.2020.100336

Mohamed MA, Hashim AF, Alghuthaymi MA, Abd-Elsalam KA (2018) Nano-carbon: plant growth promotion andprotection, Nanobiotechnology applications in plant protection. Springer, pp. 155–188. https://doi.org/10.1007/978-3-319-91161-8_7

Micháleková-Richveisová B, Frišták V, Pipíška M, Ďuriška L, Moreno-Jimenez E, Soja G (2017) Iron-impregnated biochars as effective phosphate sorption materials. Environ Sci Pollut Res 24:463–475. https://doi.org/10.1007/s11356-016-7820-9

Mukherjee S, Sarkar B, Aralappanavar VK, Mukhopadhyay R, Basak BB, Srivastava P, Marchut-Mikołajczyk O, Bhatnagar A, Semple KT, Bolan N (2022) Biochar-microorganism interactions for organic pollutant remediation: Challenges and perspectives. Environ Pollut 119609. https://doi.org/10.1016/j.envpol.2022.119609

Naghdi M, Taheran M, Brar SK, Kermanshahi-Pour A, Verma M, Surampalli RY (2019a) Fabrication of nanobiocatalyst using encapsulated laccase onto chitosan-nanobiochar composite. Int J Biol Macromol 124:530–536. https://doi.org/10.1016/j.ijbiomac.2018.11.234

Naghdi M, Taheran M, Brar SK, Rouissi T, Verma M, Surampalli RY, Valero JR (2017) A green method for production of nanobiochar by ball milling-optimization and characterization. J Clean Prod 164:1394–1405. https://doi.org/10.1016/j.jclepro.2017.07.084

Naghdi M, Taheran M, Brar SK, Verma M, Surampalli RY, Valéro JR (2015) Green and energy-efficient methods for the production of metallic nano-particles. Beilstein J Nanotechnol 6:2354–2376. https://doi.org/10.3762/bjnano.6.243

Naghdi M, Taheran M, Pulicharla R, Rouissi T, Brar SK, Verma M, Surampalli RY (2019b) Pine-wood derived nanobiochar for removal of carbamazepine from aqueous media: adsorption behavior and influential parameters. Arab J Chem 12:5292–5301. https://doi.org/10.1016/j.arabjc.2016.12.025

Nath B, Chaliha C, Kalita E (2019) Iron oxide Permeated mesoporous rice-husk nanobiochar (IPMN) mediated removal of dissolved arsenic (As): chemometric modelling and adsorption dynamics. J Environ Manage 246:397–409. https://doi.org/10.1016/j.jenvman.2019.06.008

Nekrasova G, Ushakova O, Ermakov A, Uimin M, Byzov I (2011) Effects of copper (II) ions and copper oxide nano-particles on Elodea densa Planch. Russ J Ecol 42:458–463. https://doi.org/10.1134/S1067413611060117

Oloumi H, Mousavi EA, Nejad RM (2018) Multi-wall carbon nanotubes effects on plant seedlings growth and cadmium/lead uptake in vitro. Russ J Plant Physiol 65:260–268. https://doi.org/10.1134/S102144371802019X

Pascoli M, Jacques MT, Agarrayua DA, Avila DS, Lima R, Fraceto LF (2019) Neem oil based nanopesticide as an environmentally-friendly formulation for applications in sustainable agriculture: an ecotoxicological perspective. Sci Total Environ 677:57–67. https://doi.org/10.1016/j.scitotenv.2019.04.345

Pradhan S, Patra P, Das S, Chandra S, Mitra S, Dey KK, Akbar S, Palit P, Goswami A (2013) Photochemical modulation of biosafe manganese nano-particles on Vigna radiata: a detailed molecular, biochemical, and biophysical study. Environ Sci Technol 47:13122–13131. https://doi.org/10.1021/es402659t

Qian L, Chen L, Joseph S, Pan G, Li L, Zheng J, Zhang X, Zheng J, Yu X, Wang J (2014) Biochar compound fertilizer as an option to reach high productivity but low carbon intensity in rice agriculture of China. Carbon Manag 5:145–154. https://doi.org/10.1080/17583004.2014.912866

Rajapaksha AU, Vithanage M, Zhang M, Ahmad M, Mohan D, Chang SX, Ok YS (2014) Pyrolysis condition affected sulfamethazine sorption by tea waste biochars. Bioresour Technol 166:303–308. https://doi.org/10.1016/j.biortech.2014.05.029

Rajput VD, Minkina T, Ahmed B, Singh VK, Mandzhieva S, Sushkova S, Bauer T, Verma KK, Shan S, van Hullebusch ED (2022) Nano-biochar: a novel solution for sustainable agriculture and environmental remediation. Environ Res 210:112891. https://doi.org/10.1016/j.envres.2022.112891

Ramadan MM, Abd-Elsalam KA (2020) Micro/nano biochar for sustainable plant health: present status and future prospects. In: Carbon nanomaterials for agri-food and environmental applications, pp 323–357. https://doi.org/10.1016/B978-0-12-819786-8.00016-5

Ramanayaka S, Tsang DC, Hou D, Ok YS, Vithanage M (2020a) Green synthesis of graphitic nanobiochar for the removal of emerging contaminants in aqueous media. Sci Total Environ 706:135725. https://doi.org/10.1016/j.scitotenv.2019.135725

Ramanayaka S, Vithanage M, Alessi DS, Liu W-J, Jayasundera AC, Ok YS (2020b) Nanobiochar: production, properties, and multifunctional applications. Environ Sci Nano 7:3279–3302

Richard S, Rajadurai JS, Manikandan V (2016) Influence of particle size and particle loading on mechanical and dielectric properties of biochar particulate-reinforced polymer nano-composites. Int J Polym Anal Charact 21:462–477. https://doi.org/10.1080/1023666X.2016.1168602

Sajjadi B, Chen W-Y, Mattern DL, Hammer N, Dorris A (2020) Low-temperature acoustic-based activation of biochar for enhanced removal of heavy metals. J Water Process Eng 34:101166. https://doi.org/10.1016/j.jwpe.2020.101166

Sarma H, Narayan M, Peralta-Videa JR, Lam SS (2022) Exploring the significance of nanomaterials and organic amendments—prospect for phytoremediation of contaminated agroecosystem. Environ Pollut 308:119601. https://doi.org/10.1016/j.envpol.2022.119601

Saxena M, Maity S, Sarkar S (2014) Carbon nano-particles in 'biochar' boost wheat (Triticum aestivum) plant growth. Rsc Adv 4:39948–39954. https://doi.org/10.1039/C4RA06535B

Shah UV, Wang Z, Olusanmi D, Narang AS, Hussain MA, Tobyn MJ, Heng JY (2015) Effect of milling temperatures on surface area, surface energy and cohesion of pharmaceutical powders. Int J Pharm 495:234–240. https://doi.org/10.1016/j.ijpharm.2015.08.061

Shah V, Belozerova I (2009) Influence of metal nano-particles on the soil microbial community and germination of lettuce seeds. Water Air Soil Pollut 197:143–148. https://doi.org/10.1007/s11270-008-9797-6

Singh R, Kumar R, Babu JN, Srivastava P, Singh P, Raghubanshi AS (2015) Multifaceted application of crop residue biochar as a tool for sustainable agriculture: an ecological perspective. Ecol Eng 77:324–347. https://doi.org/10.1016/j.ecoleng.2015.01.011

Singh A, Sharma R, Pant D, Malaviya P (2021) Engineered algal biochar for contaminant remediation and electrochemical applications. Sci Total Environ 774:145676. https://doi.org/10.1016/j.scitotenv.2021.145676

Singh M, Ahsan M, Pandey V, Singh A, Mishra D, Tiwari N, Singh P, Karak T, Khare P (2022) Comparative assessment for removal of anionic dye from water by different waste-derived biochar vis a vis reusability of generated sludge. Biochar 4:1–17. https://doi.org/10.1007/s42773-022-00140-7

Song B, Chen M, Zhao L, Qiu H, Cao X (2019) Physicochemical property and colloidal stability of micron-and nano-particle biochar derived from a variety of feedstock sources. Sci Total Environ 661:685–695. https://doi.org/10.1016/j.scitotenv.2019.01.193

Song Z, Lian F, Yu Z, Zhu L, Xing B, Qiu W (2014) Synthesis and characterization of a novel MnOx-loaded biochar and its adsorption properties for Cu^{2+} in aqueous solution. Chem Eng J 242:36–42. https://doi.org/10.1016/j.cej.2013.12.061

Sonkar SK, Roy M, Babar DG, Sarkar S (2012) Water soluble carbon nano-onions from wood wool as growth promoters for gram plants. Nanoscale 4:7670–7675

Sun Y, Gao B, Yao Y, Fang J, Zhang M, Zhou Y, Chen H, Yang L (2014) Effects of feedstock type, production method, and pyrolysis temperature on biochar and hydrochar properties. Chem Eng J 240:574–578. https://doi.org/10.1016/j.cej.2013.10.081

Tan I, Ahmad A, Hameed B (2008) Adsorption of basic dye on high-surface-area activated carbon prepared from coconut husk: equilibrium, kinetic and thermodynamic studies. J Hazard Mater 154:337–346. https://doi.org/10.1016/j.jhazmat.2007.10.031

Tan S, Zhou B, Wang Q (2016a) Effects of nanocarbon on the hydraulic parameters and the solute transport process for disturbed loessial soil. Arab J Geosci 9:1–13. https://doi.org/10.1007/s12517-015-2018-x

Tan X, Liu Y, Zeng G, Wang X, Hu X, Gu Y, Yang Z (2015) Application of biochar for the removal of pollutants from aqueous solutions. Chemosphere 125:70–85. https://doi.org/10.1016/j.chemosphere.2014.12.058

Tan X-f, Liu Y-g, Gu Y-l, Xu Y, Zeng G-m, Hu X-j, Liu S-b, Wang X, Liu S-m, Li J (2016b) Biochar-based nano-composites for the decontamination of wastewater: a review. Bioresour Technol 212:318–333. https://doi.org/10.1016/j.biortech.2016.04.093

Tang J, Lv H, Gong Y, Huang Y (2015) Preparation and characterization of a novel graphene/biochar composite for aqueous phenanthrene and mercury removal. Bioresour Technol 196:355–363. https://doi.org/10.1016/j.biortech.2015.07.047

Taran NY, Gonchar OM, Lopatko KG, Batsmanova LM, Patyka MV, Volkogon MV (2014) The effect of colloidal solution of molybdenum nano-particles on the microbial composition in rhizosphere of Cicer arietinum L. Nanoscale Res Lett 9:1–8. https://doi.org/10.1186/1556-276X-9-289

Tiwari D, Dasgupta-Schubert N, Villaseñor Cendejas L, Villegas J, Carreto Montoya L, Borjas García S (2014) Interfacing carbon nanotubes (CNT) with plants: enhancement of growth, water and ionic nutrient uptake in maize (Zea mays) and implications for nanoagriculture. Appl Nanosci 4:577–591. https://doi.org/10.1007/s13204-013-0236-7

Tripathi M, Sahu JN, Ganesan P (2016) Effect of process parameters on production of biochar from biomass waste through pyrolysis: a review. Renew Sustain Energy Rev 55:467–481. https://doi.org/10.1016/j.rser.2015.10.122

Tripathi S, Sonkar SK, Sarkar S (2011) Growth stimulation of gram (Cicer arietinum) plant by water soluble carbon nanotubes. Nanoscale 3:1176–1181. https://doi.org/10.1039/C0NR00722F

Tu Y, Peng Z, Huang J, Wu X, Kong L, Liang Z, Yang L, Lin Z (2020) Preparation and characterization of magnetic biochar nano-composites via a modified solvothermal method and their use as efficient heterogeneous fenton-like catalysts. Ind Eng Chem Res 59:1809–1821. https://doi.org/10.1021/acs.iecr.9b04590

Wang B, Gao B, Fang J (2017) Recent advances in engineered biochar productions and applications. Crit Rev Environ Sci Technol 47:2158–2207. https://doi.org/10.1080/10643389.2017.1418580

Wang C, Wang R, Huang Z (2018a) Effects of nano-carbon water-retaining fertilizer on yield and nitrogen and phosphorus utilization efficiency of tuber mustard. Asian J Agric Res 10:62–65. https://doi.org/10.22004/ag.econ.281779

Wang D, Zhang W, Hao X, Zhou D (2013) Transport of biochar particles in saturated granular media: effects of pyrolysis temperature and particle size. Environ Sci Technol 47:821–828. https://doi.org/10.1021/es303794d

Wang H, Gao B, Wang S, Fang J, Xue Y, Yang K (2015a) Removal of Pb (II), Cu (II), and Cd (II) from aqueous solutions by biochar derived from $KMnO_4$ treated hickory wood. Bioresour Technol 197:356–362. https://doi.org/10.1016/j.biortech.2015.08.132

Wang M, Sheng G, Qiu Y (2015b) A novel manganese-oxide/biochar composite for efficient removal of lead (II) from aqueous solutions. Int J Environ Sci Technol 12:1719–1726. https://doi.org/10.1007/s13762-014-0538-7

Wang S, Gao B, Li Y, Mosa A, Zimmerman AR, Ma LQ, Harris WG, Migliaccio KW (2015c) Manganese oxide-modified biochars: preparation, characterization, and sorption of arsenate and lead. Bioresour Technol 181:13–17. https://doi.org/10.1016/j.biortech.2015.01.044

Wang S, Gao B, Zimmerman AR, Li Y, Ma L, Harris WG, Migliaccio KW (2015d) Physicochemical and sorptive properties of biochars derived from woody and herbaceous biomass. Chemosphere 134:257–262. https://doi.org/10.1016/j.chemosphere.2015.04.062

Wang S, Gao B, Zimmerman AR, Li Y, Ma L, Harris WG, Migliaccio KW (2015e) Removal of arsenic by magnetic biochar prepared from pinewood and natural hematite. Bioresour Technol 175:391–395. https://doi.org/10.1016/j.biortech.2014.10.104

Wang S-L, Nguyen AD (2018) Effects of Zn/B nanofertilizer on biophysical characteristics and growth of coffee seedlings in a greenhouse. Res Chem Intermed 44:4889–4901. https://doi.org/10.1007/s11164-018-3342-z

Wang T, Rodriguez-Uribe A, Misra M, Mohanty AK (2018b) Sustainable carbonaceous biofiller from miscanthus: size reduction, characterization, and potential bio-composites applications. BioResources 13:3720–3739

Wang X, Gu Y, Tan X, Liu Y, Zhou Y, Hu X, Cai X, Xu W, Zhang C, Liu S (2019) Functionalized biochar/clay composites for reducing the bioavailable fraction of arsenic and cadmium in river sediment. Environ Toxicol Chem 38:2337–2347. https://doi.org/10.1002/etc.4542

Wang X, Han H, Liu X, Gu X, Chen K, Lu D (2012) Multi-walled carbon nanotubes can enhance root elongation of wheat (Triticum aestivum) plants. J Nanopart Res 14:1–10. https://doi.org/10.1007/s11051-012-0841-5

Wang Y-Y, Liu Y-X, Lu H-H, Yang R-Q, Yang S-M (2018c) Competitive adsorption of Pb (II), Cu (II), and Zn (II) ions onto hydroxyapatite-biochar nano-composite in aqueous solutions. J Solid State Chem 261:53–61. https://doi.org/10.1016/j.jssc.2018.02.010

Weber K, Quicker P (2018) Properties of biochar. Fuel 217:240–261. https://doi.org/10.1016/j.fuel.2017.12.054

Wu J, Li Z, Huang D, Liu X, Tang C, Parikh SJ, Xu J (2020) A novel calcium-based magnetic biochar is effective in stabilization of arsenic and cadmium co-contamination in aerobic soils. J Hazard Mater 387:122010. https://doi.org/10.1016/j.jhazmat.2019.122010

Wu L, Wei C, Zhang S, Wang Y, Kuzyakov Y, Ding X (2019) MgO-modified biochar increases phosphate retention and rice yields in saline-alkaline soil. J Clean Prod 235:901–909. https://doi.org/10.1016/j.jclepro.2019.07.043

Xiumei L, Fudao Z, Shuqing Z, Xusheng H, Rufang W, Zhaobin F, Yujun W (2005) Responses of peanut to nano-calcium carbonate. Plant Nutr Fertitizer Sci 11:385–389

Xu X, Cölfen H (2021) Ultracentrifugation techniques for the ordering of nano-particles. J Nanomater 11:333. https://doi.org/10.3390/nano11020333

Yang J, Chen M, Yang H, Xu N, Feng G, Li Z, Su C, Wang D (2020) Surface heterogeneity mediated transport of hydrochar nano-particles in heterogeneous porous media. Environ Sci Pollut Res 27:32842–32855. https://doi.org/10.1007/s11356-020-09482-w

Yao Y, Gao B, Chen J, Yang L (2013) Engineered biochar reclaiming phosphate from aqueous solutions:mechanisms and potential application as a slow-release fertilizer. Environ Sci Technol 47:8700–8708.https://doi.org/10.1021/es4012977

Yao Y, Gao B, Fang J, Zhang M, Chen H, Zhou Y, Creamer AE, Sun Y, Yang L (2014) Characterization and environmental applications of clay–biochar composites. Chem Eng J 242:136–143. https://doi.org/10.1016/j.cej.2013.12.062

Yu Z, Zhou L, Huang Y, Song Z, Qiu W (2015) Effects of a manganese oxide-modified biochar composite on adsorption of arsenic in red soil. J Environ Manage 163:155–162. https://doi.org/10.1016/j.jenvman.2015.08.020

Zhao L, Sun Y, Hernandez-Viezcas JA, Servin AD, Hong J, Niu G, Peralta-Videa JR, Duarte-Gardea M, Gardea-Torresdey JL (2013) Influence of CeO_2 and ZnO nano-particles on cucumber physiological markers and bioaccumulation of Ce and Zn: a life cycle study. J Agric Food Chem 61:11945–11951. https://doi.org/10.1021/jf404328e

Zhao Y, Qamar SA, Qamar M, Bilal M, Iqbal HM (2021) Sustainable remediation of hazardous environmental pollutants using biochar-based nanohybrid materials. J Environ Manage 300:113762. https://doi.org/10.1016/j.jenvman.2021.113762

Zhou L, Huang Y, Qiu W, Sun Z, Liu Z, Song Z (2017) Adsorption properties of nano-MnO_2–biochar composites for copper in aqueous solution. Molecules 22:173. https://doi.org/10.3390/molecules22010173

Zhou Y, Gao B, Zimmerman AR, Fang J, Sun Y, Cao X (2013) Sorption of heavy metals on chitosan-modified biochars and its biological effects. Chem Eng J 231:512–518. https://doi.org/10.1016/j.cej.2013.07.036

Applications and Future Perspectives of Agricultural Waste Biochar and Its Nanocomposites

Astha Dixit, Nikhil Senger, Pratik Bhoj, Rajeev Parmar, and Mangesh Lodhe

Abstract

Many biomasses found in abundant quantities in nature are the type of agricultural waste materials. This plethora of biomass is an environmental nuisance. Biochar is a black carbon produced from the pyrolysis of biomass at higher temperatures in the absence of oxygen. Biochar performs multifunctional role due to its strong physico-chemical characteristics such as adsorption, specific surface area, porosity, and crystallinity of carbon structure. The physical, chemical, and biological modifications of biochar lead to improved surface and adsorption properties of nanocomposites. The processed biochar has lots of importance due to the improved properties of pristine biochar. This chapter focuses on the application of agricultural waste biochar and its nanocomposites to waste water treatment, microbial fuel cell (MFC), and ceramic nanocomposites. The chapter is also intended to deliver the process, characterization, and mechanism of engineered biochar from pristine biochar. Further emphasis is given on applications of biochar in the development of advanced materials like graphene, carbon nanotubes (CNTs), and silicon carbide nanowires.

Keywords

Carbon nanotubes • Graphene • Microbial fuel cell • Silicon carbide nanowires

A. Dixit · N. Senger · P. Bhoj · M. Lodhe (✉)
Department of Materials and Metallurgical Engineering, Maulana Azad National Institute of Technology, Bhopal, 462003, India
e-mail: mmlodhe@manit.ac.in

R. Parmar
Department of Chemical Engineering, Maulana Azad National Institute of Technology, Bhopal, 462003, India

1 Introduction

Earth has 2.5% of freshwater resource, and out of which 30% is groundwater (Biswas 2003). This groundwater will be utilized for mining, industry, manufacturing, and various other applications which damages the quality of water. The water will have contaminants like lead, fluoride, arsenic, antimony, cadmium, etc., which are hazardous to the mankind (Amen et al. 2020). Biochar is a porous carbon material derived from the various biomasses either vegetable or animal wastes. The biological carbon derived from biomass possesses very high specific surface area and will have negative surface functional groups (carboxyl, hydroxyl, amino, etc.). These functional groups serve the purpose of absorbent to remove dangerous heavy metals from water (Amin et al. 2016; Rodriguez-Narvaez et al. 2019; Jin et al. 2020). Biomass or feedstock for Pb removal from water can be coconut fiber, rice husks, rice-straw, sugarcane bagasse, celery waste, peanut hull, and oil palm (Inyang et al. 2016; Rodriguez-Narvaez et al. 2017; Wang et al. 2017). To enhance the capabilities of biochar, besides treatments given during production of wine, some post-treatment can also be given like acid/alkali treatment. Adsorption mechanism involves of both physisorption as well as chemisorption (Jin et al. 2020).

With the rapid industrial growth, the depletion of conventional energy sources is also increased. The combustion of conventional fuels produces carbon dioxide, which is a major greenhouse gas and has revealed alarming consequences. In such cases, a microbial fuel cell can significantly fulfill the demand of future energy requirement without causing harm to the environment. In anaerobic anode compartment, microbial fuel cell (MFC) uses an active biocatalyst for production of bioelectricity. Different cathode catalysts are used to enhance the performance of MFCs. The cost of this cathode catalyst is of the major concern nowadays (Md Khudzari et al. 2019; Chakraborty et al. 2020a).

D. Mishra et al. (eds.), *Biochar-Based Nanocomposites for Contaminant Management*, Advances in Science, Technology & Innovation, https://doi.org/10.1007/978-3-031-28873-9_11

Carbon nanomaterials, especially graphene and carbon nanotubes, are widely studied by researchers all over the globe due to extraordinary electrical, thermal, and mechanical properties (Zhou et al. 2022). To produce cost-effective carbon nanomaterials, extensive research is going on biomass products. Average sugarcane bagasse contains 40–50% cellulose, 20–25% lignin, 20–30% hemicellulose, and it also contains 1.5–3.0% ash. This bagasse can also be used as source of raw material for the production of nanotubes (Yahya et al. 2015; Kandpal et al. 2021). Silicon-carbide (SiC) nanowires have highly ordered structure which offers high strength, chemical attack due to silicon-carbon bond and resistance to radiation. Various techniques like chemical vapor deposition, laser ablation, high-frequency induction heating, sol–gel and carbothermal reduction have already been reported for synthesizing SiC nanowires (Zhou et al. 2008; Attolini et al. 2010). However, these processes are energy-intensive, time-consuming, and expensive. Biomass obtained from agricultural waste contains dispersed silica and carbon and is an environmentally friendly resource to synthesize SiC whiskers (Lodhe et al. 2016a; Lodhe and Balasubramanian 2022).

Several carbon materials like activated carbon (AC), biochar (BC), expanded graphite (EG), and carbon nanotubes (CNTs) are used to develop high thermal performance composite phase change materials (PCMs) (Atinafu et al. 2021a). PCMs are the materials which can release or absorb energy during phase transition to provide heating or cooling. At isothermal working temperature, PCMs can store or release large amounts of heat energy. Some organic PCMs include fatty acids, glycols, paraffin, and alcohols. Besides their high thermal storing capacities, they offer low supercooling phenomenon, high durability, and corrosion resistance (Atinafu et al. 2021a). But, the practical application of pure PCMs is limited because of the challenging problems like low thermal, leakage, flammability, and conductivity. Seepage problem can be overcome by using supporting sold scaffolds. PCMs along with solid scaffold can make a composite. Composites can be fabricated encapsulating large amount of PCM with various scaffolding materials such as CNTs, metal–organic frameworks, porous carbons, carbon fibers, biochars, melamine foam, silica matrices, and EG, which can stabilize the shape of composites fabricated (Atinafu et al. 2021b).

In this chapter, we present the application of biochar nanocomposites for the waste water treatment, microbial fuel cell, graphene, and SiC nanowires preparation. The current techniques for the synthesis and characterization of graphene, carbon nanotubes, and silicon carbide nanowires from the agricultural waste materials are discussed in detail.

2 Waste Water Treatment

Activities like mining, domestic waste, municipal waste, industrial and agricultural discharge are responsible for the contamination and introduction of toxic heavy metals (e.g., copper, cadmium) into the water bodies. The concentration of Cu and Cd should be below 0.003 and 2 mg/l, respectively; above this concentration, they make aqueous solutions toxic as long-time exposure to these metals may cause liver damage, heart problems, kidney dis-functioning, etc. (Amen et al. 2020). Therefore, they must be removed. There are various remediation methods for heavy metals like chemical oxidation, reduction, ion exchange, electrochemical recovery, reverse osmosis, and adsorption. The adsorption technique is considered fast, efficient, low cost, and environment friendly. Adsorption of heavy metals by using biochar is being adopted nowadays. The origin and production conditions significantly influence the characteristics of biochar. Aromatic carbons dominate the chemical behavior of plant-originated biochar, while mineral or ash compounds dominate animal manure originated biochar. Lignin-rich plant feedstocks are more macroporous, and cellulose-rich plant feedstocks are more microporous.

Tharanga Bandara and his team put on the mechanism of removing heavy metals from the water bodies by agricultural derived biochar (Bandara et al. 2020). Their feedstocks comprised of poultry litter, lucerne shoot, vetch shoot, sugar-gum wood, wheat straws, canola shoot and derived biochar named as PBC, LBC, VBC, SBC, WBC, and CBC, respectively. The feedstock was collected, air-dried then pyrolyzed using mobile pyrolyzer through slow pyrolysis process (2 °C/min) at a temperature of 550 °C, holding for about 45 min. The pyrolyzed product was then quenched followed by grounding and sieving (<1 mm). All agricultural wastes-derived biochars are alkaline (pH: 9.71–10.48) due to the formation of $CaCO_3$ and $MgCO_3$ and some small acidic functional groups to neutralize primarily acidic contaminated water. All the biochars prepared by Bandara et al. (2020) show higher adsorption capacities for Cu(II) than Cd(II) because of higher electronegativity of Cu(II), smaller ionic radius, and the existence of N–, $PO_4(^{2-})$, and $CO_3(^{2-})$ containing functional groups on biochar. The mechanism of heavy metals adsorption includes precipitation, cation exchange, chemical reduction, complexation, and electrostatic interaction. The biochar is enriched with $CO_3(^{2-})$ and $PO_3(^{2-})$ and shows higher adsorption of Cu(II). This research reveals that poultry litter biochar effectively immobilizes heavy metals from mine water even under well competitive conditions with high concentrations of Al, Zn, and Fe. The removal efficiency for Cu(II) from mine water is PBC

(99.9%) > LBC (99.8%) > VBC (99.7%) > CBC (95.0%) > WBC (0.55%) > SBC (0.34%) and that for Cd(II) is PBC (99%) > LBC (82%) > VBC (55%) > CBC (20%) > WBC (6.5%) > SBC (6%). X-ray diffraction (XRD) peaks are observed at 2θ position of 20–30° showing presence of aromatic graphite structure of biochar. This aromatic graphite structure is lowest in PBC and highest in SBC.

Lead (Pb) is another significant heavy metal, which is considered as highly responsible for contamination of soil and water. The existing Pb removal techniques include flocculation, ion-exchange, chemical precipitation, and redox membrane separation. Jin et al. (2020) worked on the production of biochar, derived from grape pomace (GP), for the removal of Pb from the aqueous solutions (Jin et al. 2020). Grape pomace mainly consists of seeds, grape skins, some stems, and is a main by-product of wine making. It weighs about 20–30% (w/w) of the original grape weight. Grape pomace (GP) was collected from a winery, dried, and then milled to a fine powder (<0.85 mm). Meanwhile, the secondary waste generated during the wine making was subjected to various treatments like extraction, pressurized organic solvent, enzyme hydrolysis, and alkali pre-treatment to enhance the physical and chemical properties of GP secondary waste. Pyrolysis of GP was performed using tube furnace with continuous nitrogen purging at different temperatures (300, 500, 700 °C) with the residence time of 2 h. Finally, the product was pulverized with a motor, sieved (<1 mm). GP secondary wastes produced are of three categories, viz., GP(FOP): obtained after hexane water and ethanol extraction, GPL1: obtained after alkali dissolution of GP(FOP) followed by precipitation, GPL2: obtained after enzymatic hydrolysis of GPL1. Primary GP contains more extractives than GP(FOP). The amounts of extractives were further decreased in GPL1 and GPL2. The Pb removal from water bodies follows GPL2 (pyrolyzed at 700 °C) > GP (pyrolyzed at 500 °C) > GP (pyrolyzed at 700 °C). The effectiveness of GPL2 is due to the higher surface area (485 m^2/g), more porous morphology than its counterparts (Jin et al. 2020).

Performance of biochar highly depends on carbonization process. Five methods of carbonation, viz., direct carbonation (BC), hydrothermal treatment (HT), carbonation of hydrochar (HBC), $KHCO_3$ activation carbonation of hydrochar (KHBC), and $KHCO_3$ activation carbonation (KBC) can be utilized to enhance the biomass into a carbon/energy-rich material. These carbonation methods directly influence the carbon structure, morphology, functional groups, and in turn influences the adsorption capacities of biochar. Bin Li and colleagues studied the effect of carbonization on the performance and adsorption capacities of biochar in the removal of impurities from aqueous solutions (Li et al. 2021). Discarded tea leaves and tea stalks, which are generally considered agricultural waste, were collected from tea factories. Tea leaves were repeatedly washed with ultrapure water then dried in an oven at 80 °C, crushed and finally subjected to carbonation methods. Direct carbonation is a one-step pyrolysis process in which physicochemical properties remain incomplete; therefore, a two-step pyrolysis was performed, i.e., pyrolysis treatment then secondary pyrolysis. Hydrothermal treatment converts biomass into homogeneous materials which are rich in oxygen and carbon functional groups with low degree of atomization. To further improve the adsorption capabilities, environmental-friendly activation reagent like $KHCO_3$ was used. $KHCO_3$ offers low secondary pollution, low corrosion resistance, and high efficiency. Results showed that the combination of two step pyrolysis process and activation treatment with $KHCO_3$ in a wide range of pH (3.0–9.0) is considered to be ideal method to prepare "Engineered" biochar having adsorption capabilities 40 times higher than those of traditional biochar. Crystalline mineral structure was analyzed by XRD measurements. The diffraction peaks at 23° and 43° correspond to amorphous carbon (002) and graphite carbon (100), respectively. KBC and HKBC morphology displays well-organized pore structure as a 3D honeycomb structure caused by thermochemical reaction between tea components and $KHCO_3$ activation.

3 Microbial Fuel Cell

With the rapid industrial growth, the depletion of convention energy sources is also increased. The combustion of convention fuels produces carbon dioxide, which is a major greenhouse gas and has revealed alarming consequences. In such case, a microbial fuel cell can significantly fulfil the demand of future energy requirement without causing harm to the environment (Ok et al. 2015). Bioelectricity is being prepared from the microbial fuel cell where microorganisms are used as a catalyst in anaerobic anode compartment. Different cathode catalysts are used to increase the performance of microbial fuel cells (MFCs). The cost of this cathode catalyst is of the major concern nowadays (Md Khudzari et al. 2019).

The low-cost activated carbon was synthesized and was denoted as ABC-900 (Chakraborty et al. 2020b). It was produced by thermal activation of algal biochar at 900 °C. Using X-ray photoelectron spectroscopy, it was found that the activated charcoal was intrinsically doped with phosphorous and nitrogen. Cyclic voltammetry revealed that the ABC-900 was an active electrocatalyst with good oxygen reduction reaction (ORR) activity. Scanning electron microscopy results of ABC-900 show the presence of pore of irregular sizes at small intervals making it a porous structure. They compared three different MFCs having ABC-900, Pt-C, and bare carbon black coated cathode as electrocatalysts (Chakraborty et al. 2020b). The efficiencies for the removal of chemical oxygen demand (COD) of these three MFCs were found to be 79.5, 79, and 73.5, respectively. The use of activated biochar as electrocatalyst was advantageous as its cost was only 0.3% of the cost of Pt-C electrocatalyst. The maximum power density was 12.86 W m^{-3} for ABC-900 and 13.52 W m^{-3} for Pt-C cathode. The fabrication of such an activated biochar which was algae based required very less technical expertise making it easier to use.

Hu-Cheng Chang and his co-workers have fabricated the air cathode using balsa wood biochar which was free from many binder or catalysts (Chang et al. 2020). These air cathodes were used in microbial fuel cells. The novel biochar air cathode was formed by the pyrolysis of balsa wood at 800 °C and was obtained only in a single step. In the Raman spectra of the balsa wood biochar, the D and G bands were found to be 1350 cm^{-1} and 1587 cm^{-1}, respectively. The XRD characteristic peaks of the cathode were found at 45° and 24.3° for the graphite (001) and (002) lattices, respectively. The structure of the cathode was porous in nature due to which it had a large surface area and ultimately very high ORR activity. This binder free biochar air cathode was tested in single chamber microbial fuel cell and had got the maximum power density of around 200 mW/m^2. The power density of this biochar was approximately similar to that of air cathodes which are made up of other biochar. It was also observed that the performance of soil-based MFCs varies with the biochar chip thickness. The highest power density of around 72 mW/m^2 was found in a chip of thickness 3.5 mm (Chang et al. 2020).

Biodegradable waste materials are now widely used in microbial fuel cells (MFCs) as cathode materials due to their low cost (Pepè Sciarria et al. 2020). Metal-free activated biochar was used as an oxygen reduction reaction (ORR) catalyst in single chamber MFCs. Activated biochar was produced from olive mill waste (OMW) which was obtained from an olive oil plant. Two samples were taken; out of which one was pre-treated with supercritical CO_2 and the other was used as it is. Similarly, two salted pistachio nutshell (PNS) samples were also taken; one activated with CO_2 and other with KOH. The OMW samples were heated at 600 °C. The CO_2-activated pistachio sample was pyrolyzed at 450 °C, and the KOH activated PNS at 550 °C. Time taken for pyrolysis was 30 min which was similar for all the samples. The OMW samples were found to have granular profiles with porous structures whereas the same properties were less evident for pistachio (PNS) samples. This would have been due to the different feedstock nature and pyrolysis temperature. Activation by KOH caused lowering of surface area of OMW samples as compared to the CO_2 pre-treated one (Pepè Sciarria et al. 2020). On the other hand, non-CO_2-treated PNS samples have highest surface area (1970 m^2/g). FTIR results show the presence of C-H bonds along with nitrogen and oxygen functional groups. The catalytic behavior of activated biochar toward ORR was evaluated by cyclic voltammetry. It was found that the maximum catalytic activity for ORR was in CO_2 pre-treated and KOH activated OMW sample and without CO_2 pre-treated CO_2-activated sample. The number of electrons exchanged and the density of active locations were also maximum for OMW samples. The maximum power density of the biochar-activated MFC (271 ± 34 mW/m^2) was found to be 15 times more than that of widely used carbon black.

The cost of the formation of microbial fuel cells (MFCs) depends largely on the cathode as well as on the membrane. The membranes which are generally used in MFCs are Nafion 117 and proton exchange membrane (PEM). Chakraborty et al. (2020b) synthesized PEM by using food waste derived biochar (Chakraborty et al. 2020c). Food waste materials were pyrolyzed at 600 °C to produce biochar which was then further sulfonized. To produce MFC as a final product, poly-vinyl alcohol matrix was used. This PEM was abbreviated as SBC-600. Energy dispersive X-ray spectroscopy (EDS or EDX) with elemental analysis showed the presence of sulfur in sample (around 5–7%). Ion transport number of SBC-600 was coming as 0.891 which was slightly less than that of Nafion membrane (0.976) but was larger than that of b-PVA membrane (0.611). Proton conductivity was found by Nyquist plot by using Nova 1.11 software and was around 0.077 S cm^{-1} which was close to that of Nafion 117 (0.09 S cm^{-1}). Water uptake capacity of SBC-600 was slightly less than that of b-PVA. Other membrane properties like oxygen mass transfer coefficients of SBC-600 were found to be 8.08×10^{-3} as compared to that of b-PVA membrane (2.92×10^{-3}). Two MFCs, MFC-N and MFC-SBC, were examined in the three days feed cycle for total 75 days, and it was found that in COD removal the performance of the MFC-SBC was slightly lesser than that of MFC-N. Their work showed that MFC-SBC has comparable oxygen matter removal rate as that of Nafion membrane MFCs (Chakraborty et al. 2020c).

Fatma Allam and her co-workers (Allam et al. 2020) had used water hyacinth biochar (WHB) as a cathode electrocatalyst in single chamber microbial fuel cell (MFC). This WHB is eco-friendly in nature. The WHB weeds from canal of Alexandria drinking water company, Egypt, were pyrolyzed and then characterized. EDX results of WHB showed the presence of high Ca (35.83 wt%), K (12.65 wt%), and Si (11.95 wt%) after pyrolysis. Si was effectively used in the cathodes for measuring ORR activity whereas K was found to be useful in finding the catalytic activity in MFCs. Electrochemical activity of WHB weeds was determined by cyclic voltammetry (CV) in N_2 saturated phosphate buffer solution (PBS) electrolyte, and WHB was found to be electrochemically stable over an entire potential range

(−0.556 to 0.444) (Allam et al. 2020). The surface area was found to be around 25.924 m^2/g. The power density was 12.3 mW/m and 24.7 mW/m for WHB-based MFC and Pt-C-based MFC, respectively. The power generated by the Pt-C was approximately double than that of the WHB containing MFC.

The biochar was fabricated by chicken manure (CB), wheat straw (SB), and wood sawdust at 600 °C (Li et al. 2019). The electrochemical impedance spectrum was performed under the frequency range of 100 kHz over an amplitude of 10 mV. The cathode and anode were connected as the reference electrode and working electrode, respectively. The power is generated by connecting them with the voltage source, and the resistance was also involved in soil MFC. After that they analyze biologically the potential of biochar in the biodegradation and bioelectricity generation and by making some alterations in the soil. They found that after 223 days, the 15–17% of higher efficiency was observed in SB and CB than without biochar. The high N content and aromaticity in biochar is better for the degradation of alkanes instead of high surface area.

The low-cost biochar from corncob was produced by the pyrolysis method in temperature range of 250–750 °C, and it acts as the ORR catalyst in MFC (Li et al. 2018). The biochar was produced from corncob by making it sun dried for 5 days and then converting it into powder. After that the powder was heated in nitrogen atmosphere in the temperature range of 250–750 °C for 2 h in the furnace. The single chamber was prepared in which carbon felt acts as anode and carbon cloth acts as cathode electrode and consists of 3 layers, i.e., catalyst layer, carbon base layer, and diffusive layer. It is observed that the higher degree of graphitization improves the electrical conductivity of biochar during the process of electrocatalysis. The obtained biochar at 650 °C (CC-650) can be used as a catalyst in MFC because it has better electrocatalytic activity. The high content of graphitic nitrogen and pyridinic nitrogen in pyrolysis converts the ORR pathway from 2 to 4 electron pathways and increases the speed of electron transfer (Li et al. 2018).

The cost-effective catalyst was produced by pyrolysis of watermelon rind (Zhong et al. 2019). The higher C-N bond, pyridinic N, and graphitic N gives the good electrochemical active area of 658.90 m^2/g. The activated watermelon rind was fabricated by washing it in deionized water and held it for drying at 40 °C for 72 h, after that it was soaked in 1M HCl for 5 h. HCl-activated watermelon rind after activation at 700 °C is marked as WRC-700. The biochar was derived using environmentally friendly step of carbonization of watermelon, and watermelon rind carbon (WRC) has high electrochemical surface area and good performance as ORR catalyst (Li et al. 2018).

The applicability of biochar as a substitute of the standard carbon felt anode in microbial fuel cells was carried out (Md Khudzari et al. 2019). As in the carbon felt anode, methane emission is higher as compared to plant MFC and soil MFC with biochar anode. MFC helps to decrease the emission of methane gas and the effect of global warming. The power density released by biochar anode, graphite, and carbon was 457–532 mW/m^2, 566 ± 5 mW/m^2, and 674 ± 10 mW/m^2, respectively. These power densities were much comparable. The electrical resistivity observed in biochar was 14.8 kΩ-mm, and in carbon anode, it was found to be <30.0 Ω-mm. The experimental results showed that the biochar with MFCs was appropriate to reduce methane emissions. MFCs with biochar anodes produce less methane gas than the PMFCs with carbon felt anodes, by 19% for SMFC and 39% for PMFC. It was also observed that, after the experiment, the emission of methane gas was reduced without decreasing or changing the effect on biomass yield.

4 Graphene and Carbon Nanotubes

Pyrolysis of sugarcane bagasse (SCB) with zeolite catalyst was used to produce multiwalled carbon nano-tubes (CNTs) (Aboul-enein et al. 2021). This agricultural waste was pyrolyzed using a two-step process. Growth of CNTs was done by decomposition of products using Co-Mo/MgO catalysts. Crystal structure was determined using XRD method. Temperature programmed reduction was used to evaluate the reducibility of Co-Mg/MgO catalyst. The yield of CNTs was significantly affected by pyrolysis temperature, zeolite type, and SCB/ZSM-5 ratio (Ateyya et al. 2021). It was found that optimum yield of 24.9% was achieved at 500 °C using HZSM-5 as a catalyst with ratio keeping 6. Carbon nano-onions and bamboo-like CNTs were observed. It was also found that largest diameter distribution CNT was produced at SCB/ZSM-5 ratio of 6 (Aboul-enein et al. 2021).

Yucheng Zhou and co-workers (2022) have summarized the various synthesis methods of high-quality graphene and CNTs. Graphene has excellent photocatalytic properties and has ability to be widely used for the production of hydrogen gas (Zhou et al. 2022). The various steps involved in the synthesis of graphene from natural biomass by graphitization method are shown in Fig. 1. Graphene is mostly used as an energy storage devices due to their large surface area (Atinafu et al. 2021b). It is also reported that graphene is also very much effective for water treatment purposes because of nano-particle sizes (Li et al. 2011).

Microwave pyrolysis can be used to enhance large-scale poly-generation of carbon nanomaterials, high-value fine chemicals, and fuel gases utilizing specified biomass feedstock and reaction parameters. The active bio-component responsible for the synthesis of CNTs in the microwave has been identified as cellulose. These carbon nanomaterials are

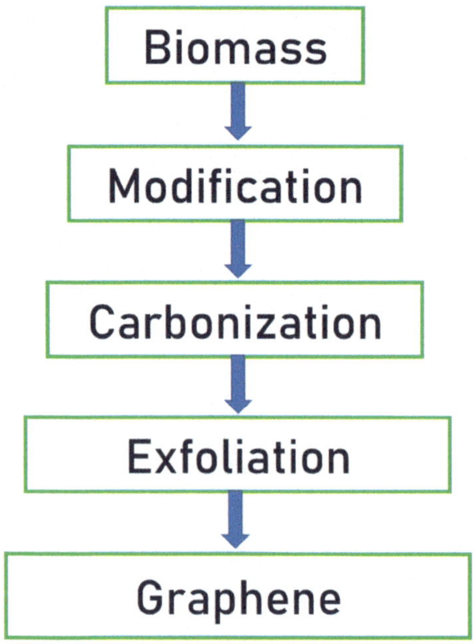

Fig. 1 Various steps for deriving graphene from natural biomass by graphitization method

generally made using three methods: (i) arc discharge, (ii) chemical vapor deposition (which uses hydrocarbon sources like CO and methane), and (iii) laser ablation (graphite application) (Omoriyekomwan et al. 2021a). Pyrolysis, template techniques, flame synthesis, and electrolysis are just a few of the technologies that have been successfully developed to manufacture carbon nanostructures over the years. Carbon nanomaterial has also been produced from biomass using the cyclic oxidation process. This approach was used by Kang et al. (2005) to produce CNTs from grass (Kang et al. 2005). The grass was first roasted in the air at 250 °C before being dried and pulverized for the experiments. After that, the sample was heated at 600 °C for 20 min under oxygen. The heating in the presence of oxygen was repeated around 50 times, resulting in hollow multiwalled CNTs with a diameter of 30–50 mm. Three primary parameters were found to be influenced by the process: (i) low-temperature pre-treatment prior to cyclic oxidation, (ii) biomass feedstock, and (iii) multiple re-duplicative oxidations under oxygen.

The advantages of microwave heating over traditional heating methods have sparked a lot of attention. Microwave frequencies with wavelengths of 0.01–1.0 m, comparable to frequencies between 0.3 and 300 GHz, are found in the infrared and radio bands of the electromagnetic spectrum. The most often utilized frequencies are 2.45 GHz and 915 MHz, with 2.45 GHz being used by most residential microwave ovens. Microwave output power level, reaction temperature, biomass particle size, volatile matter residence time, biomass properties, and reactor design are all factors that influence biomass microwave pyrolysis. Microwave heating has improved the yields of solid char fractions in the liquid phase. The reaction temperature is the determining factor in the distribution of major pyrolysis products. At higher temperatures, the liquid yield declined dramatically, although the light gas fraction yield increased. Microwave and conventional pyrolysis produce biochar with different morphologies, according to the research. Microwave-assisted pyrolysis is considered to be a potential method for producing carbon nanotubes in less time than traditional procedures. The microwave can be used to efficiently heat vast volumes of biomass. To make carbon nanotubes, microwave chemical vapor deposition (CVD) is a popular method. A catalyst, a carbon source, and catalyst support are all required for the process. Catalysts provide active sites for hydrocarbon breakdown and serve as a template for CNT development. The catalyst support aids in keeping the catalyst's activation efficiency. Carbon is usually obtained from CO_2, CO, and hydrocarbons like methane, ethane, benzene, and naphthalene. Because these species are generated during biomass pyrolysis, it is a good way to get carbon sources for CNT production (Hidalgo-Oporto et al. 2019; Zhang et al. 2021).

To describe the synthesis of CNTs using microwave-assisted pyrolysis, Omoriyekomwan et al. (2021b) proposed a "self-extrusion process." They theorized that hotspot on the char surface served as nucleation sites for CNT production. The wall structure of synthesized CNTs frequently has flaws, which limits their applicability (Omoriyekomwan et al. 2021a). As a result, reducing or eliminating these faults is critical to realizing the full potential of these materials. It has been established that post-treatment of CNTs removes amorphous carbon and contaminants from the structure of CNTs. This can be accomplished using the traditional high-temperature thermal annealing method. Single-walled CNTs must be synthesized at temperatures above 1200 °C in a vacuum, while multiwalled CNTs must be synthesized at temperatures above 1900 °C. The synthesis of CNTs from biomass using microwave pyrolysis could be interesting, with prospective applications in the electrochemical sector, due to a number of advantages (Makgabutlane et al. 2021).

5 Nanocomposites

SiC whisker/nanowire-reinforced SiC composites (SiC_W/SiC) were processed from the synthesized SiC nanowires and particles from the mixture of rice husk, coconut shell, and polycarbosilane (Lodhe et al. 2016b). SiC_W/SiC composites were prepared by the spark plasma sintering technique, and their mechanical properties were studied (Lodhe et al. 2018). Gonçalves et al. (2022) reported that biochar can be produced from valorization of chitosan (CTO) and brewed coffee (COFF). Photocatalytic properties of ZnO particles can be enhanced by incorporating biochar in different amounts by co-precipitation method (Gonçalves et al. 2022). Enhancement in interactions of ZnO particles and reduction of hydrophobicity was achieved by adding activated biochar. The addition of proper amount of biochar leads to the increase in efficiency and outperforms the pristine ZnO by 20%. Composite having 3.4% carbon (ZnO/2-COFF) had most efficient phenol degradation. At lower concentration of carbon (<4 wt%), ZnO/2-COFF samples had better photocatalytic properties compared to ZnO/CTO while vice-versa was observed for high carbon concentrations (>6 wt%). Also, poorest phenol removal was observed at carbon concentration <20 wt%. Radical excavangers presence indicated that non-oxidized carbon products improve the charge separation by acting as electron reservoir. It was found that ZnO/2-COFF samples had best performance and removed pollutants at higher efficiencies with only 0.05 g/l catalysts. Thus, the catalytic properties of ZnO can be enhanced by addition of biochar which can be used for water treatment (Gonçalves et al. 2022).

Tons of industrial and agricultural wastes such as fly ash, bagasse ash, and coconut shell ash (CSA) are discarded daily leading to various environmental issues. Research work was going on for utilizing these wastes as reinforcement with metal matrix to form metal matrix composites (MMCs). Kandpal et al. (2021) have summarized the MMCs reinforced with various agro/industrial wastes (Kandpal et al. 2021). Post-analyzing these waste products, it was found that fly ash has varied applications including road and soil-based stabilizations, etc.; many countries are using coconut shells for producing energy via boiler and casting industries. Coconut shell ash (CSA) contains various oxides primarily including SiO_2, Al_2O_3, Fe_2O_3, MgO, etc. Coconut shell ash has excellent high temperature stability and can be used up to temperatures levels of 1500 °C. Coconut shell ash density was found to be 2.05 gm/cm^3. Various other agro/industrial wastes including rice husk ash (RHA), palm oil clinger, bread fruit seed ash, sugarcane bagasse, egg shell, etc., are also used as reinforcements in MMCs. These MMCs are fabricated through various processes, based on operating temperature, such as solid liquid, liquid phase, in situ, and deposition techniques. It was reported that mechanical properties (including strength, elongation toughness) increase on increasing percentage of reinforcement in the composite. Interestingly, density was found decreasing on addition of agro/industrial wastes. Agricultural/industrial waste-reinforced MMCs are generally corrosion resistant in nature; some MMCs even show wear-resistant properties (Kumar et al. 2021; Dixit and Suhane 2022).

Aluminum alloys (mainly AA6061) are widely used in automobile industries. But, due to less wear resistance of Al alloys, researchers are exploring various alternatives. Devanathan et al. (2019) studied the tribological and mechanical properties of stir-casted metal matrix hybrid composites (MMHCs) (Devanathan et al. 2019). They used Al as matrix material and fly ash, coconut shell ash, and SiC (45–50 μm particle size) as reinforcements. Four test samples were casted using liquid stir casting method having different volume percent of reinforcements. Brinell hardness tester was used for hardness measurement, and it was reported that the sample with 20% fly ash, 10% coconut shell ash, and 10% SiC had highest Brinell hardness number (589.61 BHN) among all 4 samples. Tensile testing was done using universal testing machine (UTM). SEM micrographs of these samples showed uniform distribution of reinforcement in the matrix and presence of discontinuities in matrix. On increasing coconut ash% in MMHC, hardness and percentage elongation was increased significantly. Also, tensile strength of composite was found to be increasing with increase in coconut shell ash (CSA).

Generally, high strength materials are costly; therefore, the cost-effective ones are always preferred for engineering applications. Working in this domain, Siddharth Sarkar and his team fabricated silicon carbide (SiC) and rice husk ash (RHA)-reinforced aluminum alloys (Al-6061) and assessed their mechanical properties (Sarkar et al. 2018). Three Al matrix composites were developed having ratios of RHA and SiC as 0:1, 1:4, and 2:3, respectively. Weight percentage of reinforcements (RHA and SiC) in the composite was kept 8%. Stir casting method was used to prepare Al matrix composites. Hardness measurement was done by Brinell hardness tester, and the hardness values of composites were reported to be higher than that of the Al-6061 alloy. But interestingly, hardness of composites was found to be decreasing on increasing the RHA content. Fracture analysis indicated toward the enhanced tendency of brittle fracture on increasing SiC and ductile fracture on increasing RHA. Porosity of these composites was found to be as low as 2.86% which made these composites suitable for various applications. Tensile strength of these composites was reported to be lower than that of only SiC-reinforced Al composites but higher than that of aluminum alloys which are non-reinforced.

The composite biochar-CNT-NiO was used as an anode material in the lithium ion batteries or LIBs, which are great energy storage device. Carbon-based materials especially CNTs offer great structural properties like large surface area, increased pore volume, high conductivity, and high aspect ratios. Thus, CNT-NiO composite demonstrates superior electrochemical performance. CNT-NiO composites are synthesized via microwave-assisted chemical vapor deposition (MCVD) method as it includes direct, rapid, internal, and volumetric heating with reduced costs (Zhang et al. 2021). Biochar being an excellent absorber of microwaves, it allows in situ heating of reaction system and facilitating formation of CNTs. Ni acting as a catalyst, catalyze MCVD method, and air oxidation of Ni to NiO can increase battery performance by reducing processing time and production costs. Biochar-CNT-NiO unique structure (tube-like and multiwalled structure with diameter of around 85 nm) reduces change in volume of NiO and facilitates a pathway for electron transfer, improving LIBs capacity. It is found that NiO, rather than encapsulating inside, is uniformly distributed on surface of CNTs. Nanostructured NiO in biochar-CNT-NiO provides shorter path for Li^+ transfer by expanding effective surface area in electrolyte, thus, enhancing battery capability. Numerous macrospores and mesopores are observed in the composite. Surface area of biochar-CNT-NiO is larger than that of biochar-NiO because of more mesopores which can facilitate better Li^+ diffusion and electrolyte penetration. The charge transfer resistance for biochar-CNT-NiO is less. Improved reaction kinetics and excellent structural stability of the biochar-CNT-NiO is responsible for their outstanding performance.

6 Silicon Carbide Nanowires

Rice husk, coconut shell, and polycarbosilane mixture was pyrolyzed at 1400 °C in an inert atmosphere. The vapor-phase mechanisms were found to be responsible for the formation of SiC nanowires. The high aspect ratio nanowires with diameters ranging from 25 to 100 nm were successfully synthesized and are shown in Fig. 2. At high magnification, SiC nanowires show alternate bright and dark bands, which resembles to the defects like stacking faults and twins (Fig. 3).

The synthesis of SiC nanowhiskers from cornstalk cellulose biochar by microwave heating at 1200–1400 °C was studied (Omoriyekomwan et al. 2021a). SiC nanowhiskers formed varied in structure some are straight rod like while

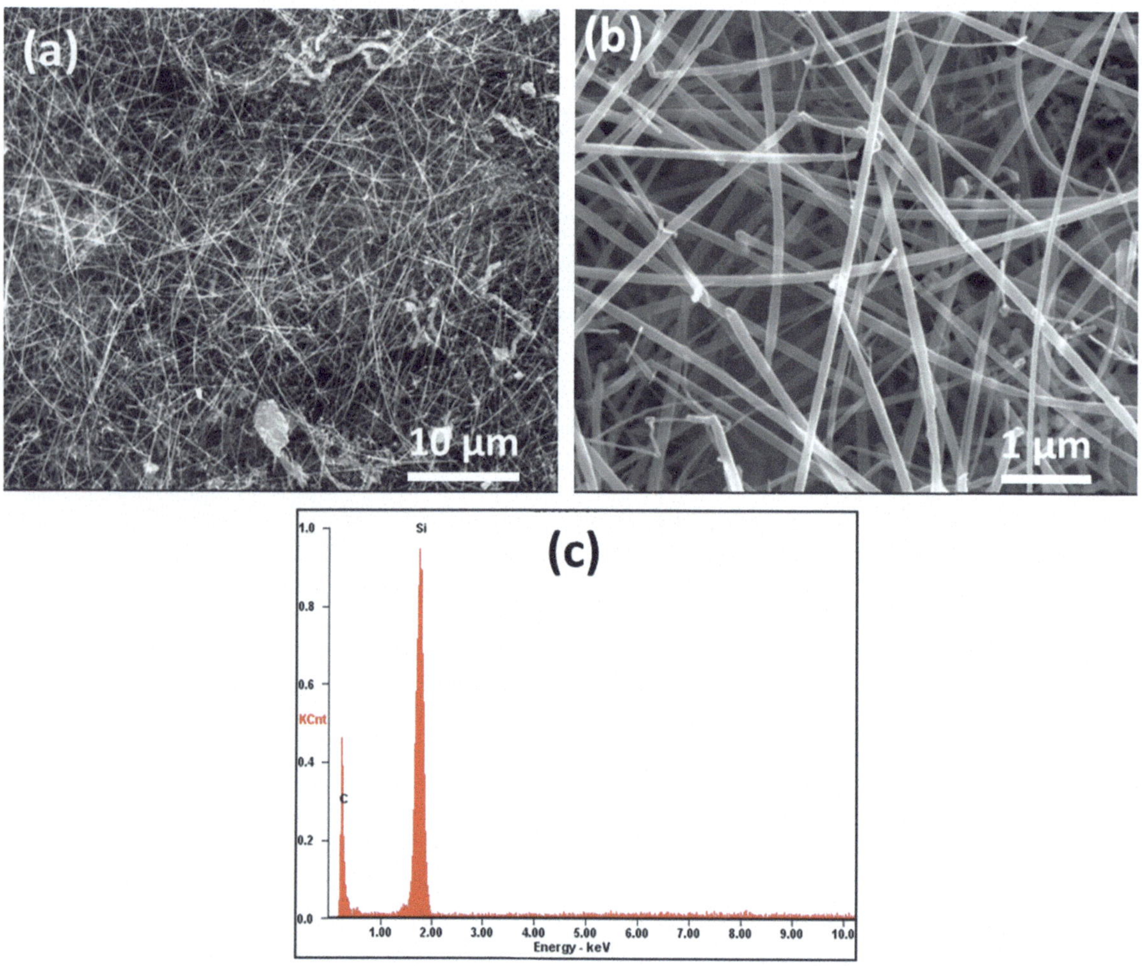

Fig. 2 Scanning electron micrographs of synthesized nanowires. **a** Low magnification image, **b** high magnification image, and **c** its EDS spectrum

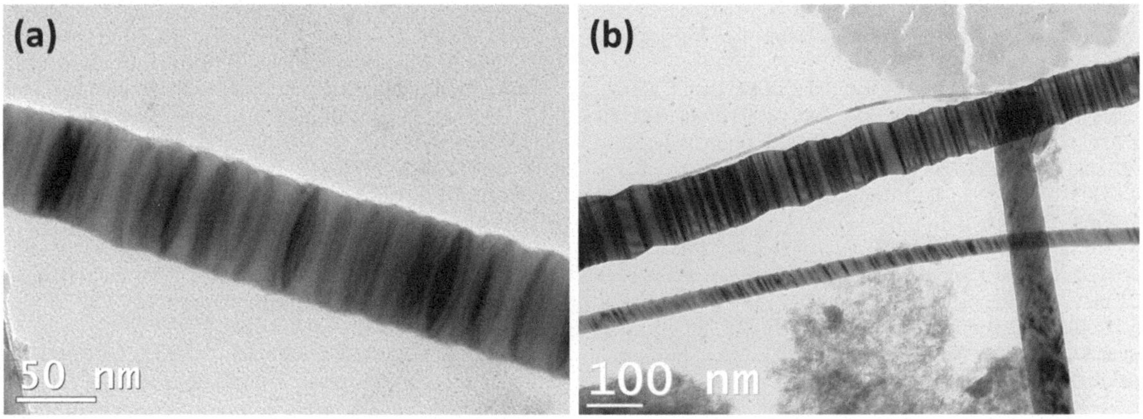

Fig. 3 Transmission electron micrographs of synthesized SiC nanowires

some entangled and twisted wire like structures. SiC nanowhiskers at 1400 °C became longer and thicker. Average length and diameter of nanowhiskers are 5 mm and 100 nm, respectively, and are primarily governed by treatment temperature. Nanowhiskers also revealed spherical droplets of Fe at their tips. SiC nanowhiskers with spherical droplets at the tip show hexagonal structure whereas 3C–SiC exhibits a cubic structure showing growth of nanowhiskers on (111) plane (Lodhe and Balasubramanian, 2022). Formation of SiC nanowhiskers in biochar is achieved at all the temperatures. At higher temperatures, graphitic carbon increased showing some amorphous carbon is graphitized, and silica is converted to β-SiC at elevated temperatures. SiC nanowhiskers also formed via vapor–liquid-solid or VLS mechanism (Milewski et al. 1985; Dhiman et al. 2011; Xia et al. 2012). Two major factors govern the formation of the SiC nanowhiskers from a biochar through microwave heating are amount and distribution of SiO_2 in biochar and high local temperatures.

7 Conclusion and Future Perspectives

Biomass is an agricultural waste material, and this plethora is an environmental nuisance. Biochar is a porous carbon material derived from the various biomasses either vegetable or animal waste. Biochar and biochar-based nanocomposites can be used for various applications which help in remediation of environmental contaminants. This chapter has summarized the application of agricultural waste biochar to wastewater treatment, microbial fuel cell (MFC), and ceramic nanocomposites. The major conclusions and future perspectives of this chapter are as follows:

- The physical, chemical, and biological modifications of biochar lead to improved surface and adsorption properties. In anaerobic anode compartment, MFC uses an active biocatalyst for production of bioelectricity.
- The SiC nanowires were synthesized from the mixture of rice husk and coconut shell.
- Advanced materials like graphene, carbon nanotubes, and silicon carbide nanowires have found many functional and structural applications.
- Future research should more focus on the extraction techniques to separate these materials from the unwanted phases, which will help us to use these materials directly for modern applications.
- Application of biochar-derived materials to advanced technologies is limited. Efforts must be given for the simplified synthesis of these materials and to enhance the yield of the products.

References

Aboul-enein AA, Awadallah AE, El-desouki DS, Aboul-gheit NAK (2021) Catalytic pyrolysis of sugarcane bagasse by zeolite catalyst for the production of multi-walled carbon nanotubes. Ranliao Huaxue Xuebao/J Fuel Chem Technol 49:1421–1434

Allam F, Elnouby M, El-Khatib KM, El-Badan DE, Sabry SA (2020) Water hyacinth (Eichhornia crassipes) biochar as an alternative cathode electrocatalyst in an air-cathode single chamber microbial fuel cell. Int J Hydrogen Energy 45:5911–5927

Amen R, Bashir H, Bibi I, Shaheen SM, Niazi NK, Shahid M, Hussain MM, Antoniadis V, Shakoor MB, Al-Solaimani SG, Wang H, Bundschuh J, Rinklebe J (2020) A critical review on arsenic removal from water using biochar-based sorbents: the significance of modification and redox reactions. Chem Eng J 396:125195

Amin FR, Huang Y, He Y, Zhang R, Liu G, Chen C (2016) Biochar applications and modern techniques for characterization. Clean Technol Environ Policy 18:1457–1473

Atinafu DG, Yun BY, Wi S, Kang Y, Kim S (2021a) A comparative analysis of biochar, activated carbon, expanded graphite, and multi-walled carbon nanotubes with respect to PCM loading and energy-storage capacities. Environ Res 195:110853

Atinafu DG, Wi S, Yun BY, Kim S (2021b) Engineering biochar with multiwalled carbon nanotube for efficient phase change material encapsulation and thermal energy storage. Energy 216:119294

Attolini G, Rossi F, Fabbri F, Bosi M, Salviati G, Watts BE (2010) Cubic SiC nanowires: growth, characterization and applications. Nanowires 1–23

Bandara T, Xu J, Potter ID, Franks A, Chathurika JBAJ, Tang C (2020) Mechanisms for the removal of Cd(II) and Cu(II) from aqueous solution and mine water by biochars derived from agricultural wastes. Chemosphere 254:126745

Biswas AK (2003) Water availability and use. In: Water resources of North America. Springer Berlin, Heidelberg

Chakraborty I, Sathe SM, Dubey BK, Ghangrekar MM (2020a) Waste-derived biochar: applications and future perspective in microbial fuel cells. Bioresour Technol 312:123587

Chakraborty I, Bhowmick GD, Ghosh D, Dubey BK, Pradhan D, Ghangrekar MM (2020b) Novel low-cost activated algal biochar as a cathode catalyst for improving performance of microbial fuel cell. Sustain Energy Technol Assessments 42:100808

Chakraborty I, Das S, Dubey BK, Ghangrekar MM (2020c) Novel low cost proton exchange membrane made from sulphonated biochar for application in microbial fuel cells. Mater Chem Phys 239:122025

Chang HC, Gustave W, Yuan ZF, Xiao Y, Chen Z (2020) One-step fabrication of binder-free air cathode for microbial fuel cells by using balsa wood biochar. Environ Technol Innov 18:100615

Devanathan R, Ravikumar J, Boopathi S, Christopher Selvam D, Anicia SA (2019) Influence in mechanical properties of stir cast aluminium (AA6061) hybrid metal matrix composite (HMMC) with silicon carbide, fly ash and coconut coir ash reinforcement. Mater Today Proc 22:3136–3144

Dhiman R, Johnson E, Morgen P (2011) Growth of SiC nanowhiskers from wooden precursors, separation, and characterization. Ceram Int 37:3759–3764

Dixit P, Suhane A (2022) Aluminum metal matrix composites reinforced with rice husk ash: a review. Mater Today Proc 62:4194–4201

Gonçalves NPF, Lourenço MAO, Baleuri SR, Bianco S, Jagdale P, Calza P (2022) Biochar waste-based ZnO materials as highly efficient photocatalysts for water treatment. J Environ Chem Eng 10:107256

Hidalgo-Oporto P, Navia R, Hunter R, Coronado G, Gonzalez ME (2019) Synthesis of carbon nanotubes using biochar as precursor material under microwave irradiation. J Environ Manage 244:83–91

Inyang MI, Gao B, Yao Y, Xue Y, Zimmerman A, Mosa A, Pullammanappallil P, Ok YS, Cao X (2016) A review of biochar as a low-cost adsorbent for aqueous heavy metal removal. Crit Rev Environ Sci Technol 46:406–433

Jin Q, Wang Z, Feng Y, Kim YT, Stewart AC, O'Keefe SF, Neilson AP, He Z, Huang H (2020) Grape pomace and its secondary waste management: biochar production for a broad range of lead (Pb) removal from water. Environ Res 186:109442

Kandpal BC, Johri N, Kumar N, Srivastava A (2021) Effect of industrial/ agricultural waste materials as reinforcement on properties of metal matrix composites. Mater Today Proc 46:10736–10740

Kang Z, Wang E, Mao B, Su Z, Chen L, Xu L (2005) Obtaining carbon nanotubes from grass. Nanotechnology 16:1192–1195

Kumar MS, Vasumathi M, Begum SR, Luminita SM, Vlase S, Pruncu CI (2021) Influence of B4C and industrial waste fly ash reinforcement particles on the micro structural characteristics and mechanical behavior of aluminium (Al–Mg–Si–T6) hybrid metal matrix composite. J Mater Res Technol 15:1201–1216

Li B, Zhang Y, Xu J, Mei Y, Fan S, Xu H (2021) Effect of carbonization methods on the properties of tea waste biochars and their application in tetracycline removal from aqueous solutions. Chemosphere 267:129283

Li M, Zhang H, Xiao T, Wang S, Zhang B, Chen D, Su M, Tang J (2018) Low-cost biochar derived from corncob as oxygen reduction catalyst in air cathode microbial fuel cells. Electrochim Acta 283:780–788

Li Q, Guo B, Yu J, Ran J, Zhang B, Yan H, Gong JR (2011) Highly efficient visible-light-driven photocatalytic hydrogen production of CdS-cluster-decorated graphene nanosheets. J Am Chem Soc 133:10878–10884

Li X, Li Y, Zhang X, Zhao X, Sun Y, Weng L, Li Y (2019) Long-term effect of biochar amendment on the biodegradation of petroleum hydrocarbons in soil microbial fuel cells. Sci Total Environ 651:796–806

Lodhe M, Balasubramanian M (2022) Polycarbosilane facilitated growth of SiC nanowires from biowaste coconut shell. Adv Appl Ceram 121:39–45

Lodhe M, Selvam A, Udayakumar A, Balasubramanian M (2016a) Effect of polycarbosilane addition to a mixture of rice husk and coconut shell on SiC whisker growth. Ceram Int 42:2393–2401

Lodhe M, Chawake N, Yadav D, Balasubramanian M (2016b) On correlation between $\beta \rightarrow \alpha$ transformation and densification mechanisms in SiC during spark plasma sintering. Scr Mater 115:137–140

Lodhe M, Logesh G, Balasubramanian M (2018) Twin induced fracture toughness in SiCW/SiC composites processed by spark plasma sintering. Mater Sci Eng A 730:280–283

Makgabutlane B, Nthunya LN, Maubane-Nkadimeng MS, Mhlanga SD (2021) Green synthesis of carbon nanotubes to address the water-energy-food nexus: a critical review. J Environ Chem Eng 9:104736

Md Khudzari J, Gariépy Y, Kurian J, Tartakovsky B, Raghavan GSV (2019) Effects of biochar anodes in rice plant microbial fuel cells on the production of bioelectricity, biomass, and methane. Biochem Eng J 141:190–199

Milewski JV, Gac FD, Petrovic JJ, Skaggs SR (1985) Growth of beta-silicon carbide whiskers by the VLS process. J Mater Sci 20:1160–1166

Ok YS, Chang SX, Gao B, Chung HJ (2015) SMART biochar technology-A shifting paradigm towards advanced materials and healthcare research. Environ Technol Innov 4:206–209

Omoriyekomwan JE, Tahmasebi A, Dou J, Wang R, Yu J (2021a) A review on the recent advances in the production of carbon nanotubes and carbon nanofibers via microwave-assisted pyrolysis of biomass. Fuel Process Technol 214:106686

Omoriyekomwan JE, Tahmasebi A, Dou J, Tian L, Yu J (2021b) Mechanistic study on the formation of silicon carbide nanowhiskers from biomass cellulose char under microwave. Mater Chem Phys 262:124288

Pepè Sciarria T, de Oliveira MAC, Mecheri B, D'Epifanio A, Goldfarb JL, Adani F (2020) Metal-free activated biochar as an oxygen reduction reaction catalyst in single chamber microbial fuel cells. J Power Sources 462:228183

Rodriguez-Narvaez OM, Peralta-Hernandez JM, Goonetilleke A, Bandala ER (2017) Treatment technologies for emerging contaminants in water: a review. Chem Eng J 323:361–380

Rodriguez-Narvaez OM, Peralta-Hernandez JM, Goonetilleke A, Bandala ER (2019) Biochar-supported nanomaterials for environmental applications. J Ind Eng Chem 78:21–33

Sarkar S, Bhirangi A, Mathew J, Oyyaravelu R, Kuppan P, Balan ASS (2018) Fabrication characteristics and mechanical behavior of rice husk ash-silicon carbide reinforced Al-6061 alloy matrix hybrid composite. Mater Today Proc 5:12706–12718

Wang B, Gao B, Fang J (2017) Recent advances in engineered biochar productions and applications. Crit Rev Environ Sci Technol 47:2158–2207

Xia M, Ge C, Yan Q, Guo H, Yue L (2012) Ti-assisted β-SiC nanowhiskers by pyrolysis of PTFE: synthesis and mechanical properties. Appl Phys A Mater Sci Process 107:777–782

Yahya MA, Al-Qodah Z, Ngah CWZ (2015) Agricultural bio-waste materials as potential sustainable precursors used for activated carbon production: a review. Renew Sustain Energy Rev 46:218–235

Zhang J, Tahmasebi A, Omoriyekomwan JE, Yu J (2021) Microwave-assisted synthesis of biochar-carbon-nanotube-NiO composite as high-performance anode materials for lithium-ion batteries. Fuel Process Technol 213:106714

Zhong K, Li M, Yang Y, Zhang H, Zhang B, Tang J, Yan J, Su M, Yang Z (2019) Nitrogen-doped biochar derived from watermelon rind as oxygen reduction catalyst in air cathode microbial fuel cells. Appl Energy 242:516–525

Zhou W, Zhang Y, Niu X, Min G (2008) One-dimensional SiC nanostructures: synthesis and properties. One-Dimens Nanostruct 17–58

Zhou Y, He J, Chen R, Li X (2022) Recent advances in biomass-derived graphene and carbon nanotubes. Mater Today Sustain 18:100138

The manufacturer's authorised representative in the EU is Springer Nature Customer Service Centre GmbH, Europaplatz 3, 69115 Heidelberg, Germany. If you have any concerns regarding our products, please contact ProductSafety@springernature.com

Printed and bound by CPI Group (UK) Ltd, Croydon, CR0 4YY
28/09/2026
02224097-0005